Social Movement to Address Climate Change

Social Movement to Address Climate Change

LOCAL STEPS FOR GLOBAL ACTION

EDITED BY
Danielle Endres, Leah Sprain, and Tarla Rai Peterson

Politics, Institutions, and Public Policy in America series
Series editors: Scott A. Frisch and Sean Q. Kelly

CAMBRIA
PRESS
AMHERST, NEW YORK

Requests for permission should be directed to:
permissions@cambriapress.com, or mailed to:
Cambria Press
20 Northpointe Parkway, Suite 188
Amherst, NY 14228

Library of Congress Cataloging-in-Publication Data

Social movement to address climate change : local steps for global action / edited by Danielle Endres, Leah Sprain, and Tarla Rai Peterson.
p. cm.
liographical references and index.
-60497-641-0 (alk. paper)
ments—United States. 2. Climatic changes—United States.
environmentalism—United States. I. Endres, Danielle.
M. III. Peterson, Tarla Rai. IV. Title.

13 2009
22

To Ian, Jamie, Jason, Jon, Kelly, Lauren, Lindsay, May, Phil, Will, and all the dedicated Americans who organized SIU events in their communities

Table of Contents

List of Figures and Table

Foreword

Sometimes it's better not to know how little you know.

We began Step It Up in January of 2007, after six weeks of meetings around dining hall tables at Middlebury College. We'd just organized a successful march across Vermont to demand action on global warming, and we figured, why not try this on a national scale? Never mind that we had essentially no money, and no real organization. We had zeal, we had the Web, and we had energy to spare. (By "we," I mean me and six kids about to graduate from Middlebury: Will Bates, Jon Warnow, Jamie Henn, May Boeve, Jeremy Osborn, and Phil Aroneanu). And we had the conviction that there needed to be a real climate movement in the United States, something more than the superstructure of economists and scientists and policy specialists that had been talking about the issue for a long time with little result.

And so we started sending out e-mails, asking people to organize rallies and demonstrations three months hence, in April of 2007. Now, that's a lot to ask—we hoped that maybe we'd find a hundred places ready and eager to participate. Instead, the thing took off like some kind of virus. It turned out there were people all over America who'd seen Al Gore's movie [*An Inconvenient Truth*], who'd watched the ruination of New Orleans, and who wanted to do something more than change their lightbulbs. And so they stepped up to the challenge—by April, people in 1,400 communities organized simultaneous demonstrations. The rallies took place in all fifty states, and they showed a depth of creativity that blew our minds—scuba divers with underwater demonstrations off endangered coral reefs, skiers descending fast-melting western glaciers.

Not only that, but they worked. Our demand—80 percent reductions by 2050—had been radical when we began, but twelve weeks later, by the time we'd finished, it had been embraced by both Barack Obama and Hillary Clinton, who were vying for the Democratic nomination for president at the time. We hadn't succeeded yet, but we'd helped jump-start a movement.

Or so it seemed to us—we had a mostly bird's-eye view of the whole process, trying hard to help people spread across the nation but never really getting to do much on-the-ground organizing ourselves. I spoke in cities across the country; everyone else spent those three months hunched over laptops sending out information, coordinating press releases, answering questions. For us, this book is a particular treat—it lets us see more clearly what was going on in the communities where the actual organizing was underway. In a sense, we did nothing more than set a date—it was like hosting a potluck dinner, and the people described in these pages, and tens of thousands more, did the cooking.

Beyond providing a window in the local events that made up SIU, this book also advances social movement scholarship. The book's analysis of the digital nature of the movement yields contributions to how we think about the relationship between social movements and the Internet. It highlights the challenges and opportunities afforded by organizing a movement mainly through a Web site and theorizes about the implications of movement building in the digital age. Through an investigation of the open-source DIY model of the movement-building, the book also reveals the tensions that many movements face and the prospects for successfully organizing DIY movements. Each chapter in the book discloses a set of theoretically informed tactics for social activism that scholars can continue to examine and movement builders can incorporate into their repertoires. Finally, this book importantly challenges social movement scholars to make their research accessible and relevant to the people who are carrying out the on-the-ground work of social activism. We need to have this information if we are going to do better next time—and given that we're still losing, there is going to have to be a next time.

Now, our team is trying to do the same thing globally. Again there's an unmet need—as the world prepares for the Copenhagen summit in December of 2009 that will decide on a framework for dealing with climate change, the plans of our leaders are much too timid. And there's no activist movement in much of the world to push them harder.

So we're taking the latest science—the climatologists' assessment, after the great Arctic melt of 2007, that 350 parts per million carbon dioxide is the most we can safely have in the atmosphere—and trying to use a wide variety of tactics to ram it home. We want 350 to be the most well-known number in the world, the one thing that people know about global warming if they know nothing else.

Again we're underfunded, and without some big organization behind us. But that's OK, because we have the Web—check out http://www.350.org—and because we have a world's worth of creativity. In the few weeks since we launched the drive, we've already heard from farmers in Cameroon planting 350 trees on the edge of their village, and from bike riders in Salt Lake City, 350 of them, circling the state capitol. Many churches have already rung their bells 350 times; students in China and India have marched under the 350 banner. It's growing.

If it grows large enough, it will matter—it will set the bar for success or failure in Copenhagen and move the debate a little further in the direction of the science. It's the most effective way we can think of to make a dent in the biggest problem the world faces, and in the short time the physics and chemistry allow. So visit the site, dream up something cool, and help us spread the message around the planet.

—Bill McKibben, scholar in environmental studies,
Middlebury College

In addition to his foreword, Bill answered a few questions for us about the SIU project.

WHAT WAS THE IMPETUS FOR CREATING THE SIU CAMPAIGN? WHAT BROUGHT YOU TO IT?

Well, I had dreamed up this march across Vermont the summer before. It went well—a thousand people by the end, five days of crossing the state's roads, real political impact. We wanted to know if it would be possible to do the same thing outside the funky confines of Vermont, so I assembled my talented crew, and we gave it a try.

In a deeper sense, I'd long thought that the climate-change movement was top-heavy with scientists, economists, policy types—and lacking the *movement* part that might give it some real heft.

HOW DID YOU COME UP WITH THE 80 PERCENT BY 2050 SLOGAN?

As I recall, that came from some group consensus that we needed a number to hold the politicians accountable. Former Vermont senator Jim Jeffords had introduced a bill that went nowhere with those numbers, so we used it during our march, and then nationally. When we began, it was seen as quite radical; by the time we were done, Hillary Clinton and Barack Obama had both signed on. It was kind of fun to listen to Nancy Pelosi lead thousands of college kids chanting "80 by 50" at the Power Shift conference in November of 2007.

WHAT WAS YOUR REASONING FOR UNITING ALL OF THE LOCAL ACTIONS WITH ONE COMMON MESSAGE? BASED ON YOUR OBSERVATION OF THE SIU EVENTS, HOW DID THIS STRATEGY PLAY OUT?

I think it played out extraordinarily well. Our sense was that people could make the case best in their own community—that they'd get far more attention from media and local politicians than if they'd all come for one more march on Washington. And we knew we had this new tool, the Web, that could make the rallies more than the sum of their parts. We also thought people would be better able to figure out how their neighbors could best understand the issue—underwater in Key West, along the Battery in Manhattan, and so on. We began with the image in our minds of pictures streaming in from iconic places around America, and so it happened.

You have used the terms *DIY* [*do-it-yourself*] and *open-source* to describe SIU. What led you to use these terms?

We lacked both the resources and the desire to control this thing. All we wanted was to give people an opportunity to engage with the biggest problem of our time. Heretofore, people had mostly been consigned to changing their lightbulbs, and they knew that wouldn't solve the problem. We offered a strategy with a reasonable chance of affecting larger politics, and people responded.

How much did you work with local organizers on local actions?

The six young people who organized the thing did endless great work with some local organizers—and others did their thing entirely on their own. We did our best to be supportive, to provide materials and advice, to help with press, and to link it all together so it really mattered.

What was your end goal for the SIU campaign? Do you think you achieved that goal?

Yep. We upped the stakes, moved the debate to the left, and helped create a bunch of great organizers. We were more than pleased.

Acknowledgments

The idea for this book emerged through conversations between three friends. In January 2007, Leah had the idea to write an academic article that would reflect on the recently announced Step It Up (SIU) "do-it-yourself" movement. She wanted to do more than a case study of one SIU rally in one location; a meaningful analysis of SIU necessitated consideration of several rallies across the country. Leah took her idea to Danielle, who was immediately interested, and Leah and Danielle starting brainstorming people who might also be supportive of this article idea. Soon after this conversation, Danielle pitched the idea to Tarla on their way to a dinner meeting. Tarla was also immediately interested and quickly raised the stakes by suggesting that what we had was not just a journal article but an edited book. In addition to emerging through friendship, this book also emerged through our shared desire to do something about the climate crisis and to engage in praxis-based research (see Endres, Sprain, and Peterson 2008). And so, the three of us undertook this project.

We have many people to thank. First, we are grateful for everyone who participated in the project. Of course, this includes all of the authors of chapters in this volume. We appreciate their adherence to our vision for the project, their contributions to a truly collaborative process of chapter development, and their willingness to be pushed to write for academics *and* practitioners. But it also includes others who participated in conference calls and project development and helped with data collection: Pete Bsumek, Leah Ceccarelli, Susan Seneca, Steve Schwarze, Chris Wells, and Lea Werbel participated in early development of the project but did not end up working on chapters, in one case because their local SIU event was cancelled. We are also grateful for those who attended events with us to help with data collection, from video recording to audio recording to taking photos to collecting interviews and field notes: Vanessa Au, Bopa Cheng, Sheryl Cunningham, M. Wayne Davis, Tabitha Hart, Vanessa Hull, Jay Hmielowski, Celia Hunko, Michelle Poff, Justin Reedy, Craig Rich, Michael Salvador, Katherine Vickery, and Rulan Wood. We also want to thank Samantha Senda-Cook for all her help on the manuscript. This book would not have happened without everyone who contributed to the project.

Second, we want to thank all of the activists, organizations, and people who were part of the national and local SIU events. The national SIU organizing team supported our project, made time for conference calls, and listened to our constructive criticisms. In particular, we thank Bill McKibben, Will Bates, and Phil Aroneanu. In addition to the national organizing team, we also want to thank the local organizers, not only for their efforts in creating SIU actions but also for allowing us to conduct research at their events. In particular, Paul Birkeland and Randy Jewart, the Austin Green Art director, were especially helpful. We had the opportunity to present our findings and receive valuable feedback from several local SIU organizers, including Catherine Ashton, Adele Bealer, Vicki Bennett, Paul Birkeland, Kerri Cechovic, Meagan Dixson, Jessica Eagle, Carol Firmage, Paul Grindrod, Nicholas Webster, and Bryan Wallis.

Third, we consider ourselves lucky to be part of a community of scholars and friends interested in environmental communication. We discussed some of our initial findings of this project at the 2008 Conference on Communication and the Environment (COCE) at DePaul University in Chicago. We are thankful for the reviewers for selecting our panel for presentation at the conference, Barb Willard for scheduling our panel as a plenary session, and the audience for providing excellent critical questions and feedback. We also had the opportunity to discuss this project in an article in *Environmental Communication: A Journal of Nature and Culture* (vol. 2, 2008, pp. 237–245) under Stephen Depoe (editor), Barbara Willard (guest editor), and Phaedra Pezzullo and Steve Schwarze (praxis section coeditors). Although the article is significantly different from the material in this book, we have received permission from the journal to reprint portions of the article. The journal can be accessed at http://www.informaworld.com.

Fourth, we want to thank our family and friends, who personally supported us through this project.

Finally, we are grateful to all activists and advocates engaged in personal and collective efforts to curb climate change.

Social Movement
to Address
Climate Change

Introduction

A National Day of Climate Change

Leah Sprain, Danielle Endres, and Tarla Rai Peterson

> We have all the parts of a movement except the movement itself.
>
> —Bill McKibben
>
> in the online magazine *Grist* on January 7, 2007

This observation was one result of his walk with students and citizens across the state of Vermont, raising awareness about global warming. They wondered what it would take to organize a national effort to curb climate change. Step It Up 2007 (SIU) was one answer. McKibben offered it as a way to "do something" other than change out lightbulbs: be part of "the first nationwide do-it-yourself mass protest." Individuals and local groups were invited to sign up on a Web site to host local actions that called for political action on climate change at iconic natural places. Just assemble your friends, hoist a banner calling for 80 percent carbon cuts by 2050, and take a picture.

In the three months between McKibben's first musings and the National Day of Action on April 14, 2007, the national organizers used a Web site to organize local actions. Word-of-mouth and media coverage led people to the Web site, where they could join the national effort. For example, on April 9, 2007, "no-impact man" mentioned SIU on *The Colbert Report*; several days later, the SIU Webmaster reported that their site had never received more traffic (Step It Up 2007). On the Web site, individuals could read about the campaign and sign up to host a local action. Information for these actions would then be posted, and site visitors could sign up to attend an action in their community. To support local actions, the national organizers posted materials—banner templates, action ideas, media guides—that could be used to plan events and develop new resources. After a local organizer asked for help about how to attract media attention, the national organizers wrote a sample press release, e-mailed a copy to the local organizer who requested help, and then posted a copy on the Web site for everyone to use. Later, the team added a function that enabled local organizers to send a mass e-mail to all of the participants who signed up to attend a particular action. The Web site was the center of SIU; it made the call for action, offered people a way to join "the movement," and provided support for local organizers. By April 13, 2007, over 1,400 actions were scheduled to take place across the fifty states. Though the events varied in size, complexity, and location, one message linked all the rallies: "Step it up, Congress! Enact immediate cuts in carbon emissions, and pledge an 80 percent reduction by 2050."

In the early morning hours of April 14, thirty Middlebury College students in Vermont created a photograph of the phrase "Step It Up" using flashlights in the dark. This image was uploaded to the national index site, where it became the anchor for a dynamic digital slideshow of images from actions. Photographs displayed bike riders in Portland, Oregon; people bundled in coats at a morning rally at Mendenhall Glacier in Alaska; an aerial of people spelling "Step It Up" and "we hear your call" in Inuit in Park City, Utah; and scuba divers displaying posters underwater off Key West, Florida. In the following days, over 850 action reports describing local events and sharing photos were uploaded to the national site.

Considering SIU: A Case of Movement Building

The Web site provides one account of SIU, one way to see and experience the national day of action. Rather than relying on the Web site alone to tell the story of SIU, the contributing authors to this book attended local actions as participant researchers. We heard speeches about the importance of action on climate change—the power of individual action, the promise of legislation, the potential of collective action. We asked participants questions about why they came to SIU actions, what they thought SIU was, and what the next steps should be. We interviewed organizers to gain insight into how actions were organized and what they were attempting to accomplish. We wrote our observations about what we saw at local actions and what we sensed was missing from them.

We did all of these things because we believed that SIU was an intriguing moment in the national discussion about climate change. We attended SIU actions because we wanted to be part of a potentially historic effort to address one of the most pressing environmental problems of the twenty-first century. SIU represented an overt attempt to spark a movement, to push for political action, to organize people across the nation, to demand that politicians step it up. It attempted to do all these things with a Web site and three months' time. We wanted to understand SIU. We wanted to see how it worked, whom it attracted, whether it succeeded and why. We approached SIU as researchers because we thought that it could provide insight into efforts to build social movements on climate change and other social issues. Taking the opportunity to study a movement from the beginning, we sought to learn about the implications of using new media technologies to organize a decentralized movement.

This book focuses on SIU's effort to create a national movement for climate-change action through over one thousand events that took place across the fifty states. Drawing on this case, we explore the question, what does it take to build a social movement in the twenty-first century?

Although SIU focuses on generating a movement about climate change, our analysis speaks to the broader effort to spark political action. As we chronicle and analyze SIU as an exemplar of how one contemporary campaign emerged out of and interacted with today's political and economic pastiche, we aim to offer academically rigorous yet accessible advice to activists, movement builders, and citizens interested in constructing future social movements. For this reason, each chapter concludes with concrete suggestions for people engaged in advocacy and movement building. The next section sketches the three theoretical threads that characterize movement building: rhetorical strategies, modes of organizing, and practices of citizenship. Finally, we preview the essays in this book by discussing how they relate to these threads.

Research Approach

Shortly after McKibben announced the SIU campaign on *Grist*, we initiated a collaborative national research project on SIU. As local organizers were developing their actions, we were planning how to best study SIU. Mirroring the attempt to start a movement through coordinated local actions, we sought out researchers across the country to attend and observe local SIU actions. In the end, we had eight research teams examine seventeen local actions in California, Idaho, Massachusetts, Michigan, North Carolina, Texas, Utah, and Washington state.

Through several conference calls, research teams across the country developed a common approach to studying local actions. Members of the research teams would attend local actions, take field notes of their observations, record speeches, take photographs, and conduct interviews with people at the events. The interviews included several shared questions[1] across the sites, and individual research teams were encouraged to ask additional questions based on their research interests. Several teams gathered additional information, such as collecting paper handouts from informational booths or conducting brief surveys. In addition to gathering materials from local actions, we examined the national Web site, making electronic caches of the site, including the organizer discussion board and the blog. Using LexisNexis, we gathered local and national mainstream-media coverage of SIU. Research teams also looked for nontraditional media coverage of their local events, such as coverage in podcasts, blogs, and online networking sites. Finally, we interacted with local and national organizers, including interviewing local organizers before and after the events, attending organizing meetings before the events, and attending a postevent meeting with a congressman. We also held conference calls with several of the national organizers after the day of action to learn their assessment of SIU and the ideas that informed their leadership of the national event. Finally, we talked to McKibben directly about SIU and its place in the ongoing climate movement.

All of the above materials were available to chapter authors through a shared database. Some chapters focus on a particular geographical area (e.g., Texas in chapter 1 and Salt Lake City in chapter 4) and use all of the available materials pertaining to their site. Other chapters analyze multiple locations (e.g., chapter 10 on identity construction and chapter 11 on new social movements), drawing on particular types of data (e.g., speeches and interviews). In the chapters, quotations from interviews or speeches are identified by location and the type of speaker (e.g., the mayor of Austin), but individuals are not named to protect confidentiality. Unless otherwise cited, all of the materials come from our database of interviews, field notes, and speech transcripts.

Theoretical Foundation

This book centers on a fundamental question: what does it take to build a movement in the twenty-first century? We examine SIU as a case to answer this question, considering how SIU provides a guide for what to do and what not to do. The following chapters are grounded in our empirical observations of what happened at SIU. Many chapters begin with observations from local actions, quotations from interviews, claims made by national organizers, or photographs from events. However, this book does not aim to simply describe and catalogue SIU. Instead, we seek to use SIU as a touch-point for understanding the processes of movement building—whether the topic is climate change, environmental destruction, equal rights, worker treatment, immigration policy, race relations, or welfare reform. Given this motive, we considered SIU from a theoretical foundation that helped guide us to focus on key aspects of SIU that would help us better understand it and explain movement building. Our theoretical foundation consisted of three threads—rhetorical strategies, modes of organizing, and practices of citizenship—that together, we argue, are fundamental to movement building in the twenty-first century. Each of the chapters in this book draws on one or more of these threads.

We are not alone in examining movements through the lens of rhetorical strategies, modes of organizing, and the practices of citizenship. In particular, there are well-established academic traditions of examining social movement rhetoric and the organizational dynamics of movements. This book draws on these traditions in communication, sociology, and political science. Each chapter explains relevant theories and research in an accessible way, and then uses them to help make sense of SIU and movements. By highlighting these three threads, we are attempting to pay attention to phenomena that are particularly important to contemporary movements. To explain what these perspectives bring into focus, we briefly outline each piece of our foundation.

Rhetorical Strategies

A focus on rhetorical strategies in social movements is a concern with how the tactical use of words, symbols, and images contributes to efforts for social change (Morris and Browne 2006). Rhetorical tactics can be used instrumentally to achieve certain goals and can be constitutive of new ways of thinking, adhering, and acting in relation to climate change or other pressing social issues. Rhetorical theory can highlight inventive resources—materials for crafting new, compelling, jarring, or controversial messages and arguments—such as metaphors, values, condensation symbols, myths, and images that uniquely draw people together, agitate for change, or transform public opinion. This perspective offers insight into how future movement builders might craft messages that fit particular audiences and situations. In addition to a focus on inventing messages, rhetorical theory can also be used for criticism and evaluation. Rhetoricians draw on theory to criticize the rhetorical choices of a particular speaker or movement.

The chapters in this book use various forms of rhetorical criticism to consider the specific dynamics of texts (i.e., speeches, images, protest chants) in order to explain how they succeed or fail to meet certain objectives. This type of criticism results in both further sophistication and understanding of the rhetoric of social movements and insight into how advocates might best approach similar communication situations.

Modes of Organizing

Within social-movement research, many theorists attend to how people and resources are organized. Doug McAdam and W. Richard Scott (2006) explained the reason for this: "resources must be mobilized and momentum maintained for movements to be successful, and both tasks require instrumental activities and coordination of effort: in short, organization" (5). Given these practical objectives, organizational theory considers the structuring of movements—from the relationships between organizers, community groups, and participations to the institutionalization of movement groups into established organizations—and how different structures shape movement outcomes. A movement's principles of organization (e.g., a democratically controlled movement where all members have equal voice in decision making) can lead to common dilemmas or benefits. An organizational perspective can also be used to evaluate the success of a movement to achieve its most practical goal: organizing people and resources to create symbolic and material change.

We focus on the methods and principles of organizing SIU, tracking SIU from idea to day of action to potential organization. Analyzing modes of organizing directs our attention to different levels of movement—from local actions and their organizers to the national leaders to the affiliated organizations and groups to the social networks of participants—and the relationships between them. Such analyses should enable us to explore specific elements and processes that characterize the institutional changes that accompany social movements. They also offer a window on impression-management strategies that both establishment and social-movement leaders use to maintain and/or gain legitimacy. As with rhetorical analysis, organizational analysis offers guidance for more effective advocacy among would-be agents of change.

Practices of Citizenship

Traditionally, political participation has been associated with activities like voting and, to a lesser extent, petition signing, letter writing, and political campaigning (Verba and Nie 1972). In the twenty-first century, the ways in which citizens can participate in governance has expanded to include participation in public forums, online deliberation, and new practices of political organizing, such as face-to-face meet-ups organized online (like SIU). At the same time, Lance Bennett (1998) and others have argued that people are moving away from expecting the government to solve problems and toward solving social problems through civic associations and lifestyle choices, a move he calls "lifestyle politics" since people are making political choices through lifestyle decisions such as group membership. Although some celebrate these changes as indicating that opportunities for citizens to act politically are increasingly open, others caution that they also contribute to powerful exclusionary forces (Peterson et al. 2006). Building a movement requires playful struggle with notions of how individuals can act as citizens in relation to the movement.

Our analysis of citizenship practices considers how SIU navigated the shoals of contemporary politics by using electronic media to create new roles for citizens and by providing safe venues where citizens could try out new identities. At the same time, by focusing on communication as a political act, we suggest ways advocates for change can more effectively marshal their resources by treating communication as a political act, as a practice that contributes directly to a successful movement.

Volume Overview

The book is divided into three sections corresponding to the three theoretical threads described above. Each of the chapters draws primarily on one of the theoretical perspectives to highlight the dynamics of SIU and offer practical suggestions for movement building based on this analysis. Between the sections, we have included texts that capture moments from SIU, including interviews with organizers and speeches from local actions. Together, the analytical chapters and the narrative interludes between the sections provide a multilayered description of SIU that results in both theoretical and practical insights into movement building.

Rhetorical Strategies

The four chapters on rhetorical strategies each approach rhetoric from a different perspective. Chapters 1 and 2 consider the rhetorical resources available to advocates who want to craft messages that bring people together into a movement. Chapter 1, "How Identity and Myth May Impact a Movement," uses events in Texas to explore how a strong sense of place and community ethos can be used to draw people together. The chapter offers strategies for how movement builders can invoke identity and myth to plan events and appeal to potential participants. Chapter 2, "Calling All Artists: Moving Climate Change From My Space to My Place" focuses on another rhetorical resource available to movement builders: creating participatory art experiences that constitute the local and personal implications of climate change. Using three illustrations of green art—environmentally themed cultural activism that is political, activist-oriented, and community-based—from SIU, this chapter argues that the coconstruction of green art can uniquely explain and connect participants to abstract environmental crises such as climate change due to their participatory, experiential nature.

Whereas these chapters explore rhetorical resources available to advocates, the next two chapters analyze the rhetorical choices made at local SIU sites. Reflections from these analyses can enable activists to make informed choices about how to manage common rhetorical challenges.

Chapter 3, "Demonstrative Protest Rhetoric and the Boston Step It Up Campaign," analyzes a Boston event to see whether and how the physical presence of people at the event sent a political message about climate change. Unfortunately, Prelli argues, SIU did not make a new movement visible to outside publics or politicians. Given this conclusion, he offers strategies that would more effectively use demonstrative rhetoric to draw media attention and build support for a movement. Chapter 4, "Step What Up? Rhetorical Framing and Dialectical Tensions in Salt Lake City's Step It Up Events," considers the tensions and disconnects between organizers and participants at rallies; for example, the chapter identifies a disconnect between organizers who advertised a "concert" and some participants who sought a "protest." Noting these tensions, the authors suggest how different rhetorical framing could help manage these tensions and more effectively draw diverse publics to future events.

Modes of Organizing

Each of the chapters in this section focuses on a distinct aspect of organizing. In chapter 5, "Organizing Step It Up 2007: Social Movement Organizations as Collective Resistance," the authors consider the dynamics between national and local SIU organizers. By encouraging local organizers to take control over their actions, SIU avoided some common problems that established environmental organizations face. Yet, the lack of control created new dilemmas typical of loosely organized movements such as SIU. Reminding readers of the goals behind collective resistance, the authors suggest key themes that organizers must consider in structuring their efforts—even when advocates do not want to replicate traditional organizations.

Chapter 6, "Toward Just Climate-Change Coalitions: Challenges and Possibilities in the Step It Up 2007 Campaign," builds on the discussion of the national-local dynamics to advocate that national organizers need to explicitly build coalitions with climate-justice groups nationally and create strategies for establishing connections with existing local organizations beyond an open invitation on the Internet. This advocacy stems from the authors' observations that climate-justice organizations, environmental-justice values, and underrepresented populations were missing from SIU actions across the country. The authors suggest that collaborations with faith-based organizations might offer a social network that could be mobilized to advocate for justice in the climate movement.

Finally, chapter 7, "A Social Movements Success Story? Assessing a Self-Identified Movement for Climate Action," applies an organizational lens to the same question explored in chapter 3 and comes to a somewhat different conclusion. The authors tackle the question of whether efforts to accomplish the ultimate organizational challenge—building a movement—were successful. They use a theoretical framework of the persuasive functions of movements developed by Stewart, Smith, and Denton (2007) to characterize SIU, situate SIU in ongoing efforts to address climate change, and discuss what types of mobilization will be necessary to address climate change.

Practices of Citizenship

In the final section, each of the chapters highlights a new way to practice citizenship in the twenty-first century. Examining the movement "cookbook" that national organizers offered potential grassroots organizers, chapter 8, "New Media, New Movement?" moves beyond the previous chapters on organizing to focus on how new media technologies, in particular, enabled and constrained action. Shifting from how local actions facilitate citizen action, this chapter considers the potential for relying on new media technologies to organize and practice activism. The analysis demonstrates that new media enabled a shift to local power that opened up the ways that individuals could participate in climate action.

Chapter 9, "Step It Up and Image Politics in the Pacific Northwest" applies the theory of image events—spectacles designed to gain media attention and disseminate visual messages (DeLuca 1999)—to examine actions in the Pacific Northwest. Citing research that creating image events to get media attention is an effective means of changing dominant discourse, this chapter offers practical lessons on how organizers might use image events effectively through the possibilities and challenges exemplified by SIU actions.

Whereas the previous chapter focused on participation through image events, chapter 10, "'Smells Like Folk Life': Participants' Identity Construction at Step It Up," demonstrates how participants are themselves shaped through their participation in advocacy and public events. Specifically, participant identities are renegotiated to fit the activities they are involved with, including advocacy and activism. The chapter concludes by discussing how this shift for individuals can be a potential resource for campaigns.

Next, chapter 11, "Environmentalism 2.0: New Forms of Social Activism," argues that SIU represents a new system of activism in which organizational tactics and rhetoric differ from establishment environmentalism. This new system of activism makes the roles of participants differ too. In environmentalism 2.0, individuals are offered an open invitation to contribute to the movement, whereby their participation creates a constantly evolving and changing organization. After describing the new system of advocacy, the authors discuss the practical and critical implications of these changes.

Finally, chapter 12, "An Ecologist's Response to the SIU Campaign: The Role of Natural Sciences in Environmental Movements," examines the roles of natural science and scientists in SIU as the basis for discussing the challenges scientists face in participating in campaigns and movements. Explaining why the philosophy of natural science does not automatically lend itself to activism, the author suggests how advocates and activists might approach scientists with requests for data and their interpretation of data, which can contribute to advocacy efforts and enable scientists to act on their values.

Drawing on theory from rhetoric, organizational studies, and political science, these chapters tell the story of SIU from multiple levels and perspectives. But they do more than capture a past event; they use SIU as grist for discussion about what it takes to build a movement. Together, they provide a set of strategies for advocates and activists interested in building a movement in the twenty-first century. This volume builds on the efforts of SIU to create a movement to curb climate change and other environmental crises.

Endnote

1. The shared interview protocol included the following questions: Why did you come today? How did you first hear about Step It Up? What is Step It Up? What do you think is the most important message? What do you think should happen next?

REFERENCES

Bennett, Lance. 1998. The uncivic culture: Communication, identity, and the rise of lifestyle politics. *Political Science and Politics* 31:740–761.

DeLuca, Kevin M. 1999. *Image politics: The new rhetoric of environmental activism*. New York: Guilford.

McAdam, Doug, and W. Richard Scott. 2006. Organizations and movements. In *Social movements and organizational theory*, ed. Gerald F. Davis, Doug McAdam, W. Richard Scott, and Mayer N. Zald, 4–40. Cambridge: Cambridge University Press.

McKibben, Bill. 2007. Step It Up: Introducing a brand-new, mass-protest climate movement for 2007. *Grist: Environmental News and Commentary*. http://www.grist.org/comments/ dispatches/2007/01/08/mckibben/ index.html (accessed September 6, 2008).

Morris, Charles E. III, and Stephen Howard Browne. 2006. Introd. in *Readings on the rhetoric of social protest*, 2nd ed., ed. Charles E. Morris III and Stephen Howard Browne, 1–3. State College, PA: Strata Publishing.

Peterson, Tarla Rai, M. Nils Peterson, Markus J. Peterson, Stacey A. Allison, and David Gore. 2006. To play the fool: Can environmental conservation and democracy survive social capital? *Communication and Critical/Cultural Studies* 3:116–140.

Step It Up. 2007. No-impact man features Step It Up on *The Colbert Report*. Step It Up 2007 blog. http://april.stepitup2007.org/article.php ?list=class&class=20&offset=80 (accessed September 6, 2008).

Stewart, Charles J., Craig Allen Smith, and Robert E. Denton. 2007. *Persuasion and social movements*. Long Grove, IL: Waveland.

Verba, Sidney, and Norman Nie. 1972. *Participation in America: Political democracy and social equality*. New York: Harper and Row.

Chapter Interlude

Speech by Mayor Bernero of Lansing, Michigan

Transcribed by Micheal Vickery

Jessica Pociask, Step It Up (SIU) organizer: All right, well, I'm going to introduce Virg Bernero. This is the mayor of Lansing. He's going to come out today and talk about his Go Green, Go Lansing initiative.

One of the things that he's doing right now is he actually has just signed the Mayor Climate Protection Act. This is a component of the Kyoto treaty. This is very, very important because we have this gentleman, who is in an automotive-ruled state capital, he is shaking things up. He is taking the initiative. And I think he is a leader that is stepping up and outside the normal bounds of what we can expect, but he is setting an example for many people to follow. So I just want to give him a round of applause when he comes up here to speak.

Virg Bernero, mayor of Lansing, Michigan: Thank you, thank you, and you're too kind. What I really want to say is congratulations to

Jessica and to Molly for hosting this, for bringing us all out, for bringing us together.

I'm a late bloomer in all this green stuff, really. And I'm not ashamed—I'm a little embarrassed—but I'm not ashamed to admit that I've been a follower on this. You guys are the leaders. People, you guys that have been in the trenches on this, I was busy doing other things that I thought were more important. And I'm making this public admission to you to let you [know] that what Jessica said is just exactly correct: you can make a difference in this. You can make a difference.

I ran for mayor, and signing the Kyoto Protocol was not at the top of my list. It wasn't. I had a bunch of things I wanted to do. I had a big list of priorities, and signing the Kyoto Protocol wasn't even on that list of priorities. That's how uninformed I was going into office. Thanks to people like Jessica and many of you standing here, I got educated, I got informed, and I did the right thing. That's because you led the way.

[*applause*]

So, I tell you this—it's a painful admission, it's painful to admit your own ignorance, but I grew from that. And we can be educated. Even at forty-three, we can be educated, maybe even older. I hope I can keep learning. But we can make a difference. Where we stand today, we can make a difference if we choose to. You students, you parents, you children, you can make difference today, where you stand. For your kids, for your grandkids, you can make a difference today by reaching out, by taking the extra steps that Jessica talked about.

Don't assume that the state senators, the state representatives, the council people, and the mayors know all that they need to know. Even about our, protecting our planet. It's just a little matter of protecting and preserving our planet. We have other priorities, you see. Government is busy with so many things that to take time out to save the planet, you know, it doesn't always make the top ten. Unless we get involved and make a difference.

And I'm glad Jessica brought up *An Inconvenient Truth*. For my birthday on March 31, which also happens to be Al Gore's birthday, we had an Al Gore party at my house. I had my staff over, and we watched the movie in honor of Al Gore. And I told them to forego presents and write a check to the Mid-Michigan Environmental Action Council. We had a great celebration, a little bit depressing. [*coughs*]

But it's a challenge to us, you know, it's a challenge. You can sit there and look at the glass half empty and say, "My God, it's over, we can't do anything," or you can make a difference starting today. And when you see the statistics, you realize how quickly we could make a difference, if we choose to. And we've chosen to do that in Lansing.

And so, again [*removes notes from jacket pocket*], leadership is also about followership. When you find people doing the right things, you line up, you listen, and then you do the right thing. So I want to thank you for being in the driver's seat on this one. And we're happy to follow.

We're happy to be in cooperation with Michigan State. I see John is here from Michigan State. Thank you for your leadership. The Land Policy Institute at Michigan State University has been a phenomenal partner to us. Continues to be a partner. We're excited about what we're going to be able to do. The Board of Water and Light is working with us to become greener. They are, right now, let's face it, you know, they're a polluter, but we're all polluters because we want the energy that they provide, which is mostly through coal. They're looking to become cleaner, and we're working with them. We're working with Lansing Community College in partnership. They want to join on and do this.

So, I've done some things in Lansing. I, I, just to let you know, I'm not bragging about it, because, like I say, you, you led the charge. But we, I issued an executive order for an RPS. I learned what an RPS is: renewable portfolio standard. Which means we're looking at our energy use and making it renewable. And we've established a goal of 10 by '10, 15 by '15, and 20 by '20. The 20 percent [*applause*], thank you, is more aggressive than a lot of communities. And it will require the city to purchase energy-efficient products and use biofuels and we're working on all that.

I'll tell you an exciting thing. It's amazing how one thing leads to another in this whole Go Green—we call it our Go Green, Go Lansing initiative. And it's about to be expanded with Mayor Sam Singh, my brother mayor. We're working together on a lot of things. And we're going to be working—I'll let him talk about it—but we're going to be working together now on this green initiative. Just by announcing it, and by setting the bar—you know, part of what we have to do as human beings, you know, we've got to stretch ourselves. If you've noticed, the times when you grew, the times when you've learned, the times when you've made real change, was when you had to stretch a little bit. We get comfortable, we get in our comfort zones day-to-day, but then we have to stretch. And so this goal—we met with our management team, and they said, "Gosh, I don't know, 10 by '10, '15, man, that's a lot, you know, how do we…" Ah, but, we said, "We think it's manageable." We look at what other communities are doing and what we can do. I signed the U.S. Mayors' Climate Protection Agreement. We took the EPA Energy Star Challenge. In Lansing, Michigan, we were the second city in the country to sign the EPA Energy Star Challenge, which says we will reduce our energy use by 10 percent. [*applause*]

And just by making these announcements, we look at things differently. Our managers, our department heads in city government now understand. They don't just [say], "Oh, well, this one's cheaper. This bulb is cheaper. We'll just buy the cheapest because that's how it works. You know, otherwise we'll get pounded if we buy something different." No. They know now that just because something costs a little more on the front end, they can now—they're encouraged to make the case to us. We want to know, we want to look at where we can make strategic investments. We're not saying, "Oh, we're in a deficit situation, so we're always going to do the cheapest thing, always, right now." No. We're going to make strategic investments, to become more efficient, to reach this goal.

It changes the tenor. It changes the dialogue. It gets people thinking creatively. I have people that have been with the city for years, in the engineering department, here and there, that are thinking in new ways, that are looking for new products.

And now, because we announced in the State of [the] City this goal, and we introduced our new electric vehicle, because we did those things that seemed symbolic, we actually have one of the leading alternative-energy companies. I can't say, but—because it's still secret—but we're in negotiations with a company that wants to build the new operation in Lansing that would be all green. It would be green technology. And it would be for a new biofuel-related product. Which would be phenomenal. They will, there will be global demand for this product. And it may be headquartered, based, created in Lansing, Michigan. Because they watched the speech. And they said, "Here's a city that's committed to green; we want to be in a city like that." And they grew up in Lansing. They graduated from MSU. They have operations in California, and they could go anywhere in the world. But they're thinking about Lansing because of what we did.

So, not only is it the right thing to do for our future. It's the right thing to do, morally, for our children. There's economic development value in saying, "We're going green." Because in the future, that's where it's at. [*"Right on," from the crowd.*] It's, it's about sustainability. [*applause*] It's about sustainability.

To think—I won't go on; I know we've got other people to hear from. I'm excited about this. I'm excited about this. People who know me know that it may take me a while, you know, I can be dense, it may take me a while to get there, but once I'm there, I'm committed. And we are committed to sustainability.

I'm convinced. Before, I thought I could get economic development. See, I was obsessed with economic development. I was too busy to really consider all this stuff. The reality is, sustainable economic development is where it's at. And, you do the right thing and, and it makes good sense economically. It will attract the right people. It will attract the creative class. It's the growth model for the future. So, it will pay off. It will pay dividends.

And anyway, I'm very excited about it. I'm grateful for the leadership, the grassroots leadership, the depth that we have in this community. The experts, the folks [who] have been toiling in the fields. At Fenner Nature Center, at Michigan State University, at Lansing Community College, in different environmental groups, the Mid-Michigan Environmental Action Council, and so many others, the Sierra Club, that really lead the way on this. And we're working with these groups now in our Go Green initiative. So, I just want to encourage you, talk to ten more people. You know, if you think sometimes you're talking to the wall when you tell people about this global warming and the things that are happening, and there are people in denial. And then there are people that are, again, that are distracted. And I thought I was doing economic development and that was it.

You don't have time not to do this. We don't have time not to consider it. In fact, the time that we're wasting is time we can't afford to waste for our future.

So, I'm just so excited and grateful to be a part of this. I'm excited about what we're going to be able to do together. Sam and I, just by coming together today, we work together on so many projects, too many to mention, for our community, because we recognize, we're not in competition, Lansing versus East Lansing, it's Lansing and East Lansing together in mid-Michigan and what we can accomplish. Now we're talking about doing some creative new things on the environment. Because if I, I just signed the Kyoto Protocol, and we're just ramping up, as I say, to make sure that we're making the right decisions. Sam's about to, which is very exciting. We're talking about working together and doing some regionally so that other regional government leaders will see that, "Well, if Lansing and East Lansing [are] doing it, gosh, I better jump up."

So we can always, always look at what more we can do. If we're involved, that's terrific. We can do more, and spread the word and make a big impact. So, congratulations to all of you. Thank you.

Part I

Rhetorical Framing

Chapter 1

Step It Up in the Lone Star State

How Identity and Myth May Impact a Movement

Andrea M. Feldpausch-Parker, Damon Hall, Cristi Horton, Jodi Minion, Anna Munoz, Israel Parker, and Tarla Rai Peterson

On April 12, 2007, thirty Step It Up events were scheduled for Texas, the second-largest state (both in terms of space and human population) in the United States. On April 14, only fourteen of the thirty took place. In this chapter, we explore three of the SIU events in Texas to discover how identity and myth may enable successful organizers of social movements to reach diverse audiences and motivate them to action. We also discuss some of the challenges posed by identity and myth, which may have contributed to this relatively high attrition rate across Texas.

The SIU events in Austin, San Antonio, and the Dallas/Fort Worth metroplex demonstrate how appeals to identity and myth can contribute to, as well as challenge, the development of social movements. Austin, Plano (a Dallas/Fort Worth suburb), and San Antonio all project a strong Texas-centric identity. To varying degrees, organizers in all three locations drew on local identity and its connection to the Texas myth to rally people around the cause of global-warming activism. The Austin event occurred on the steps of the Texas capitol, which provided a symbolic backdrop representing the seat of Texas government. The city of Plano promoted Live Green in Plano at the Cox Playhouse in recognition of the National Day of Climate Action; without an iconic setting, it was less successful in terms of numbers in attendance and raising local awareness of climate change. San Antonio rallied at the Alamo, the most revered site in Texas. Sam Houston's famous battle cry, "Remember the Alamo," which is widely credited with filling his Texas soldiers with the necessary rage to beat Santa Anna at the battle of San Jacinto on April 21, 1836, was only the beginning of this mythology.

We first describe how identity and myth contribute to community building, specifically as applied to social movements. To clarify our points, we briefly summarize the mythology of Texas. We then explore how SIU organizers negotiated the challenges and opportunities posed by strong identification with the Texas myth, and use our experiences as participant-researchers to suggest lessons other organizers and activists may take from the Texas events.

IDENTITY AND MYTH

Identity plays a crucial role in how we view ourselves in society. As members of a community, we present ourselves to others based on our occupations, extracurricular activities, upbringing, and a multitude of other experiences. Our identities define us and enable us to differentiate ourselves from others. As such, we feel the need to guard our identity from incursion (Rothenbuhler 2005). Kenneth Burke described identity as "an entity in itself and by itself, a demarcated unit having its own particular structure" (1969, 21). It is the necessary yet fragile niche each person holds within a social network (Mol 1976). And yet, as Burke also noted, "if [humans] were not apart from one another, there would be no need...to proclaim their unity...Identification is affirmed with earnestness precisely because there is division. Identification is compensatory to division" (1969, 22). Communicative efforts to achieve community arise out of an effort to transcend the very separation we take such pains to guard. Nowhere is this paradox more pronounced than in social-movement rhetoric, which must motivate people to identify simultaneously with other movement participants and against mainstream culture outside the movement.

People use rhetoric to link their own thoughts and feelings to recognizable identities, such as Americans' belief in freedom and democracy. It provides a means to awaken shared thoughts and values, allowing for cooperation with their fellows (Peterson 1997). Organizers of Texas SIU events used identity to link the message of climate action to the state and its cities in various ways. Austinites care about climate change because they have identified themselves as *the* progressive city in Texas, including on issues related to the environment. The Dallas/Fort Worth region has historically identified with oil and cattle and currently identifies with the energy industry. As a result, there are mixed feelings about climate change and what responding to it could mean for energy production and the Texas economy. San Antonio is home to the iconic Alamo, a symbol of valiant struggle in the face of insurmountable odds. Though not necessarily insurmountable, climate change provides San Antonio residents the opportunity to demonstrate the fighting spirit central to the Alamo legend. In all of these settings, framing the impacts of climate change as a threat to Texan identity offers a powerful catalyst for action.

Like identity, myth plays a role in the development and characterization of self, providing a framework for interpreting reality and sanctifying experience (Mol 1976). "Mythic vision is a powerful source of identification because it is rooted in a group and affects group action...Myth provides the means whereby social relations are sanctified" (Peterson 1990, 9). One example of this sanctification is the American faith in individualism, with its roots in the frontier myth, where settlers were reputed to have carved their homes out of the forbidding wilderness. The Texan identification with rugged individualism, accompanied by sanctification of private-property rights, can be traced back to the frontier myth.

As an expression of a collective mentality, myths reinforce our identities by embodying our "deepest inclinations" (Livingston and Harrison 1981, 98). In Austin, one speaker unself-consciously incorporated the myth of Texas independence into his speech when he compared Texas with "other nations." Every student who goes through the public-education system in Texas is thoroughly aware that when Texas became part of the United States, it retained the legal right to secede if it so chose. All students learn that Texas has fought for its independence on numerous occasions and was an independent nation when it voluntarily joined the United States.

Social-movement organizers face the challenge of linking to deeply held values within traditional mythology at the same time as they engage dimensions of that mythology that highlight the need for new practices. They must use existing mythology as justification for new patterns of human action. The United States has a history of responding to environmental problems when human health and well-being are threatened (Cox 2006). For example, the demonstration of direct links between human health and air pollution, water pollution, pesticide and heavy-metal accumulation in the food chain, and nuclear fallout contributed to the establishment of the Clean Air Act in 1963, the Land and Water Conservation Act in 1964, the National Environmental Policy Act, Occupational Safety and Health Act, and United States Environmental Protection Agency in 1970, and the Superfund in 1980. Rachel Carson (1962) successfully tapped into the American myth of scientific ingenuity in the service of humanity to suggest that unregulated use of pesticides was not only harmful to birds, but was decidedly un-American in its disregard for human health. With its petition to Congress for more stringent regulations on carbon dioxide emissions, SIU was well positioned to take advantage of this pattern of political activism framed as benefiting human health and well-being.

A Texas State of Mind

Texans are known for their state pride in the production of oil, natural gas, coal, and wind energy to fuel American vehicles and light, heat, and cool American homes. Texans brag about their longhorns and rodeos, bluebells and big trucks, Tex-Mex food and Texas country music (Middleton 1990). The big trucks and SUVs traveling Texas highways frequently sport bumper stickers proclaiming the driver a "native Texan." Texas rhetoric is thick with revolution, independence, and boundary conflicts. Road signs, T-shirts, and commercials are littered with the message "Don't Mess with Texas" (the Texas litter campaign since 1986). His own interactions with Texans led John Steinbeck to write, "Texas is a state of mind, but I think it is more than that. It is a mystique closely approximating a religion" (1962, 203). The name, location, and sensation of Texas have come to symbolize individuality, as a hyperextension of the American frontier myth (Champagne and Harpham 1987). As Tom Pilkington noted, "no other state of the Union has such a full-blown and extravagant mythology" (1998, xii). Its history of territorial conflicts, cowboys, and oil production is continually reinvented through its mythology.

Texas has a long history of struggle against foreign occupation, with Spain, France, Mexico, and the United States molding its boundaries as well as its culture. In 1836 Texas won its independence from Mexico at the conclusion of the Texas Revolution and then remained a sovereign nation until 1845, when it joined the United States as the twenty-eighth state (Texas State Historical Association 2001). During its time as an independent nation and after its voluntary annexation by the United States, boundary conflicts continued, some of which contributed to the Mexican-American War (1846–1848) and the addition of territory to Texas. The territorial imperialism of the nation of Texas was a comfortable fit with the economic and cultural imperialism of the United States.

Texas seceded from the United States in 1861 to join the Confederate States of America until its collapse in 1865.

As with any myth, the power of the Texan story emanates from people's ability to weave a believable and appealing narrative from their experience and imagination. The self-reliant and self-assured Texas frontier hero is embodied in the hardscrabble cowboys and oil producers of legend. Ranching activities such as cattle drives up the old Chisholm Trail from South Texas through Oklahoma Territory to the new railheads in Kansas spawned an entire genre of novels, television shows, and movies. Land speculator Sam Maverick's lax cattle-branding practices gave us the term *maverick*, a television show, and a movie. The owners and managers of Texas' historic King Ranch (established 1853) spent decades struggling to develop the Santa Gertrudis breed of beef cattle, which could thrive in the hot, muggy climate of southern Texas (Graham 2003). The Texas oil industry came into its own at the beginning of the twentieth century with the discovery of the Spindletop Gusher in 1901, and Texas oil fueled U.S. efforts throughout World War II (Olien and Olien 2002). Oil reserves gave new life to economically marginal cattle ranches and enabled a broad industrial base that stretched from exploration through drilling to refining. The oil boom, which lasted a little over fifty years, built huge cities, attracted blue-collar workers from throughout the nation, and generated wealth to build highways, parks, and universities. Like all booms, the Texas oil boom eventually went bust, leaving Texas-sized bankruptcies in its wake. Due to a decline in Texas oil production combined with a rapidly increasing population, Texas was a net energy importer by 2007. And the role of agriculture (i.e., ranching and farming) continues to decline with concurrent increases in urbanization.

Although agriculture and oil production have waned, the Texas myth has flourished. The Texas myth of hearty individualism has proven resilient, outlasting cowboys and cattle barons, roughnecks and oil reserves. In the 1980s, as geologists saw the state's oil reserves dwindling, Hollywood paid homage to the wealthy Texas oilman via an over-the-top cowboy oil family, the Ewings, in the long-running soap opera *Dallas*

(Goodwyn 1996; Pilkington 1998). In the twenty-first century, Texas has turned to alternative energy, overtaking California as the top wind-farm state in 2006. In February 2008, the Texas land commissioner proclaimed, "We're No. 1 in wind in the United States and that will never change" (Krauss 2008, 1). Oil-field heroes are getting in the act too. Kati Fehrenbacher with *Earth2Tech* wrote, "Our favorite oil-baron-turned-wind-power-wildcatter, T. Boone Pickens [plans] to build the world's largest wind farm" (2008). The wind-energy boom promises to revitalize economically depressed ranching communities.

Texas mythology offers a hyperextension of the American devotion to individual freedom, while at the same time almost mandating uniformity in its valorization of tradition. These values, as a source of identity for Texans, were incorporated into SIU events through the use of iconic settings, speeches, art, music, and other activities. In the following sections, we examine how local SIU organizers used identity and myth to appeal to their audience in Austin, Plano, and San Antonio. At each location, organizers responded to, while simultaneously attempting to mold, Texas identity. Identity concerns influenced where the events were held, how they were framed, and how the climate-change message was communicated.

Austin: Green Capital of Texas

Texans view Austin, the centrally located state capital, as a progressive city. Before becoming the state capital in 1846, Austin was a village named Waterloo, surrounded by stockades because of its regular encounters with Mexican marauders and its location on the edge of Comanche lands (Hylton and Rossie 2001). In addition to having hostile neighbors, Austin was also located far west of prospering Texas towns such as Houston, Galveston, and Nacogdoches.

Frontier expansionism eventually made Austin's location a reasonable choice for the state capital. Austin still exudes the spirit of expansion, adventure, imagination, and possibility. It is a demographically young (median age 30.9) and educated city (52 percent of residents have some college education), with 680,000 people living in the metropolitan area and nearly a million people living in surrounding communities (Ashford et al. 2006). The state government, University of Texas, and several technology companies (an area of town dubbed "Silicon Hills") shape the daytime practices, while the night is host to Austin City Limits, the self-proclaimed Live Music Capital of the World. Austin is the home of Dell Computers, Whole Foods, and Willie Nelson.

The Austin SIU event both strengthened and reflected the city's shift away from identification with the fossil-fuel industry, toward more environmentally friendly energy production. Austin organizers used SIU as an opportunity to encourage Austinites to further develop this new identity by celebrating local achievements such as a recent agreement to combat climate change.

Although Austin has had some success on the environmental front, it remains a site of dissonance concerning climate change and environmental policy. Texas has just over half the population of California, while producing twice the greenhouse-gas (GHG) emissions (Witt 2007). According to a state representative's rally speech, Austin has the highest GHG emissions of any state capital in the United States. The representative told an enthusiastic crowd that "when compared to other nations of the world, Texas is the seventh-largest emitter." Despite these figures, Austin is listed by the International Council for Local Environmental Initiatives (2007) as one of the fifteen greenest cities in the world. Austin has one of the oldest green-builder programs in the United States, with the best recognized eco-rating program for houses and buildings (Seiter 1997). In 1998 Mayor Kirk Watson responded to rapid growth by developing smart-growth policies, which offered a number of environmentally friendly incentives and initiatives for water conservation, energy conservation, and the improvement of Austin's quality of life (Hylton and Rossie 2001).

Although Austin's SIU rally included the national demand for congressional action to slow global climate change, organizers framed it more as a celebration of a recent accomplishment for city leadership and the people of Austin. Less than two months earlier, the city had passed "the nation's most aggressive plan for carbon-reduction," called the Austin Climate Protection Plan (ACPP). The plan went beyond the efforts of many United States cities known for their climate-protection efforts, such as Portland, Oregon (Witt 2007). On April 14, ACPP organizers took the party to the Texas capitol. Austin's SIU rally celebrated the power of local citizens, municipal government, and professional associations working together to contribute to global change at the local scale. The rally also sent a message to other municipalities throughout the state of Texas that climate-change reversal begins at home.

Appealing to a Progressive Identity

At precisely noon on April 14, 2007, Austin SIU organizers started their rally on the steps leading to the front doors of the Texas capitol. They chose the capitol as a rally site for the same reason civil rights movement organizers chose the Lincoln Memorial in Washington, DC, for Martin Luther King Jr.'s "I have a dream" speech in August 1963, the same reason Eleanor Roosevelt responded to the refusal to allow Marion Anderson to perform in Constitution Hall by arranging for the concert to be performed on the steps of the Lincoln Memorial on Easter 1939. As Burke argued, "there is implicit in the quality of the scene the quality of the action that it is to take place within it" (1969, 7). By holding the rally on the steps of the state capitol, organizers and rally participants were taking the message of climate change directly to their state government, as well as symbolically responding to SIU's national goal of initiating official government action to reduce carbon emissions.

The Texas capitol sits in a park-like setting filled with stone monuments to the Confederacy, soldiers of various foreign wars, volunteer firefighters, and other public servants. Students on school field trips, Austinites on lunch breaks, and tourists from around the state and beyond frequent the lush grounds. Since the capitol is a popular location for tourists and lawmakers alike, it was ideal for attracting a nonthreatening crowd. A throng of more than one hundred people joined SIU Austin organizers at the state capitol on April 14, 2007.

The Austin SIU event was cosponsored by the U.S. Green Building Council and the Sierra Club, with speakers from the American Institute of Architects, Habitat for Humanity, the University of Texas Solar Decathlon Team, Austin Green Art, local green-building contractors, a state representative, and the mayor of Austin. The state representative focused on the Texas economy and its move away from identification with oil. Using the backdrop of the state capitol to his advantage during the speech, he wove a story of the Texas petroleum industry and discussed its impact on global climate change. He urged listeners to push Texas state government operating "in there" (pointing at the capitol doors behind him) to be an agent of change. He told them the conversation about climate change and energy already had changed: "I no longer have to argue that global warming is coming. Now the argument is that global-warming regulation is coming." After comparing Texas GHG emissions to emissions of other "nations," he noted that Texas is the "headquarters of the energy industry...because we just happened to be sitting on top of a lot of oil." The representative's message incorporated SIU into the Texas identity with the energy industry, arguing for Texas to "lead the new energy technology" because "sustainability is cost-effective" and "profitable." He noted that "no economy will suffer more than ours if we do not take proactive, preemptive action and lower emissions per unit of GDP."

The state representative also offered a vision rooted neither in lost resources nor accountability to future generations, but one that focused on new economic opportunities and the importance of retaining the competitive edge during the next energy boom. Looking to a future in renewable energy, he noted that Texas happens to be in a place of abundant wind- and solar-energy resources, spinning a new take on Texans' traditional identity as the U.S. center for energy production. He also mentioned bills currently on the state legislature floor, suggesting that state policy change was the next step to continue Austin's successes in combating global climate change. By showcasing Austin's progressive actions, he made action part of the social pattern of the city, a key component of myth building.

Austin's mayor spoke about how proud he was to be part of Austin municipal government efforts to reduce GHG emissions locally. An architect by training, he had been involved in the green-building community of Austin, the American Institute of Architects, and municipal government for many years before being elected mayor. He also drafted the ACPP, which he outlined to much applause from proud Austinites. The plan involved powering 100 percent of Austin's municipal facilities with clean, renewable energy sources such as solar and wind power by 2012. By 2020, the plan shifts 30 percent of the city's total residential, commercial, and industrial energy to clean, renewable energy. Changes in building codes for new homes and incentives for using high-efficiency materials to retrofit existing homes give the plan teeth, but have made some home builders wary of its economic implications (Witt 2007).

Proponents of green building were not the only voices to be heard at the Austin SIU rally. After numerous speeches from people involved with the ACPP, a young, energetic man from the Sierra Club took his turn at the microphone. Other speakers having set the groundwork for action, he distributed national information packets, tip sheets regarding how individuals can reduce their emissions, coupons to Home Depot, and, of course, free compact fluorescent lightbulbs, which seemed to be the standard parting gift for any SIU event. The Sierra Club representative's speech shifted from a focus on Austin's local successes to discussion of the Cool Cities Campaign, a nationwide campaign driven by regular meetings of U.S. mayors, who discuss climate-protection agreements. He challenged listeners to bring their municipalities into the group of 442 (at that time) cities and towns that had crafted local Cool Cities agreements. When he read the fourteen cities in Texas, members of the crowd reacted and applauded in shock at some of the historically conservative towns, such as Sugarland (a suburb of Houston), participating in the campaign. The representative encouraged people to pick up a packet if their hometown was not on his list. Then he circled back to the prior speeches, praising the mayor's efforts and the city of Austin for developing the ACPP.

Speeches were not the only activity on the docket. Austin organizers also invited resident artists to participate. In addition to hearing speeches, the crowd saw a sixteen-foot-tall art installation signifying the plastic garbage bags accumulating in the North Pacific Gyre (a convergence of a number of ocean currents where garbage has accumulated). According to the artists, the gyre is a ten-foot-deep, thousand-mile-wide circle of plastic bags. The purpose of the art installation was to boost awareness of the consequences of consumer waste by imitating the accumulation of plastic and other garbage in the North Pacific Gyre. Although the metal structure decorated with grocery bags painted teal green did not relate directly to SIU's call to cut carbon emissions, it signified Austin's move away from identification with the petroleum industry. The artists reminded viewers that petroleum is basic to the production of the plastic bags that have become so ubiquitous throughout U.S. society. The art installation, along with performances by local musicians, helped generate a festive air. By involving local artists, Austin organizers took advantage of Austin's identity as the cultural nucleus and most progressive city in Texas.

Overall, the Austin SIU rally was a celebration of what can be achieved on the community level, which, according to a sign carried by a teenaged participant, is to "Slow the Warm." Austin organizers, speakers, and artists appealed to Austinites' identity and the ongoing reconstruction of the Texas myth to offer participants a sense of self that allowed them to remain true to their traditions at the same time as they looked forward to change. Austin organizers also took advantage of the state capitol as an iconic place, along with representatives of the local government, to signify that the government is listening and positive action is possible.

PLANO: THE CHALLENGE OF SUBURBIA

Plano, which is located in Collin County, originated in the early 1840s. Situated in the heart of the Blackland Prairie, farming and livestock production provided its economic base. The arrival of the Houston and Texas Central Railway in 1872 provided new employment opportunities and gave local producers access to an outside market, thus prompting production of commercial crops such as wheat and cotton. By 1890, Plano's population numbered 1,200, and the town had two railroads, six churches, two steam gristmill-cotton gins, three schools, and two newspapers. Plano continued as a farming community until the 1970s, when its population began to nearly double every five years. Property taxes skyrocketed, contributing to the demise of agriculture. By the mid-1990s, Plano was a city of seventy-two square miles with a population of approximately 130,000. With changing demographics, Plano became an upper-middle-class city supportive of Republican state and national tickets. By 2000, the city had approximately 7,800 businesses and 222,000 citizens (Texas State Historical Association 2003). The city of Plano had designated 2007 as the Year of Living Green (City of Plano 2007b). Through this initiative, Mayor Pat Evans promoted resource conservation and healthy ecosystems (City of Plano 2007a).

Plano is one of many suburbs of the rapidly growing Dallas/Fort Worth metropolitan area. Dallas was founded in 1841 as a trading post on the Trinity River (Texas State Historical Association 2008a). After becoming the county seat in 1850, the city thrived as a hub for transporting merchandise in and out of the region. When oil was discovered in east Texas, Dallas became a financial center for the oil industry, and is now the world headquarters to companies such as ExxonMobil and ENSCO Offshore Drilling. Founded in 1849 as an army outpost, Fort Worth later became home to the world-famous Fort Worth Stockyards (Texas State Historical Association 2008b). Although the last livestock auction was held in 1992, the yards remain an economically vibrant section of town, catering now to American consumers in search of quaint little shops, restaurants, and nightclubs. The two cities are often referred to as the Dallas/Fort Worth metroplex or just the metroplex, since they eventually merged into one metropolitan area.

Holding Fast to Tradition

On the morning of April 14, 2007, the SIU Web site was full of Texas cancellations. Our research team had traveled to Plano the previous evening to attend the Fair Friends Action Day in a neighborhood park. People were encouraged to bring food and enjoy family-friendly entertainment from 4 p.m. to 10 p.m. We showed up early, hoping to interview organizers.

We had the park to ourselves all afternoon, however. Using a laptop computer and the Internet, we searched in vain for an announcement of a changed venue or cancellation. A few minutes after four o'clock in the afternoon, we gave up on the seemingly deserted suburban neighborhood of high-end homes. We logged into SIU's Web site again and discovered an SIU rally scheduled for downtown.

The event, sponsored by the city of Plano, was touted as a celebration of the National Day of Climate Action. The SIU organizer for this rally was also the coordinator for Live Green in Plano. Live Green in Plano proposes simple lifestyle changes that will enable citizens to improve and protect the environment and integrate environmental values into the local economy and everyday life. Several green events were planned, including the Environmental Awards and the Live Green in Plano Concert and Film Series (Flemmons and McGathey 2007).

On April 14, the city hosted a self-guided walk through historic downtown, addressing how each person could make a difference to climate change. To start off the event, the organizer gave a brief speech, which included the SIU challenge to "cut carbon emissions 80 percent by 2050." Participants then walked back to the Cox Playhouse for a barbecue. A documentary titled *Kilowatt Ours* was shown at 7:15 p.m., and a concert by The New Hokum Boys, a blues/jug band, concluded the evening. The SIU event served as the kickoff to the Live Green in Plano Concert and Film series.

The Cox Playhouse, tucked into one of the quiet downtown streets across from a park, seemed unlikely to draw significant public attention; no signs had been posted to direct participants. When we asked a resident for directions to the playhouse, he said he had no idea what we were talking about. We eventually found the event and joined other participants in a walk through quiet streets lined with turn-of-the-century heritage homes. The area looked prosperous, with attractive coffee shops and new condos in close proximity to the light-rail system (Dallas Area Rapid Transit, or DART for short).

The Plano rally seemed less about SIU and more about resources and activities offered by Live Green in Plano. Throughout the event, people came and went based on individual activities, and none described themselves as attending an SIU event. The eight residents who showed up for the guided tour, for instance, generally came for its historical content. One woman who had moved to Plano in 2005 came to learn more about her new community. The brochure on the concert/film series did identify April 14 as a "National Day of Climate Action," but SIU was not emphasized. As with all scheduled activities, the organizer focused on more environmentally friendly living, while minimizing the connection to climate change.

For the viewing of *Kilowatt Ours*, the crowd swelled to thirty-six people (including the four researchers), the largest number for any of the day's activities. This documentary advocated cutting fossil-fuel consumption and addressed energy policy, global warming, and pollution. The film centered on the Southeast United States. Interviews with people about where their electricity came from, clips about what scientists have to say about the problem, general discussions of air and water pollution, arguments against nuclear energy and its toxic waste, and finally, a discussion of human health and environmental justice summed up the energy/climate-change problem in the film.

Plano's SIU organizer enthusiastically explained Live Green in Plano to us and described the information packet about sustainable living that her program was sending to each Plano household. When prompted with a question about SIU, she became more tentative, responding, "We want to be in line with what the movement is about," but also noting that local politics made any reference to climate change or global warming risky. She explained that many Plano residents "don't believe in global warming":

> You know, I'm with the city of Plano, and so we're careful how much we raise the global-warming issue…Just yesterday there was a letter to the editor and it was slamming my program, which is a fairly new department. It's the department of sustainability for the city of Plano;…we've been careful not to say "global warming" in anything, in any of our literature, but then we publicized this and, of course, they noticed…his [Al Gore's] message was going to be

> there. Well, it got slammed in this letter to the editor about how we've gone wacky, what's the city doing, and it's all political. So you're going to get that, but we believe that it's important.

Since many Plano residents perceived climate change or global warming as part of the "left agenda," she feared they would simply "close their ears" to any message of sustainability if it were attached to global warming:

> We would rather they hear what we want them to do and the changes that we want them to make that will get us to the same place. We feel that if we throw out global warming immediately, they'll shut us down and then they won't hear what we're saying.

She shared with us her efforts to educate other Plano residents on climate change. For example, she said, "We [Live Green in Plano committee members] took an afternoon and went and saw Al Gore's film at the Angelica theatre when it was here...to just start whetting people's appetite and for them to be exposed to issues."

When asked again about the SIU event, she responded, "We weren't going to avoid the issue tonight...but we were going to be cautious about how we used or how we emphasized global warming." Plano's SIU organizer recognized that identifying the city's Live Green in Plano program with the "left agenda" would be political suicide in Plano, and chose to minimize the SIU message because of her reservations about the receptiveness of community members. Having already experienced negative responses to the city's tentative efforts to advance an environmentally progressive agenda via Live Green in Plano, she feared politically motivated confrontation and exercised caution by organizing a low-visibility event and scheduling politically acceptable activities, such as a tour of the historical district and a barbecue.

Organizing an SIU event in a Texas community that had not self-identified as progressive or environmentally friendly posed difficult challenges. Where Austin's identification with the oil industry had already transformed itself into identification with the larger energy industry, the politics of Plano had discouraged such transformation. And as a part of the Dallas/Fort Worth metroplex, Plano's residents clung more tightly than ever to the financial security they saw represented by highly profitable companies such as ExxonMobil.

In this case, the challenge of separating SIU from an economic attack on the industry many local residents viewed as central to their prosperity proved too difficult, and the organizer chose not to link SIU directly to Texas identity. Given that the organizer had already received letters expressing anger about Live Green in Plano, the Plano SIU event's message of climate-change action was stifled by fear of a political backlash against future efforts toward sustainability. This fear, which may have contributed to lack of participant recruitment, led to low attendance and a weakened message.

Our description should not be read as an indictment of the Plano event. Several SIU events in the Dallas/Fort Worth metroplex and throughout Texas were cancelled, many with no explanation offered. Houston, the nation's fourth-largest city, managed only one gathering, with twenty-three participants (Step It Up 2007). Fort Worth participants, who danced to a Caribbean beat while holding posters that demanded, "Expose Exxon," may have drawn wider attention but also may have alienated undecided Texans. In each of these cases, organizers and advocates made strategic choices about how best to negotiate a relationship between SIU and local identities.

San Antonio: Forging a New Tradition

San Antonio is the only major city in Bexar County, Texas. It represents a new form of American urbanism that blossomed in the Sunbelt following World War II. Its growth patterns owe much to the same postwar public policy that contributed to Plano's growth; these policies included generous funding for new highways and tax incentives for single-family home ownership (Fink 1993). Consumption of land increased and population density decreased as the single-family home and suburbia became the benchmark for success (Garreau 1991). San Antonio epitomizes these new urban-growth patterns, which have become pervasive throughout and beyond the American Southwest.

These urban-development patterns have had serious impacts on air and water quality. Declining levels of aquifers, increased nutrient- and contaminant-loading of streams, decreasing freshwater inflow to estuaries, loss of endemic species, and invasion of exotic species represent a few of the human-induced ecological problems with serious socioeconomic repercussions. Municipal use accounts for 71 percent of water used in the San Antonio River Basin (Texas Water Development Board 1997). Air quality has suffered as well. In 1999 local elected officials and air-quality planners responded to designation of San Antonio as a "near-nonattainment" area in regard to national ambient air-quality standards related to ozone levels by proposing a voluntary program for reaching attainment to the Texas Commission on Environmental Quality (TCEQ). It was designed to avoid regulatory action by demonstrating intent to voluntarily achieve the standards. The TCEQ signed on almost immediately and approached the EPA. In 2002 EPA Region 6 endorsed the approach and entered into an agreement with the Alamo Area Council of Governments. Since that time, the EPA has deferred the effective date of nonattainment designation several times, based on claims that the region is seriously attempting to achieve clean-air standards (Texas Commission on Environmental Quality 2005).

The area's socioeconomic heterogeneity presents significant complications to any planning effort. Agriculture, which anchored the county's economy for the past century, no longer provides sufficient focus for planning and development. Rather, the region's economy has become increasingly diverse over the past decade. Recent estimates indicate that, although trade, government, and services provide most employment opportunities, construction, manufacturing, utilities, mining, and agriculture also contribute to the county's economy. In addition to the economic diversity of the region, there is also diversity in ethnic background. In 2000, 67.7 percent of the population of San Antonio was white. Of those, 58.7 percent were Hispanic or Latino and 41.3 percent were not Hispanic or Latino. Other groups comprising more than 1 percent of the local population included "some other race" not listed on the census (19.3 percent), African American (6.8 percent), two or more races (3.7 percent), and Asian (1.6 percent) (U.S. Bureau of the Census 2007). San Antonio illustrates the volatile mix of cultural groups and economic interests that characterize rapidly growing urban areas throughout the United States.

Remember the Alamo!

There is no more revered site in the state of Texas than the Alamo, another SIU rally location. For over one hundred years, storytellers have amplified and embellished the mythology of the Alamo and what it means to Texans. Its defenders have been valorized in Texas and U.S. history as symbols of courage and sacrifice for the cause of freedom. More than 2.5 million people a year visit the 4.2-acre complex known as "the Alamo." Most come to tour the old mission where a small band of Texans held out for thirteen days against the supposedly ferocious army of General Antonio Lopez de Santa Anna. Local SIU organizers joined a long and varied parade of other activists and organizers in selecting the Alamo for the San Antonio site (Barra 2003; Flores 1998).

Similar to the Austin event, the local chapter of a nationwide green-building organization organized and sponsored SIU San Antonio. The green builders saw SIU as a means for doing public outreach for sustainable architecture and technologies. Consistent with the Texans' identification with individual volunteerism, the organizers hoped to present reducing individual carbon impacts as one way to lessen the threat of industry-wide state or federal regulation. When the lead SIU San Antonio organizer explained the relevant goals of the American Institute of Architects (AIA), she identified people within her industry as both the problem and solution to reducing carbon dioxide emissions:

> In an effort to designate the impact on our built environment, the AIA's commitment on the environment has a mission to increase awareness of sustainable design practices. The building sector alone accounts for 39 percent of the total U.S. energy consumption and 71 percent of its electricity. The AIA has its own challenge, the 2030 Challenge, which calls for an immediate energy reduction of all new and renovated buildings to one-half of the national average. By 2030, all buildings will be carbon-neutral.

The Green Builders Association focused the majority of its speeches toward the municipalities, concentrating on the issues of expansion and energy use. Speakers also addressed individual homeowners, usually with an appeal to economics. As in the organizer's statement in the interview, the speeches identified the citizens of San Antonio as part of the solution, if they were willing to take the necessary steps to change.

Like most of the SIU rallies, local politicians took the opportunity to share their visions of the future. A Texas state representative called on Texan identity and pride when he shouted from the speaker's platform, "Texas is forty-ninth in the union for clean air, are you proud of that!?" The crowd responded instantly, "No!" Then he asked, "Texas consumes more energy than any other, are *you* proud of *that*?!" Again, the crowd roared back, "No!" After presenting a series of similar questions and receiving equally resounding negative responses, he then explained that only by "stepping up to the plate" on climate action would participants live up to the hardy, strong-willed Texas image represented by the Alamo in the backdrop.

Speeches were given by the lead event organizer, other members of the local green-builders organization, the chair of the Alamo chapter of the Sierra Club, and local politicians. Although few were able to hear him, a ninety-four-year-old local renewable-energy guru and legendary business maverick drew energetic applause from the crowd. Most of the speeches drew on the iconic site of the rally as a symbol for action, noting the Alamo's bloody history in the quest for freedom. With the Alamo directly behind her, the city of San Antonio's environmental director told the audience,

> I was very *moved* and humbled that we'd be here in front of the Alamo, and I was thinking about the very ground our forefathers who battled so courageously for their liberty and sovereignty over this land. But today we're in a different battle—we're in the battle over our environment. Over the prevention of global warming... to save the air we breathe, the water we drink, and the *very land* that sustains us...Let's honor our forefathers with a battle here at the Alamo and for future generations to come by doing our part to prevent global warming.

Nothing could be more traditional, more Texan, than battling to save the "land that sustains us."

Of all the speakers at SIU San Antonio, the Bexar County commissioner was the most successful at tapping into the Texas myth, identification with the Alamo, San Antonio's centrality in the Texas history of action, and its potential significance to the future of climate-change activism:

> Hopefully we won't say *we* didn't have the *guts* to do anything about it [global warming]. 'Cause this thing right behind me [*turning and gesturing toward the Alamo*] in 1836 was the place, was the lightning rod of the Texas Revolution. You know, *we* used to be the capital of Texas...*we* were the work horse and the war horse of the Texas Revolution. And that's the message I want to bring to you. Individually, you may think, "Oh, I can't do anything much..." [Y]ou can do *something*. "I can't do a whole lot." You can *do* something. We can ask *our* Congress to step up to the plate. It's not too much to ask that we breathe clean air. *Is it?*

The commissioner worked his crowd with the skill of a camp revivalist. Texas pride, and especially Bexar County pride, emanated from his words and movements: "If an army marches on its *feet*, then *a democracy* marches on its *education* and *information*, and I submit to you that that is what we are doing here today!" By reframing words such as *education* and *information* as the energy to fuel "an army," he called up Texans' identity as people of action, as the nation's best personification of the rugged frontier myth. He explicitly drew on the presence of the Alamo to link Texas identity to climate-change activism:

> The same spirit of the Texas Revolution right here behind me, is the same spirit that we need to take a revolution of thinking about how we get our energy. The leader of the free world cannot be proud to import 60 percent of its oil;...his country has 60 percent of its oil coming from foreign entities that are distant, uncertain, and hostile. And I am asking you to join with me in an effort to create a revolution of thinking in how we use energy. Let us conserve, let us be energy efficient, let us use renewables that are clean and *free*!

By holding the rally at the Alamo and incorporating its symbolism into their talks, SIU San Antonio speakers appealed to the Texas version of the frontier myth, encouraging everyone to build on the tradition embodied by the Alamo. The seemingly insurmountable challenges associated with climate change were something for true Texans to embrace. By the time everyone gathered around for the group picture, they were ready to take on climate change along with all other enemies that came their way.

EXPLOITING THE POTENTIAL OF IDENTITY AND MYTH: LESSONS LEARNED

SIU's national coordinators recognized the importance of identity in SIU events. Local organizers were encouraged by the national coordinators to meet at iconic locations and incorporate residents' identities into the delivery of the message for action. Since these iconic locations are sustained by myth and are often linked to a collective identity, they offer the potential of imbuing any event held on the site with power. The San Antonio and Austin SIU rallies exemplified how organizers and leaders can tap into that potential: the Alamo and the capitol provided material contexts that encouraged participants to interpret SIU as consistent with being a Texan. Organizers and advocates did not, however, stop at this. Instead, they maximized the potential by referring directly to the structures and their stories at strategic points in prepared speeches as well as in unscripted interactions with participants. In both cases, local organizers drew on the power of iconic places to induce action in participants through connections with identity and myth.

Aside from the availability of an iconic location, identification with contemporary local culture and industry can also help an event connect more directly to participants. Austin's organizers took advantage of local culture with their art installation and entertainment provided by local musicians. San Antonio's organizers highlighted the diversity of their city by pairing the tradition of the Alamo with a performance by local belly dancers. Plano's organizer drew on potential curiosity about the community's historic district. National coordinators encouraged local organizers to build on local identity. In an interview held four months after the April events, one of the national organizers explained,

> I'm sure there were plenty of people that were involved with Step It Up who really saw themselves [as]...being part of something that was more local in nature...We talked to a number of groups who intentionally created a more local action identity, like...in Portland...They wanted to develop their own identity, their own brand around this idea, their own organizational name.

The freedom to present SIU in locally recognizable terms gave event organizers a chance to attract local residents to the movement while at the same demanding that the federal government cut carbon emissions across the nation. Through appeals across time and place, organizers at the Austin and San Antonio rallies provided their participants with a shared identity from which to build a unified voice. These appeals were exhibited most strongly in Austin's focus on opportunities to regain and surpass Texas' past position as the U.S. leader in energy production and San Antonio's focus on the heroic tradition of defending the land and "stepping up to the plate" for freedom.

The Plano event began at a local historic site, but one that has not attained anything close to iconic standing. The organizer also chose not to play up potential connections between SIU and local culture and industry. Plano's spatial proximity to the Dallas/Fort Worth metroplex and the oil industry, along with its general political tone, led the organizer to de-emphasize climate change, even at an event that was supposed to promote climate-change action. The Plano event succeeded to the degree that it happened, and that it may have served as an information-sharing venue. However, the weak links between SIU, the local identity and myth, and the place may not have been the most successful way to promote the main message of the campaign. The Plano case provided an opportunity to explore how social-movement leaders, organizers, and advocates might promote flexibility in mythic boundaries. In this case, for example, organizers might expand Plano's existing identification with the fossil-fuel industry to include the energy industry. They could use the frontier image to encourage residents to retain or regain their (self-proclaimed) role as the "national energy leader" through embracing a more diverse and ecologically friendly assortment of energy technologies that directly attack climate change.

These Texas SIU events suggest possibilities for mythic revision that social-movement leaders and event organizers may find useful in a variety of instances. Possibilities include:

- Learn both the details and the main storyline of local mythology, and then pull from cultural resources present in the community to make your message familiar to the local citizenry.
- Use mythic symbols to keep the message positive and action-oriented. People are more comfortable with positive messages that suggest constructive action.
- Use mythic symbols to rally the citizenry, increase saliency, and build momentum for the cause.
- Take advantage of living myths' flexibility. Use this characteristic to suggest new interpretations that revise and expand the existing traditions in response to contemporary problems. Audiences want to believe that their mythology can evolve to incorporate new challenges.
- Provide multiple options for participants to align their identities with the expanded definition of myth.

As Burke pointed out,

> Rhetoric…is not rooted in any past condition of human society. It is rooted in an essential function of language itself, a function that is wholly realistic, and is continually born anew; the use of language as a symbolic means of inducing cooperation in beings that by nature respond to symbol. (1969, 43)

The Bexar County commissioner demonstrated sophisticated skill at influencing rhetoric's continual rebirth as he wove his audience's reverence for the Alamo into climate-change activism. His listeners learned that by demanding federal action to reduce carbon emissions, they were simply carrying on tradition. Austinites could congratulate themselves on having chosen to cooperate with national efforts such as SIU at the same time as they revived and revised the future of energy production. Both reconstructions of the Texas myth illustrate Peterson's argument that by "making mythic images more explicit, we create opportunities to consider breaching old boundaries" (1991, 305). In the case of Austin, organizers reshaped the boundaries of the Texas oil myth to include alternative energies such as solar and wind power. In San Antonio, climate change displaced Santa Anna's army as the major threat to freedom. Rather than rejecting existing Texan identity, speakers and organizers reshaped the myth within which that identity flourishes to include climate-change activism. Although every local action faces different specific challenges, all social movements share the need to revise existing patterns of identification and myth. Both identity and living mythology present a constantly shifting landscape (see chapter 10). Organizers of both the San Antonio and Austin SIU events strategically guided that fluidity to individual and community-level actions that inculcated the fight against climate change into the Texas identity and myth.

REFERENCES

Ashford, Kate, Carolyn Bigda, Tara Kalwarski, Sarah Max, Donna Rosato, Asa Fitch, Michelle Kalkhoff, Shelly Ng, and Ingrid Tharasook. 2006. Best places to live. *Money* 35 (8): 95–108.

Barra, Allen. 2003. Trying for the truth about the Alamo: A report from the field on the attempt to make a historically accurate movie about an event that has always been both highly controversial and wildly romanticized. *American Heritage* 54 (6): 50–55.

Burke, Kenneth. 1969. *A rhetoric of motives*. Berkeley: University of California Press.

Carson, Rachel. 1962. *Silent spring*. New York: Houghton Mifflin Co.

Champagne, Anthony, and Edward J. Harpham, eds. 1987. *Texas at the crossroads: People, politics, and policy*. College Station: Texas A&M University Press.

City of Plano. 2007a. http://www.plano.gov (accessed December 20, 2007).

———. 2007b. Live Green in Plano. http://www.plano.gov/Departments/Environmental+Services/GreenLiving/ (accessed December 20, 2007).

Cox, Robert. 2006. *Environmental communication and the public sphere*. Thousand Oaks, CA: Sage Publications.

Fehrenbacher, Kati. 2008. T. Boone Pickens: Kicking off the world's largest wind farm. Earth2Tech. http://earth2tech.com/2008/04/14/t-boone-pickens-kicking-off-the-worlds-largest-wind-farm/ (accessed May 1, 2008).

Fink, Marc. 1993. Toward a Sunbelt urban design manifesto. *Journal of the American Planning Association* 59:320–333.

Flemmons, Stephanie, and Liz McGathey. 2007. "Green" challenge to residents, businesses. *Plano Insider*, April 9, 2007. http://www.

planoinsider.net/articles/2007/04/09/ news/story3.txt (accessed December 20, 2007).

Flores, Richard R. 1998. Memory-place, meaning, and the Alamo. *American Literary History* 10 (3): 428–445.

Garreau, Joel. 1991. *Edge city: Life on the frontier*. New York: Anchor Books.

Goodwyn, Lawrence. 1996. *Texas oil, American dreams: A study of the Texas Independent Producers and Royalty Owners Association*. Austin: Texas State Historical Association.

Graham, Don. 2003. *Kings of Texas: The 150-year saga of an American ranching empire*. Hoboken, NJ: John Wiley and Sons.

Hylton, Hilary, and Cam Rossie. 2001. *The insider's guide to Austin*. 3rd ed. Guilford, CT: Globe Pequot Press.

International Council for Local Environmental Initiatives. 2007. Local governments for sustainability. http://www.iclei.org/ (accessed December 15, 2007).

Krauss, Clifford. 2008. Move over, oil, there's money in Texas wind. *New York Times*, February 23, 2008. http://www.nytimes.com/2008/02/23/business/23wind.html?pagewanted=1&_r=2&hp (accessed May 1, 2008).

Livingston, David, and Richard Harrison. 1981. Meaning through metaphor: Analogy as epistemology. *Annals of the Association of American Geographers* 71 (1): 95–107.

Middleton, Richard. 1990. *Studying popular music*. Philadelphia: Open University Press.

Mol, Hans J. 1976. *Identity and the sacred: A sketch for a new social-scientific theory of religion*. New York: The Free Press.

Olien, Roger M., and Diana Davids Olien. 2002. *Oil in Texas: The Gusher Age, 1895–1945*. Austin: University of Texas Press.

Peterson, Tarla Rai. 1990. Jefferson's yeoman farmer as frontier hero: A self-defeating mythic structure. *Agriculture and Human Values* 7:9–19.

———. 1991. Telling the farmers' story: Competing responses to soil conservation rhetoric. *The Quarterly Journal of Speech* 77 (3): 289–308.

———. 1997. *Sharing the earth: The rhetoric of sustainable development*. Columbia: University of South Carolina Press.

Pilkington, Tom. 1998. *State of mind: Texas literature and culture*. College Station: Texas A&M University Press.

Rothenbuhler, Eric D. 2005. The church of the cult of the individual. In *Media anthropology*, ed. Eric D. Rothenbuhler and Mihai Coman, 91–100. Thousand Oaks, CA: Sage Publications.

Seiter, Douglas. 1997. The city of Austin green builder program: Developing sustainability through marketing and education. Paper presented at the First Florida Sustainable Building Conference, Miami.

Steinbeck, John. 1962. *Travels with Charley: In search of America*. New York: The Viking Press.

Step It Up. 2007. http://april.stepitup2007.org /index.php (accessed August 30, 2008).

Texas Commission on Environmental Quality. 2005. Environmental quality. http://www.tceq.state.tx.us/nav/eq/ (accessed September 10, 2007).

Texas State Historical Association. 2001. Texas Revolution. *The handbook of Texas online*. http://www.tshaonline.org/handbook/online/articles/TT/qdt1.html (accessed September 21, 2007).

———. 2003. Plano, Texas. *The handbook of Texas online*. http://www.tsha.utexas.edu/handbook/online/articles/PP/hdp4.html (accessed December 20, 2007).

———. 2008a. Dallas, Texas. *The handbook of Texas online*. http://www.tshaonline.org/handbook/online/articles/DD/hdd1.html (accessed March 30, 2008).

———. 2008b. Fort Worth Stockyards. *The handbook of Texas online*. http://www.tshaonline.org/handbook/online/articles/FF/dif4.html (accessed March 30, 2008).

Texas Water Development Board. 1997. *Water for Texas: A consensus-based update to the state water plan.* Austin: Texas Water Development Board.

U.S. Bureau of the Census. 2007. State and county quick facts. http://quickfacts.census.gov/qfd/states/48000lk.html (accessed September 13, 2007).

Witt, Howard. 2007. Overheated planet seeking solutions in heart of Texas, drumbeat for green: Austin attacks the problems of climate change—right at home. *Chicago Tribune*, September 28, 2007. http://www.ci.austin.tx.us/council/downloads/mw_planet_trib.pdf (accessed December 15, 2007).

Chapter 2

Calling All Artists

Moving Climate Change From My Space to My Place

Damon M. Hall, Leigh A. Bernacchi, Tema O. Milstein, and Tarla Rai Peterson

Each year, the Gallup Poll reports that more Americans claim to understand global climate change better than they did the previous year (Carroll 2007; Saad 2007). Climate-change-communication researchers Susanne Moser and Lisa Dilling (2004) have argued that while people understand climate change, they do not yet feel sufficient urgency to take meaningful action. They and others have suggested the enormous time lags in biophysical and social systems, the fact that developed countries and the power elites within them are relatively insulated from the effects of climate change, and the overshadowing of climate change by political and economic crises (Moser and Dilling 2004; Moser 2006; Begley 2007) have combined to minimize attention to this phenomenon. Bill McKibben also observed that global warming is a background issue in the United States, where "it hasn't registered in our gut; it isn't part of our culture." McKibben emphasized his point with a set of rhetorical questions: "Where are the books? The poems? The plays? The goddamn operas?...Where are the artists in this movement?" (2005, 1). Organizers for Step It Up (SIU) sought to fill this void, placing climate change on America's center stage by incorporating art.

Guided by the assumption that "images would count most," the national SIU organizers explicitly called for a visible national movement, inviting potential hosts to "gather your friends and take a picture" (McKibben et al. 2007, 101). Over 1,400 group photographs captured local SIU events on April 14, 2007. Local organizers uploaded these photographs to the SIU home page, and national organizers assembled a running slideshow (see figure 9 in chapter 8). In addition to being seen, organizers also implicitly called for a community-based movement by encouraging local organizers to interpret the SIU policy message of cutting national emissions of CO_2 "80 percent by 2050" in whatever manner they deemed most appropriate (Barringer 2007). In this chapter, we examine artistic performances that went beyond taking a picture in Austin, Texas; Salt Lake City, Utah; and Seattle, Washington. These performances illustrate how climate-change activists can use community art to reconfigure perceptions of climate change from a distant and ambiguous spatial phenomenon to a personalized and specific danger to individual places. These community-level artistic performances suggest ways to enable tangible experience with climate change as well as other issues that present the challenge of scale mismatch. After explaining what we mean by the term *green art*, we describe each of the performances listed above, pointing out how they interact with both the local and national scene to make climate change more visible and visceral. We conclude with suggestions for how environmental organizers and other activists may use green art to achieve their goals.

WHAT MAKES ART GREEN?

John Delicath (2004) argued that cultural activism and other informal contexts for citizen protest open public-participation opportunities to those who are not comfortable in more formally constituted settings. We use the term *green art* to specify cultural activism that is not only environmentally themed, but also political, activist-oriented, and community-based. These are not mutually exclusive categories, but labels whose colors bleed into each other (Scher 2007). Our concept of political art expands on David Orr's claim that "poetry and politics are both matters of verbal persuasion" (2008, 409). In one sense, poetry, painting, sculpture, dance, and other art forms are performances, and performance implies an audience. The attempt to persuade an audience, however, is not sufficient to designate an artistic performance as political. Orr describes "political poetry" as being

> concerned with a specific political situation; rooted in an identifiable political philosophy; addressing a particular political actor; written in language that can be understood and appreciated by its intended audience; and finally, offered in a public forum where it can have maximum persuasive effect. (2008, 415)

Although art may be political without necessarily exhibiting all of these characteristics, art that does exhibit these characteristics is unquestionably political.

Green art also is oriented toward activism. Douglas described activist art as "critical intervention that is uncontained by fixed categories and insists on foregrounding power and dominance relations…[It expresses] a critical attitude toward institutions and structures" (2006, 19). For example, although Arnold Friberg's popular painting of George Washington titled *Prayer at Valley Forge* is clearly political, it is not activist art because it does not challenge the status quo. Rather than celebrating the current political regime, activist art sends a message that subverts the regime and its ideology. Activist art with an ecological theme can attempt to directly influence environmental actions and policy, or it can take an indirect route to change. When McKibben and other national organizers invited artists to participate in SIU, they guided artistic endeavors toward indirect rather than direct action. The event "cookbook" (see chapter 8) urged organizers to "provide art supplies and invite your participants to create art—maybe something that speaks of the importance of your iconic place" (Step It Up 2007).

Perhaps one reason for the indirect approach was a desire to present SIU as a safe venue in multiple communities. Because direct action often involves confrontation, it alienates many who do not see themselves as extreme (see, for example, chapter 10). Artistic performances for SIU in Austin, Salt Lake City, and Seattle all exemplified indirect action. Time's Up, a New York City-based group, on the other hand, illustrated direct action when members participated in the Bike Lane Liberation Ride. Dressed in brightly colored clown costumes, they rode through Manhattan bike lanes "searching out motor vehicles illegally parked in their lanes" (Time's Up 2007) and issued them parking tickets. Although community-based art may include both direct and indirect action, most SIU events appealed to people who indicated greater comfort with indirect action. The important point is not whether the political action is direct or indirect but that community-based art must begin where the community is; it must grow out of community needs and desires (Siskin 2006).

Community-arts practice is more about the creative process than it is about producing an object. Although both the technical expertise and aesthetic sensibilities of professional artists are important dimensions, they do not form the core of this artistic genre. Community-based art projects are "collective processes…with artists acting more as facilitators of community members' creativity" (Scher 2007, 5). Although they may be initiated by any individual or group, community-based projects must be participatory and inclusive. For example, a group of interested community members jointly created the ceremony for the Salt Lake City water ritual and then offered participants the opportunity to further develop the ceremony. The critical reflection generated by encounters with community-based art offers the potential for expanding the original community of concerned citizens and galvanizing previously unconcerned participants.

Community-based art performances and installations provide shared spaces where people can gather. Those material spaces provide focal points that enable community members to imagine themselves as part of (rather than apart from) the group that engages with the art, including those whose engagement with that space occurs at different times (Forsbert et al. 2006). By producing and displaying community-based art, professional and amateur artists can focus attention on the most important issues we face, especially those, such as climate change, that are ordinarily invisible or difficult to visualize. We now examine three examples of SIU green art that enabled attendees to become cocreators of simulated encounters with climate change.

Seattle: Envisioning Rising Sea Levels at a Local Beach

Waves of blue ribbon fluttered gently in the ocean breeze. Far above the ribbon, strips of yellow caution tape flapped wildly. On a Seattle-area beach, forty individuals marked a new shoreline with bamboo poles, ribbon, and caution tape, creating a disconcerting line well above the heads of morning beachgoers. Members of Sustainable Ballard met on April 14 to make a picture, an evocative image of change, of danger, of foreboding arrest (see chapter 9). They used their bodies to provide an opportunity for every beachgoer to experience a personal encounter with the seemingly far-off specter of climate change. One of the local organizers explained,

> People need to know that global warming isn't a theoretical construct; it is actually happening now. The blue tape is one meter. A meter is what the scientists are predicting we'll get everywhere if present trends continue everywhere. It won't happen tomorrow. It's real gradual. That's one of the problems with climate change. You don't see it...We have a couple bamboo poles that are really high that have caution tape on them. That's halfway to catastrophe. That's what Al Gore was talking about if everything melted, the Greenland Ice Shelf, the Antarctic Ice Cap. So that's

> only the halfway mark…we're talking about twenty-three feet. So we absolutely can't let that happen.

Sustainable Ballard acted on the assumption that when people can imagine how something may endanger us and our special places, we will work to fix it. Rather than presenting a reasoned argument about climate change, they invited fellow Seattleites to imagine what they risk losing if they do not act.

Participants in Sustainable Ballard's SIU action used a participatory, embodied, community-art performance to call attention to the ramifications of global warming for sea-level rise in general and particularly for the Golden Gardens Beach. The line that participants formed ran roughly parallel with the waterline at high tide. Because the yellow caution tape

FIGURE 1. Sustainable Ballard's community art.

attached to the group's bamboo poles flapped high above everyone's heads, passersby were obliged to crane their necks sharply to identify the elusive sight. While the blue ribbon connecting the entire line of poles sat at a comfortable waist level, the garish caution tape bedecking individual poles was much higher. One participant holding a bamboo pole explained the performance as an attempt to concretize our ephemeral experience with climate change:

> How do you get people concerned about this completely seemingly nonexistent problem? A problem people can't see or witness. It's almost like people need to see it to make something happen. Like, you have kids getting killed with guns, you get gun legislation. You have to get people thinking about it.

The visual image of the altered high tide provided a simultaneously dramatic and nonthreatening (at least in the immediate sense) opportunity for local residents, tourists, congressmembers, and other decision makers to imagine the effects of climate change on their everyday lives. It played on personal self-interest or productive narcissism, suggesting that climate change matters in each of our lives (to our families, the things we do, the places we love, etc.). Everyone (activist or not) who went to the beach that morning encountered this representation.

For another participant, the performance was an opportunity to offer an alternative to the dominant political discourse surrounding climate change. After arriving on his bike, flying an earth flag, he explained his frustrations:

> It's "climate crisis," not "global warming" or "climate change." It's "climate crisis." This term is meaningful because it invokes what it really is. "Global warming" is inaccurate, as some places will freeze over—and some people think "warming" is nice. And "climate change" is ambiguous...It doesn't help communicate the issue at hand. "Climate crisis" does...We're using the language brought to us by the U.S. Chamber of Commerce, the Cato Institute, as well as the oil-company-funded industries and organizations. Those groups have dictated the language used to the corporate media, who have—without thinking—just parroted the language.

For this participant, simulating the loss of his beach was one way to overcome the strategic ambiguity that results from political vocabularies designed to protect elite interests. The message was clear: climate change will destroy this beach.

Salt Lake City: Gathering the Waters in a Desert

The fourteenth of April, 2007, was a "bluebird" spring day in Salt Lake City. A light breeze rustled the giant banner that spread across picnic tables, and feeding birds rippled the surface of the human-made pond at Sugarhouse Park. Twelve students of nature writer Terry Tempest Williams created an SIU ritual to celebrate the LifePlace of Salt Lake City, the Great Basin, the Great Salt Lake, the Oquirrh Mountains, the shores of Pleistocene-era Lake Bonneville, and the snow-capped peaks of the Wasatch mountain range (Thayer 2003).

Seven watersheds, familiarly referred to as canyons, flow into the Salt Lake Valley. The streams that carved these canyons rely on annual snowpack, which is also a mainstay of the economy and culture for a state whose license plates proclaim, "Utah; the greatest snow on earth." Organizers and performers focused on the hydrologic cycle and how residents value the water. Collectively, they created four touchstones for their SIU event: (1) gather the waters of the world representatively, (2) engage a group of local residents who would promise to protect Earth's waters, (3) provide information to the attendees on actions they can take, and (4) gather signatures and public presence to influence Congress and Utah's governor to seek policy solutions for global climate change. Each artist collected water from a different watershed so that all would be represented. All attendees were invited to bring water from their communities. Cups were provided so that those who did not bring water with them could carry water from the stream running through the ceremonial site.

On the day of the performance, informational posters adorned the adobe firepits and the wooden struts of the public gazebo. As people arrived, they were greeted by an artist, invited to sign the banner asking the governor and Congress to cut carbon 80 percent by 2050, and directed to the posters and large kettles where the ceremony would be performed. Printed programs with a photograph of water and a short refrain on the cover were dispersed among attendees. A local environmental activist and graduate student then announced that the ceremony was about to begin. He called on everyone to form a circle and join in gathering

FIGURE 2. Ceremony kettles.

the waters. The resulting circle of participants was about twenty feet in diameter and encompassed one of the kettles, leaving the other on the outside.

The speaker stood on the circumference of the ragged circle, slightly apart from others, to deliver a prepared statement written by two of the organizing artists. It opened,

> Climate change is an increasingly serious and omnipresent issue, as it has the capacity to affect all life on earth. One of the major ways is that it will change patterns and quantities of precipitation. While no one is certain how the Wasatch Range will be affected, as floods or droughts are equally as likely and equally as much a part of the water cycle over time, we believe that whatever human impact could lead to such change should be recognized, studied, and decided upon, based on a multiplicity of factors (economic, ethical, ecological). We are emotional beings. We are moral beings. We are part of this place.

The audience of approximately forty white, middle-class participants was attentive as he delivered a lyrical description of the water cycle: "the winds, the disparate heating and cooling of the earth's atmosphere, push and pull, sometimes dragging up and out these great collections of water." He used familiar objects and relationships to relate water to climate change and provide something people could identify with, noting that "only the slightest change in temperature determines the shape of a cloud. Only the slightest change in temperature determines where it will go." The text of his presentation was designed to encourage his audience to recognize the connection between water and global climate change.

A group leader set the tone by clearly stating what was meant to happen during the ceremony: "Let us take this time to revere the water source we have treated unkindly. Let us take this time to celebrate what a gathering of the waters can mean, for us and for our future." After a moment for reflection, the first of eight ceremonial cycles began. Each time, the speaker stated the name of the watershed or canyon, read a four- to eight-line poem about the canyon and the neighborhoods through

FIGURE 3. Circle gathered by the kettles.

which it flowed, and then invited people to join in the ceremony. Those who had brought water from that watershed were invited to pour it into the communal vessel in the center, in response to the phrase, "Let us gather the waters of _____ Canyon." As they poured from plastic water bottles or crouched down with their pottery cups, other attendees recited text printed on the program: "These waters remind us of what is possible and what is important. As waters change, so can we; bodies of water, we flow." Then the leader would announce the next canyon. The ritual continued in this fashion until all seven watersheds of Salt Lake City had been gathered. Gradually, a rhythm settled into the group. Some participants were visibly moved to tears, and everyone contributed water to the central kettle.

Eventually, the group had gathered all of the streams that would have at some point flowed into the Great Salt Lake (or Pleistocene Lake Bonneville), where the water would have resided in brine or evaporated, bonding with sandy particulates and coalescing into clouds that would rise and catch the mountains' edges again. One participant contributed the following sample of local knowledge and place-based description:

> A swirl, spirals of salts and larvae waves.
> Brine shrimp below a sky flat surface.
> Breeding birds, flocking black, rising out of small white caps, to arc and glint in the sun.
> Seagulls bowing, pull their necks in and run, sprint at the flies, snapping devouring.
> Reeds and marshes by Bear River trace a saline edge green. A highway edges the lake and carries the migratory people back to their suburban islands.
> Islands: Antelope, Stansbury, Gunnison.
> Quiet, still, a resting place. Rooting shallow in the desert, deep with salt, Remembering past shores: Lahontan, Bonneville, Great Salt.
> Let us gather the waters of the body that gathers our local waters, Great Salt Lake.
>
> —Catherine Ashton and Leigh Bernacchi

Finally, to invoke a relationship with the global system, participants were invited to contribute water from beyond the Salt Lake Valley. One participant added metaphorical water from a recent trip to Rwanda, while others dripped precious water from tiny streams in southern Utah's desert into the kettle. Participants then repeated one last time, "Let us gather the earth's waters."

Then the leader guided participants into the second segment of the ceremony. One member of the organizing team had started a fire in the kettle that sat outside the circle. As participants turned toward the fire, the speaker described fire's ability to forge something new and strong, and water's ability to quench a fire. He then likened the fire to global climate change. He encouraged participants to consider ways to use the water they had gathered, suggesting, "Let us take this water and make a promise to this place, this planet. We will be bound by this combination of fire and of water: clouds. A pledge to the cycle." At the conclusion of this presentation, participants responded to an invitation to read a pledge from their programs: "We promise to respect this water's source. We promise to respect this water's path." For the final segment of the pledge, participants were encouraged to fill in their own individual action: "I promise to…" Each person silently rehearsed his or her promised action as he or she walked to the water kettle in the center of the circle, filled a vessel or cupped hands, and then carried water to the fire kettle. As the attendees poured water on the fire and observed the sizzling embers, the leader read a closing paragraph, and the ritual was complete.

The event's solemnity was somewhat marred when the gradually dispersing group was hurriedly called back to perform a requirement of SIU: gather your friends and take a photo. The group took pictures with everyone holding the banner, then some with just the organizers. The signed banner was never delivered to any political office, but photos of the event were posted to the SIU Web site later that day.

Despite the failure to deliver the banner, both those who created the ritual and those who participated were attempting to send a political message to their governor and to Congress. Perhaps more importantly, they experienced the immediate presence of climate change as they listened to the ritual's narrative, trickled water into a communal kettle, and reflected upon potential actions they might commit to. All of them brought to the ritual an awareness that, especially in the desert, water is life. Participation in the ritual enabled them to sharpen that awareness and link global climate change directly to their local water supply.

Austin: Taking Out the Trash

In Austin, Texas, an art installation shared the stage with several SIU speakers. From afar, the strange-looking metal sculpture resembled a giant blue-green pineapple. Closer observation revealed four metal bars affixed to a metal circular base weighed down by sandbags and rising to a pinnacle eight feet tall. Upon this point rested a swirling, propeller-like top made of four six-foot-long curved metal blades. Zip-tied and knotted to all points of the frame were teal-colored plastic grocery bags that sounded like flags flapping in the stiff Texas wind. The sculpture was a twelve-foot-high mass of moving blue-green plastic.

During a speech, the director of Austin Green Art explained the installation to the Austin SIU audience. He told the crowd of his attendance at a modern-art show in Pittsburgh, Pennsylvania, where McKibben had asked, "Where are the artists in this movement?" He responded to McKibben's question via e-mail and received immediate instructions about how to get involved with SIU. He collaborated with Austin artist Virginia Fleck to tell the story of a situation that is directly related to anthropogenic causes of climate change: the North Pacific Gyre (NPG). The phenomenon was first documented by Don Walsh, who wrote, "Since the oceans are 'downhill' from virtually all mankind's activities, the sea is the final resting place for tens of millions of tons of lost and dumped plastics" (2005, 82). The NPG is a direct result of a global economy that runs on fossil fuels, which, in turn, contribute directly to increased rates of climate change. The ocean currents pull plastic trash (a petroleum product) from around the Pacific Rim into a space that is one thousand miles in diameter and ten feet deep with plastic trash. Some of the inert and long-lived plastics are broken down into tiny plastic fibers and granules that become suspended in the upper one hundred feet (Thompson et al. 2004). A Greenpeace study of the Texas-sized site reported that more than 267 species of marine animals—including turtles, seals, sea lions, whales, fish, and 110 species of seabirds—have been entangled in or have ingested debris from the NPG (Allsopp et al. 2006; Kay 2006).

After reading about this repository of plastic consumer waste, the director of Austin Green Art said Fleck "started immediately making art out of plastic bags." In a personal interview, he described experiencing her art:

> At first, when you see it from a distance, you don't know what the material is. You just see this beautiful swirling thing, and then you get up close, and it is like "Whoah?! It's shopping bags!" And she'll have a little thing beside it that says, "Don't use shopping bags!" She feels like this a problem that has a solution.

Figure 4. View from afar of the Austin SIU art installation.

Austin Green Art used this installation to encourage audience members to move from curiosity to interest, repulsion, and finally, to action.

Fleck described the installation as a metaphor for the kind of thinking and visualizing necessary to respond to impossibly large global problems such as climate change. Like climate change, the accumulation of plastic in the NPG results from apparently invisible and gradual accumulation of waste from the fossil-fuel-based global economy. Like Earth's atmosphere, the ability of subtropical gyres to absorb and recycle this waste has been taken for granted and seen as limitless (Moore 2003). Consumed plastics are so common that people rarely think about their impacts. The sculpture deconstructed this *unseen* environmental phenomenon and reassembled it as overlaid upon the audience's everyday *seen* experiences with plastic grocery bags, the porters of consumerism.

Speaking from behind the podium, Austin Green Art's director told of countries that have adopted policy solutions, such as Bangladesh, where plastic bags are outlawed, and Ireland, which instituted a tax on bags and cut its bag use by 90 percent in one year. He concluded by encouraging attendees to take a teal plastic bag from the barrel and tie it onto the frame of the sculpture, remember the NPG, and commit to using alternatives to plastic shopping bags. He bluntly demanded that his audience "quit using shopping bags!"

FIGURE 5. Austin SIU participants tying on bags.

After all the speakers had finished, people lined up to tie bags onto the sculpture. Participants had to reach into a small barrel, pull out a bag, and tie it onto the sculpture. It was impossible to pull the bags out of the barrel without getting your hands sticky with wet paint. When asked about this, the artistic director explained that "the experience of pulling the bags from the barrel is meant to be nasty." By accepting his invitation to add to the sculpture, SIU attendees became artists, allies, collaborators, and advocates. As audience members solemnly tied bags to the frame, they relived the request to make a personal commitment to be conscious of this swirling blue-green place of plastic trash in the Pacific Ocean. By providing a tangible encounter with the vast and interrelated issues of global climate change and plastic waste, the sculpture enabled participants to close mental gaps between individual actions and their potential impacts upon Earth's biophysical systems.

All three of these performances demonstrated how green art may affect participants and communicate climate change in a unique way. All three presented climate change as a complex biophysical phenomenon that was also a discrete political issue. In the Seattle case, the artists took their performance to the larger community, presenting all beachgoers with a startling yet nonconfrontational image. Community artists in Salt Lake City and Austin performed for a more narrowly defined audience of SIU attendees. They took advantage of their venue by providing opportunities for audience members to become part of the artistic performance, whether by pouring water onto hot coals or by sticking their hands into masses of gooey plastic bags.

Practical Suggestions

In the SIU organizers' postevent handbook, *Fight Global Warming Now*, McKibben et al. spoke of creative visual expressions of climate change as a "low-cost high impact solution" that nourishes citizen understanding "in a deep and visceral way" (2007, 92). Based on the three artistic performances we analyzed, we offer the following suggestions for national and local organizers who want to use green art to encourage climate-change activism.

1. *Green art should be community-based; it should appear simple, sincere, collaborative, fun, and nonthreatening.* Community-based art differs from mass-produced national nonprofit activist messages, professionally designed high-gloss advertisements, and high-art exhibits. The construction of community-based art is made with everyday materials like ribbon, plastic bags, bottles, and cups of water. The homemade character and apparent low cost help potential participants envision themselves as cocreators.

This does not mean that professional artists should not participate. Organizers should encourage collaboration among diverse and creative groups of people. Organizers can suggest ways that professional artists might incorporate local images, events, or issues that demonstrate the connections between seemingly abstract and ephemeral biophysical processes. They also can challenge all participants to find new uses for everyday or readily available materials in image design. For example, Austin Green Art collaborated with community volunteers to create "Cup City," a sixty-five-by-eighty-five-by-fifteen-foot installation of over 30,000 recycled coffee cups, aluminum cans, and plastic bottles (Sayer 2005). The creation process as well as the display of the artistic image should be sincere and fun, for the collaborative process of producing community-based art is at least as important as the end product. Although Cup City resulted in a product (and was commissioned by Starbucks), the more important aspects of this effort were community interactivity, collaboration, and fun.

2. *Artistic endeavors should create a face for the problem, giving audience members a recognizable focal point.* This imaginary meeting acts as a starting point for conversations that are less easily dismissed than political slogans. Only if artistic performances or installations can give the problem scenario a face can the audience have a face-to-face encounter with the problem. Through these encounters, participants locate themselves in relation to the issue presented. Identification with the need to address the problem transcends the discourse of culpability and political problem solving, and the beholders face the immediacy of ecological change as it confronts their places, their homes, their families,

their neighbors, and their daily lives. In other words, art should foster meetings between the distant global and the immediate local.

Organizers can help give a face to the problem by encouraging the production of dramatic and ridiculous-looking artistic images that excite curiosity and wonder, such as the Austin installation. Organizers can prompt these artistic endeavors to emerge locally by asking challenging questions of local participants. For example, what does global climate change look and feel like where you live? How do you see climate change at home, in your town, or in your favorite place? What is the highest hurdle people must get over to see, smell, hear, touch, and taste global climate change? Organizers could call for visual, visible, tactile, audible, and other sensual experiences of the problem (or solution), with care taken to allow for artistic freedom and local nuance.

3. *Organizers should encourage designers to subject their creations to some kind of art criticism.* By art criticism, we mean criticism that focuses on aesthetics rather than politics. This encourages a thoughtful and appreciative response and enables community activists to differentiate artistic performances and installations from explicitly didactic arguments. Cup City, for example, was submitted to the Texas Society of Architects and received a state-level architecture award in 2006. As a result, the ideas it embodied became part of the conversation in multiple forums, extending beyond those traditionally focused on environmental topics. Artistic expression may be less easily dismissed than campaign slogans, declarative signage, and political platforms because it does not prescribe a structure for action but names a reality that connects to the human condition.

Because sloganeering alienates some audience members, organizers may even consider challenging designers to limit usage of overtly politicized terms, letting the agnosticism of the expressive artistic form lead. This new relationship to an issue constitutes a relatively safe space where individuals may begin to view themselves as part of a shared Earth. Artists and other educators may respond more openly to options for reclaiming the ocean around the Pacific Gyre, for example, if they read about it in the context of an award-winning art installation, the *Laguna Gyra* (Rockett 2006).

4. *Community-based art should fit the local setting and audiences.* Organizers should remind designers to consider both external (publics) and internal (planners, designers, and performers) audiences. While the Seattle shoreline display may have led to few public interactions, the forty participants who formed this shoreline experienced their own power as they stood ten feet apart on the beach, braved the cold, clutched their poles, and gazed out into the ocean for that chilly morning hour. The Salt Lake City students who created a ritual for collecting the waters from their valley experienced a new level of connection with their special place. Both of these performances may have been more significant for the performers than for an external audience. By cocreating a frightening new shoreline or measuring out water in the desert, these participants experienced the material presence of climate change.

Timing, location, and normative activities of the setting are also important considerations. For example, walking, running, or strolling along the beach is an intimate and contemplative practice, one that fits seamlessly with the opportunity to view the new sea level represented by the blue ribbon rippling between bamboo poles. Sustainable Ballard was unlikely to have offended or disrupted beachcombers. At the same time, the chilly April weather and early morning hour yielded few beachgoers to see this representation. Austinites, who are accustomed to viewing themselves as an avant-garde outpost in conservative Texas, accepted the plastic-bag sculpture as consistent with their city's sensibilities. All three performances focused attention on climate change in ways that were relatively comfortable for local attendees.

Of course, community-based art sometimes takes a more confrontational tone. Since all three of these examples were part of the SIU campaign, organizers designed performances that presented climate activism as a reasonable, safe, and relatively mainstream activity. While organizers should remind designers to honor the local community, the discourse of community should not be limited to dominant political discourse.

5. *Organizers should use artistic performances and installations to suggest possibilities for rearranging status quo understandings of political and scientific complexities.* Although the goal is change, change usually is

incremental, and it always begins in the present. Audiences seek a familiar point of connection, a way to reorganize familiar patterns into new ways of thinking or feeling. Artistic performances that begin from a point of familiarity can disrupt habitual attitudes towards projected climate-change impacts. Showing projected sea-level rise and the related changes to the look and feel of a familiar place challenges the claim that climate change is unimportant to residents of the United States. Gathering the waters of a desert valley leads to questions about just how close to the edge our human settlements are. Participants are led to imagine how the slightest increase in temperature may make their homes uninhabitable. Learning about a Texas-sized plastic-trash vortex in the Pacific Ocean interrupts the comfortable assurance that Wal-Mart is always available to recycle the plastic bags.

Imagining the local relevance of global ecological impacts empowers individuals and communities to creatively envision alternative futures and relations with the natural world that differ from dominantly held sensibilities. Organizers should invite interpretations that disrupt long-held and habitual beliefs. They can encourage designers to re-vision dominant forms of thought that prevent publics from envisioning, understanding, and experiencing the impacts of global climate change.

6. *Organizers should remember that artistic performances are multiple and open-ended. Art invites multiple interpretations and defies standardization.* These interpretations depend on contextual elements, the event itself, and the manner in which a performance is staged or exhibited. Organizers should embrace the impossibility of controlling possible interpretations, because such control would severely limit the creative power of art. At the same time, they should exert due diligence in encouraging interpretations consistent with their advocacy goals. To do this, they should surround displays, installations, and performances with appropriate supporting persons and other resources.

At the same time, they should be leery about trying to create a master narrative for the sake of a uniform artistic sensibility. Although the standardized campaign formula often has proven successful, its application to artistic performance is less well documented. SIU offers new ways of *doing* social-movement organizing (see chapter 5 and chapter 11). Rather than following the corporate model used by most nonprofit organizations (complete with magazines, mailers, meetings, and organizational charts), it formed a network of active citizens and friends. Rather than maintaining itself in a static organizational format, the network musters when called or otherwise moved. Each node has local concerns, local problems, local groups, and local goals.

SUMMARY AND CONCLUSIONS

One strength of the SIU network is its diversity of voices, faces, and talents. Although the specific art performances and installations that emerged from the many SIU events were not predictable, their diversity was. As an introduction to climate change, green art can constitute a starting point for thinking about how individuals see themselves as a part of the problem, the solution, and ultimately, as part of a shared Earth. These artistic performances were more than slogans that ended the conversation with a command to do or say something; they began conversations that continue to spin out in multiple directions, and where every new message has the potential to be an equally creative response rooted in relationships to Earth and other beings. What remains of critical importance is that the artistic performances enable an embodied encounter with something beyond the self.

The genius of the green-art pieces at SIU was their ability to personally engage audiences and participants with complex multiscaled human-environmental phenomena like climate change. They did this by providing a *local face* for the problem. W. J. T. Mitchell argued that images "present, not just surface, but a *face* that faces the beholder" (1996, 72). Any sense making or meaning making derived from an image occurs by audience members placing themselves in relation to the image—literally searching its face. From meeting the local face of global problems, a more personalized understanding may be realized. By inviting a personal encounter, green art makes abstract problems tangible, visceral, and specific to local places, neighbors, and selves.

The green art we observed did not attempt to ignore or devalue existing activities, objects, and relationships within our highly technologized society. Members of Sustainable Ballard used recognizable caution tape to signify rising sea levels. Salt Lake City residents staged their ritual of gathering the desert's water on the shores of a human-made pond in a city park maintained by a sophisticated irrigation system. The dramatic form of the plastic-shopping-bag sculpture relied on a framework of steel bars. Instead, each of these performances gave audience members an opportunity to "know the textures, rhythms and tastes of the bodily world, and to distinguish readily between such tastes and those of our own invention" (Abram 1996, x). Sustainable Ballard participants shivered through an hour's vigil on the cold, windy beach. Those who gathered the waters in Salt Lake City felt the moisture on their faces and listened to the sizzle of water disappearing into the atmosphere. The audience flinched and grimaced when they encountered the gooey residue on the plastic bags they attached to her sculpture. These green-art performances made otherwise ephemeral connections between local communities and global climate change more tangible at the same time as they encouraged audiences to register uniquely evocative meanings.

One powerful dimension of these examples of green art was that people experienced them directly with their bodies rather than as electronically mediated experiences. Audiences had the opportunity for personal experience, even becoming creative producers. Although green artists were pleased to receive mainstream-news coverage and to post photos of their performances on the SIU Web site, the face-to-face encounter was a more central objective. The more fully audience members interacted with the art, the greater the opportunity for understanding. These green-art installations were designed to be thoughtfully pondered, read, touched, criticized, taken in, and experienced.

Within the cacophony of environmental discourses and argumentation, green art offers an important option for constructing and voicing alternative meanings from the dominant political-economic voices. The artistic expressions encountered at SIU do not suggest that we apply a more accurate renaming or any ideal frame to communicate climate change. Instead, these examples of green art suggest ways to communicate climate change so that persons can locate themselves in relation to the problem, so they look into the face of climate change and sense its impacts on everyday practices, places, and people. Whether an individual interpretation resists or resembles the artists' interpretation is less important than the experience of seeing how climate change places one's object of desire (e.g., self, local beach, clean water, good health, fresh air) in jeopardy. Green art fosters a meeting between the local self and the global problem. This intimate encounter enables powerful imagination of the local impacts of global phenomena and positions the local self in relation to the global problem.

References

Abram, David. 1996. *The spell of the sensuous*. New York: Vintage.

Allsopp, Michelle, Adam Walters, David Santillo, and Paul Johnston. 2006. Plastic debris in the world's oceans. United Nations Environment Programme. http://www.unep.org/regionalseas/marinelitter/publications/docs/plastic_ocean_report.pdf (accessed September 30, 2008).

Barringer, Felicity. 2007. Renewing a call to act against climate change. *The New York Times*, March 14, 2007.

Begley, Sharon. 2007. The truth about denial. *Newsweek*, August 13, 2007.

Carroll, Joseph. 2007. Americans assess what they can do to reduce global warming. *Gallup News Service*, April 24, 2007.

Delicath, John W. 2004. Art and advocacy: Citizen participation through cultural activism. In *Communication and public participation in environmental decision making*, ed. Stephen P. DePoe, John W. Delicath, and Marie-France Aepli Elsenbeer, 255–266. Albany: State University of New York Press.

Douglas, Susan. 2006. Art + Activism = ATSA. *Women & Environments International Magazine* 72/73:10–12.

Forsbert, Liz, Leah Burns, Clare O'Connor, Laura Reinsborough, and Jill Tomac. 2006. Transforming space into place. *Women & Environments International Magazine* 72/73:44–46.

Kay, Jane. 2006. The plastic garbage pit of the Pacific. *San Francisco Chronicle*, November 6, 2006.

McKibben, Bill. 2005. Imagine that: What the warming world needs now is art sweet art. *Grist Magazine*. http://www.grist.org/comments/soapbox/2005/04/21/mckibben-imagine/ (accessed September 30, 2008).

McKibben, Bill, Phil Aroneanu, Will Bates, May Boeve, Jamie Henn, Jeremy Osborn, and Jon Warnow. 2007. *Fight global warming now: The handbook for taking action in your community*. New York: Holt.

Mitchell, W. J. T. 1996. What do pictures really want? *October* 77:71–82.

Moore, Charles. 2003. Trashed: Across the Pacific Ocean, plastics, plastics, everywhere. *Natural History* 112:46–51.

Moser, Susanne C. 2006. Talk of the city: Engaging urbanites on climate change. *Environmental Research Letters* 1:1–10.

Moser, Susanne C., and Lisa Dilling. 2004. Making climate hot: Communicating the urgency and the challenge of global climate change. *Environment* 46:32–46.

Orr, David. 2008. The politics of poetry. *Poetry* 192:408–418.

Rockett, Benne. 2006. Green Wave: Austin Green Art. *Voices of Art Magazine* 13 (2). http://www.voamag.com/index.php?option=com_content&task=view&id=71&Itemid=9 (accessed August 30, 2008).

Saad, Lydia. 2007. Did Hollywood's glare heat up public concern about global warming? *Gallup News Service*, March 21, 2007.

Sayer, Kyeann. 2005. Austin Green Art Temporary Recycled Artwork Installation. Treehugger. http://www.treehugger.com/files/2005/10/austin_green_ar.php (accessed August 30, 2008).

Scher, Abbey. 2007. Can the arts change the world? The transformative power of community arts. *New Directions for Adult and Continuing Education* 116:3–11.

Siskin, Sharon. 2006. Drawing the circle: From the community to the studio to the classroom and back again. *Women & Environments International Magazine* 72/73:6–9.

Step It Up. 2007. Action ideas. http://april.stepitup2007.org /article.php?id=70 (accessed August 30, 2008).

Thayer, Robert L. 2003. *LifePlace: Bioregional thought and practice*. Berkeley: University of California Press.

Thompson, Richard C., Ylva Olsen, Richard P. Mitchell, Anthony Davis, Steven J. Rowland, Anthony W. G. John, Daniel McGonigle, and Andrea E. Russell. 2004. Lost at sea: Where is all the plastic? *Science* 304:838.

Time's Up. 2007. Bike lane liberation ride. http://times-up.org/index.php?page=bike-lane-liberation (accessed August 30, 2008).

Walsh, Don. 2005. The plastic ocean. *U.S. Naval Institute Proceedings* 131:82.

Chapter 3

Demonstrative Protest Rhetoric and the Boston Step It Up Campaign

Lawrence J. Prelli

On April 14, 2007, about four hundred people gathered at the Parkman Bandstand at Boston Common on a sunny Saturday afternoon to participate in the first National Day of Climate Action, urging Congress to "step it up" and pass legislation to cut carbon emissions by 80 percent by 2050 (Step It Up 2007). More than 1,300 similar actions occurred at localities across all fifty of the United States. Bill McKibben and six students from Middlebury College had planned the nationwide event, with McKibben billing it a "brand-new, mass-protest climate movement for 2007" (2007b). What made the purported climate movement new was the use of the Internet to organize, build, and maintain it (McKibben 2007a, 2007c; see also chapter 8). Still, McKibben and other organizers compared the "new" mass protest with the actions of protest movements during the 1960s, especially the venerated civil rights movement. Those comparisons spurred this study of the rhetoric of Step It Up (SIU). In this chapter, I examine how SIU operated as "demonstrative" protest rhetoric at the Common, and whether that rally and others like it generated the public perspective that a new, nationwide, climate-protest movement was emerging. In what follows, I articulate three qualities of demonstrative protest rhetoric, analyze the extent to which SIU enacted or failed to enact those qualities, and delineate recommendations for future actions.[1]

DEMONSTRATIVE PROTEST RHETORIC AND THE INVENTION OF A MOVEMENT

Richard McKeon, philosopher and historian of ideas, argued for a classically grounded conception of demonstrative rhetoric that could account for the protest demonstrations he witnessed during the 1960s and early 1970s (1971, 1968). For McKeon, "demonstration" applied as much to showing feelings and commitments during street protests as it did to the process of drawing valid conclusions in logic. For him, both were forms of "demonstrative rhetoric." One of the qualities that McKeon attributed to the "new" twentieth-century demonstrative rhetoric is that it consisted of discourse and action that disclosed previously unknown, muted, or concealed perspectives toward facts, meanings, relationships, or actions. For him, this was the *inventional* quality of demonstrative rhetoric. Though McKeon did not illustrate that inventional quality with a focused analysis of its embodied, situated practice, other scholars have since furnished examples that clarify this important aspect of demonstrative rhetoric.

DeLuca's (1999b) analysis of the "image events" of Earth First! activists illustrated the inventional capacity of demonstrative protest rhetoric.[2] To protest logging operations, one activist placed herself one hundred feet above the ground in the canopy of a Douglas fir, and another had himself buried up to his neck in a logging road. DeLuca explained that both actions embodied a perspective that challenged conventional understandings of humanity's place at the top of a hierarchical chain of value vis-à-vis other living things: "In dislodging the blinders of a human-centered worldview, the protesters *bring into being* an ecocentric perspective. In identifying with the tree and the earth, the protesters invite viewers to also identify with the natural world" (DeLuca 1999b, 13–14; emphasis added). The new perspective manifested through those bodily acts can disrupt more conventional understandings of nature as a repository of "resources" ready for "harvest" to meet human needs and wants. The audience holding that conventional view is reachable, provided that the protesters encapsulate their ecocentric perspective within bodily forms that are evocative enough to engage the news media and turn up as "image events" on nightly television news or in the daily newspaper (Delicath and DeLuca 2003; DeLuca 1999a).

SIU's primary problem of invention was how to bring a climate movement into being that would disclose previously unacknowledged or muted public support for urgently needed political action. As McKibben wryly put it, "We've heard the science, the economics, even the policy proposals. The only part of the movement we haven't had is the movement itself" (quoted in Donnelly 2007). Did participants at the Common rally make this emerging movement "present" for actual and potential audiences? I shall explore this question by examining, first, the speeches that formed the core event at the rally; second, what participants at the rally thought its most significant disclosures were; and third, how the Common rally and other rallies like it were related to the nationwide SIU action in the print and television media.[3]

Speakers depicted the Common rally as part of a wider, nationwide political-protest movement. The coordinator of the Boston Climate Network and organizer of the Common event said, "We represent a tidal wave of people who want our government to take action. And our political leaders are starting to listen." A Massachusetts congressman proclaimed, "This has the potential of being one of the largest mass movements ever created, not just in the history of the United States, but of the world." The associate director of the Harvard Medical School's

Center for Health and the Global Environment presumed the existence of a widespread movement when he said,

> Great movements come from great changes—the kind we're witnessing today in the climate system. And sweeping movements like us today can generate bold leaders. Movements like the one we're part of today emerge from a convergence of concern around the world. And a vision, a shared vision, is taking hold.

Most who attended the Common rally held attitudes consistent with the "movement" perspective that speakers evoked. Many expressed that the rally demonstrated widespread public concern and caring about the climate-change issue. One person said, "It's a question of showing that a whole lot of people really care." Another remarked that the rally brought people out "to show that they recognize it [climate change] as a problem." Yet another participant said the rally showed that "it's the beginning of people [being] aware of something that's very, very important." Still another commented, "It hopes to prove that there are a lot of people who care about this issue and are worried about the future."

Closely related to this theme of public concern and caring is the view that the nationwide SIU rallies demonstrated that action was urgently needed. As one participant said, "There's public support for doing more now!" Another commented, "At some point, we're gonna have to do something, so why not start now?" Still another person addressed the same theme, saying, "It is up to everybody to get involved, not just the politicians, but every person on the planet." Thus, a clear majority of those interviewed perceived the Common rally as demonstrative of widespread public concern and the need for immediate action, a perception consistent with seeing the Common rally and rallies elsewhere as exemplars of an emerging climate movement.

Did news reportage convey to readers and viewers the perception that SIU was an emerging social movement? At first glance, it seems that it did. Reportage of the Common rally illustrates the general pattern of newspaper coverage. *Boston Globe* articles incorporated the Common action as one among several examples of rallies held across the state, region, or nation. Before the event, newspaper coverage often included organizers' depictions of SIU actions as an emerging political movement. For instance, SIU was depicted "as the largest day of citizen action on global warming in the nation's history" (Daley 2007). After the event, coverage continued this pattern, adding quotes from selected people who spoke at the Common rally (Levenson 2007). Newspaper articles and television news coverage across the country conveyed similar generalizations (for examples, see Lindberg 2007 and ABC News 2007b). Many adduced numbers to reinforce the nationwide reach of the action. ABC's *Good Morning America* announced "hundreds of events" nationwide (ABC News 2007a), while NBC Nightly News reported that "thousands of Americans" were involved in more than 1,400 events across the country (NBC News 2007; also see Anon. 2007).

Despite generally positive coverage,[4] it is doubtful that many viewers and readers perceived SIU actions as a national protest of movement proportions. Questions about verbal generalizations quickly come to mind. For example, in response to the claim that SIU is the largest citizen action on global warming in the nation's history, one might ask, "Were there others?"[5] So, too, were cited numbers questionable. How many "hundreds" or "thousands" attended events across the country? Are there reliable estimates of nationwide turnout? Based on news coverage, readers and viewers could not readily see parallels with other well-known movements, contrary to organizers' depictions.

New York Times coverage of the lower Manhattan rally expressed such doubt by minimizing bold claims made about the action's movement proportions:

> The campaign, called Step It Up, was described by organizers as the largest environmental demonstration in the country since Earth Day 1970, a gathering that helped start the modern environmental movement. Yet the many Step It Up events seemed far smaller in comparison. Some drew a few hundred, others a few dozen. In Manhattan, organizers estimated a crowd of 1,200. (Fernandez 2007)[6]

Modest turnouts like 1,200 people at lower Manhattan and 400 people at Boston Common leave room for doubt about SIU's movement proportions. The many rallies did direct attention to potential impacts of global warming on significant local places, but did not sufficiently *manifest* the magnitude of support alleged or make the movement *visibly present* through conventional news sources. Put otherwise, the movement was not *brought into being* for many who were not already participants or like-minded supporters. For contrast, recall familiar images of Martin Luther King Jr.'s speech at the Lincoln Memorial in 1963, before about 250,000 people. Those images *embodied* the civil rights movement, made its existence *visibly present*, and did so indisputably. There are no comparable images of an SIU rally or group of rallies that could suitably "stand for" an emerging protest movement of political significance on global warming.[7]

In the absence of images or other compelling rhetorical displays, SIU did not hold media attention very long after "The Big One—One Earth, One Climate Rally" took place in Boston. Today, the SIU campaign is known primarily among its participants and supporters. When I ask politically engaged colleagues or informally poll students, I discover that few have even heard of SIU. Though admittedly anecdotal, I still cannot imagine similar results had I asked about the Million Man March or the Million Mom March one year after they occurred. The invention of a protest movement requires rhetorical demonstrations that manifest its "presence" before a wide audience in some strikingly tangible way. I shall return to this point in recommendations offered later in this chapter. Next, I examine how the embodied presence of SIU relates to the second quality of demonstrative protest rhetoric: enacting a compelling proof of power.

DEMONSTRATIVE PROTEST RHETORIC AND THE PROOF OF POWER

Invention of new or previously unknown or concealed perspectives on facts, meanings, relationships, or actions hinges on the capacity to "prove" or "demonstrate" their existence through situated discourse and action.

DeLuca identified (1999b) and Delicath and DeLuca detailed (2003) the "argumentative" qualities exhibited in embodied actions framed as "image events," and Hauser (2006, 234–235) identified the logic by which embodied demonstrative rhetoric operates. One central element of that logic is disclosed through comparison of protest demonstrations and logical demonstrations: both secure audience adherence to claims based on their consistency with "purportedly irrefutable premises" (Hauser 2006, 234). Put otherwise, the *force* of a demonstrative proof is experienced when one is compelled by the rules of logic to think from premises to conclusion according to a valid reasoning pattern. Similarly, the "force" of a protest demonstration is experienced whenever one feels compelled to discourse and act in ways that are consistent with the cause's presumptive moral and political convictions. We might add that much as the *soundness* of logical reasoning presumes that premises are certain, true, or otherwise acceptable, so too does the "soundness" of demonstrative protest rhetoric depend on identification with the cause's moral and political convictions as indisputable and, thus, as constituting the legitimating grounds of specific lines of discourse or courses of action.

Jerry Blitefield's study of Saul Alinsky's conflict-based approach to community organizing illustrated how protest actions can compel discourse and action consistent with its leading moral and political certainties. Alinsky's approach sought to reverse power distribution so that "the underpowered gain power while the powerful lose power, if only temporarily," a reversal that must be demonstrated through staged public actions (Blitefield 2006, 261). Blitefield discussed one organization's action against a bank accused of racially discriminatory practices in issuing mortgage loans (263–264). The new organization, Baltimoreans United in Leadership Development (BUILD), had sixty of its account-holding members go "conduct business" at the bank. These sixty "customers" were extraordinarily clumsy with their change, confused about numbers for their accounts, or confounded by some other seemingly routine detail. As the police looked on (since no laws were broken), they returned to the back of the line after their transactions were finally completed to await their next turns. The result? The bank's president agreed to hold a previously denied meeting with the group due to pressure from frustrated customers, bank officials, and tellers (270).[8] BUILD's action illustrates how the logic of protest demonstrations works to compel actions in accord with the cause's leading premises.

The BUILD example also illustrates that the logic of proving in demonstrative protest rhetoric is confrontational or even coercive, especially when applied to the words and deeds of those who are uncommitted, indecisive, or opposed to the cause.[9] After all, whenever the certainty or rightness of premises is taken for granted, those who adhere to them see no need to *argue with* an audience about alternative premises or even their applicability in many particular cases. Thus, disruptive actions against opponents are often seen as legitimate tactics of protest demonstrations. Similarly, the uncommitted or undecided must be told about or shown (or, to use the latter-day euphemisms, "educated" or "informed" about) communicative practices that manifest adherence to the cause's leading premises.

At the Boston Common, speakers relied on four supposedly inarguable presumptions as grounds for warranting appropriate lines of discourse and courses of action. Two were elucidated as unambiguous matters of scientifically certified fact: human-induced global warming is real, and human-induced global warming is a threat to life as we know it. A third was a moral imperative: we should act now in behalf of future generations. And, finally, there was the pragmatic presumption that effective change was probable if individuals and leaders would only take appropriate action. As expected of the logic of demonstrative protest rhetoric, those presumptions were not open to question since they presumably were the very grounds of legitimate proof. Persons who question them would readily be dismissed as misinformed, self-interested, or suspect in motive. Others who speak or act in ways inconsistent with them would be perceived, again, as requiring education, as irrelevant to the cause, or both. Unsurprisingly, few among those interviewed at the Common rally even indirectly exhibited skepticism about any of the four presumptions or the specific kinds of discourse or action that speakers implied they sanctioned.

I will focus primarily on the pragmatic presumption, given its centrality to SIU's core message: Congress should take action to reduce carbon emissions 80 percent by 2050. Speakers at the Common stressed the need for their audience to pressure Congress and other elected leaders to embrace that message. Some insisted that only continued pressure from the "grass roots" and the implied threat to political power would compel elected leaders to make the right policy decisions. A journalist and author illustrated:

> Stepping up and speaking out is the key. It's not enough to just turn down your own thermostat and carpool and change your lightbulb. You have to tell lots of other people, loudly and clearly, why you're doing it. And that way, you and the people you talk to will be giving our elected officials the support and courage they need to make the really big changes that nature is telling us to make.
>
> He later added,
>
> So thousands of people around the country gathered at SIU rallies today; they are sending a loud and clear message from the statehouse to the White House: that time has run out for industry lies and political cowardice. Our planet has run out of patience, and so has our sense of justice.

Most speakers evoked a more collaborative and less confrontational relationship between citizens and leaders—especially elected political leaders. A Massachusetts state senator illustrated:

> There will be those, when we try to move forward with legislation, who will try to stop us every step of the way. And that is why myself, and other legislators and city councillors and mayors and people in elected positions that are going to step up front and say, "We need to do something now"—we're gonna need you to stand up loud. And we're gonna need you to stand up and cheer. And we're gonna need you to be at those hearings and to talk to the other legislators, and to make sure that, when we come out with that comprehensive bill to stop global warming, to meet those goals, that you are going to be with us—that you are going to stand up tall so we have the votes.

One speaker's closing remarks captured the general collaborative tenor when he said, "So finally, I'm going to steal a few rosy words from a really brave New Englander from years ago, 'Ask not what your leaders can do for you, but ask what you and your new leaders can do together.'"

The pragmatic premise that warranted these generally positive appeals draws its rhetorical potency largely from the assumption that public support for change is sufficiently strong to influence or pressure politicians' actions. Many at the Common associated their presence at the rally with the public's presumed power to influence or pressure political leaders. For them, the rallies clearly were targeted at "political leaders," "leaders in government," "the politicians," and so on. As one man put it, "Just having all these people here means that the politicians can't ignore it." Another participant offered a more elaborate analysis:

> Well, I'd say the leaders in government. They're the ones who will hopefully make some changes in their policies, but since we live in a capitalist country, they're going to try and get whatever they can out of it, if they decide to go that way. And I'm hoping they'll be a little more humane about it.

Still another captured well the connection between public demonstrations and political action when saying, "Oh, the politicians. I mean, the people who are supposedly elected in our interest who do not do anything in our interest, but only in their self-interest...I mean, nothing can be done unless it's forced."

Did the rally enact a demonstrative proof of power sufficient to compel participants to act in accord with the pragmatic presumption? Though several respondents grasped that they needed to write their representatives and local leaders or engage in other political actions, a clear majority were unclear about what *they* were expected to do to exert political pressure on decision makers. A substantial minority of respondents stressed the need to educate or inform rather than muster political action as the aim of the rallies. From that vantage, they did not include themselves among the rally's targeted audiences; instead, the rally presumably disclosed facts or appropriate actions for the "unaware." Two groups most in need were "younger people" and a general category of people variously described as the "general public," "those not here," those who are "unaware of the problem," or even "everybody." Children followed these two categories as persons most in need of the "education" presumably fostered by the rally.

Many respondents left the next political actions in the hands of elected leaders (e.g., "I just hope the people, like the local politicians and the national politicians, actually do do something."). Others spoke in vague or confused generalities (e.g., "I don't know. We should do something about the environment. I don't know how that happens, but I guess through everything—politics and art and all that. Eventually, people will take action and do something about it."). One politically astute respondent perceived this lack of clarity and thought that the organizers needed to direct rally participants' subsequent actions:

> I think there should be a continuation of rallies like this, and communication between the people who are supporting the event. The organizers should contact me and others who have attended and inform us about other rallies and demonstrations and help us organize to not only make changes in our personal lives but also on a policy level at the statehouse and in our local communities, local governments, and in our schools. I think…a lot of people need help organizing and getting together.

Based on confusion about expected, specific actions, it is not surprising that only two hundred people attended the second SIU rally at the Cathedral Church of St. Paul in Boston on November 3, 2007 (Peters 2007). Organizers blamed the low turnout on bad weather, but one neophyte protester, whose protest experience was limited to the two Boston actions, had a less optimistic appraisal: "Sadly, we don't respond to a crisis until it hits us right between the eyes," he said. "I expect, ironically, as things get progressively worse, we'll get more people at events like this" (quoted in Peters 2007). Evidently, few at the Common grasped that they, as members of an emerging and significant political cause, were *expected* to turn out.

Nor did the Common rally register as a protest capable of enacting a proof of power that could, in turn, confront political decision makers or otherwise compel them to action. Some local and state politicians seemed already in step with the SIU campaign message before the rallies even were held. For example, the Boston mayor had expressed his administration's support of SIU earlier in the week, but it is clear that he had previously enacted a green agenda (City of Boston n.d.; also see Menino 2007). However, I did not find one arguable example of a local or state political decision maker who changed position on global-warming issues due to SIU efforts to exert pressure.

The Boston rally could very well be exemplary of the nationwide rally's failure to enact proof of power for congressional leaders, though SIU did successfully inject the highest capping standard of 80 percent emission reductions by 2050 into public discussions. Some congressional leaders appealed to SIU, but did so in support of their own legislative initiatives. Vermont Senator Bernie Sanders used the SIU action as evidence of public concern about the global-warming problem and endorsed the "80 percent by 2050" standard as "consistent with the legislation" he and California senator Barbara Boxer had previously introduced (Sanders 2007). Several other legislative efforts also were already underway before the SIU rally. Indeed, California representative Nancy Pelosi made global warming a priority for the 110th Congress when she became Speaker of the House in January 2007.[10] Overall, then, it also seems doubtful that congressional leaders were influenced, let alone compelled, by SIU's nationwide action to adopt the SIU standard or take positions on the climate-change issue that they otherwise would not have taken.[11]

As this chapter was written, the presidential and congressional campaigns were underway for the 2008 election. Both presidential candidates had staked out positions on global warming, a clear indicator of its salience as an issue with voters. Though the electorate recognizes climate change as an important issue, they still rank it as the least important among all issues recognized as important, with the economy and the Iraq war included among the top issues (Pew Reports 2008). In the months leading up to the election, high gas prices and failing financial companies only whispered of implied environmental catalysts and concerns. Even in the aftermath of Hurricane Katrina, the unexpectedly high rate of Arctic ice melting, Al Gore's *Inconvenient Truth* and subsequent Nobel Prize, and media coverage that conveyed these and related concerns, there still is no special sense of public urgency about global warming. And, I must add, I suspect that there was little sense of urgency among those in attendance at the nationwide SIU rallies, at least insofar as they resembled those who attended the Common rally.

Whether in Boston or viewed nationwide, SIU did not enact a demonstrative protest rhetoric that established the urgency of responding to the global-warming issue because it did not exhibit a convincing proof of power that would motivate supporters and confront politicians with the need for immediate action. Indeed, one reason is that the SIU action turns out not to be much of a *protest* rhetoric at all. A protest movement, as a protest movement, must *demonstrate* urgency through enacting proof of power that constrains both demonstrators' subsequent words and deeds and political leaders' subsequent decisions to remain consistent with the cause's leading moral and political presumptions. This, SIU clearly did not do. I shall return to this point in the final section, but next, I argue that the SIU rallies at Boston Common and elsewhere did not disclose a climate movement with sufficient proof of power to bring about political change, due at least partly to the performance of its demonstrative rhetoric.

Demonstrative Rhetoric and the Performance of Protest

For Kenneth Burke, politics is drama (1973, 310) and drama is politics (129–131), in that both symbolically evoke attitudes toward problematic political or social situations. Demonstrative protest rhetoric is a kind of enacted performance that anticipates the presence of both participating and witnessing audiences. For example, Blitefield, following Alinsky, viewed protest demonstrations as carefully staged dramas that symbolically set in-group performers apart from witnessing out-groups, inducing simultaneously "in-group solidarity and power" and "out-group recognition of that power" (2006, 264). The performance of protest demonstrations, as with any symbolic drama, is amenable to analysis as rhetoric since we can gauge whether it induces audience attitudes that adequately encompass or sum up the political or social situation. In this section, I will examine SIU's rhetorical performance of protest both at the Boston Common and nationwide (as represented through the news media), disclose the attitudes those performances evoke, and assess the adequacy of those attitudes toward the climate-change situation.

Speakers dramatized the climate situation through their verbal depictions. That drama involved nothing less than "saving the planet," "preserving life," "reclaiming our future." The Union of Concerned Scientists' president said in his speech, "Global warming is the greatest threat to our planet's ability to sustain life that we have known since the thermonuclear standoff at the height of the cold war." A Massachusetts congressional representative similarly said, "We're either going to live together or we're going to die together as a planet. We are either going to know each other or we're going to exterminate each other as a planet."[12]

A dramatic situation so urgent calls for heroes who must prevail over the villains who resist needed change. Some speakers acknowledged that change would not come easily, but still stressed that supporters should take confidence from the extraordinary people on their side. These heroic persons often came in the form of general types. Indeed, one could readily infer that each speaker represented a heroic type: scientists, schoolchildren, students, activists, clergy, and city, state, and federal Democratic Party politicians. Speakers praised young people and students as courageous change agents. A lawyer and a scientist, along with several Democratic Party politicians, were singled out and lauded for their exemplary actions. Based on speakers' depictions, heroes possess the virtues of knowledge, courage, commitment, and moral vision. The villainous coal and oil industries and the politicians who defend their interests oppose the heroes by obstructing needed change. President

Bush and his administration were singled out for special scorn based on their indifference, ignorance, and folly.[13] The villainous were depicted as displaying the vices of avarice, dishonesty, cowardice, and moral blindness, betrayed by their evident indifference about how their actions will impact the lives of children and future generations. One speaker characterized the opposition as "dark forces…that are leading us towards a precipice."

Audience members also were characterized as actors in the drama. Speakers depicted them as change agents and praised their attendance at the rally as an important accomplishment (again, often alluding to political strength in numbers). For example, one speaker told audience members that they were "sending a loud and clear message from their statehouse to the White House," and that they represented "the thousands of people around the country gathered at SIU rallies today." Speakers also exhorted listeners to perform a range of specific actions. Throughout the event, the audience was called upon to recycle, carpool, use energy-efficient lightbulbs, eat less fast food, turn down thermostats, join supportive organizations, donate in support of environmental initiatives, and perform other specific acts. And, after the speeches concluded, the gathering was asked to take the SIU "final pledge" to own responsibility for changing how they lived and for contacting political leaders to counter the growing climate-change crisis.[14]

Speakers articulated elements of a high-stakes drama that "summed up" the climate situation in words that disposed hearers to change some daily habits and to contact political leaders. Those responses seem a rather weak denouement to the drama they had characterized, literally, in earth-threatening terms. But other dramatic elements at the Common militated against that threat and diffused a sense of urgency. As a result, the overall performance did not dispose the audience to respond in proportion to the grave situation the speakers had depicted.

One reason for a lack of urgency is that the performance at the Common seemed bereft of conflict. There was very little specificity in delineating "them" as opposed to "us." Who, specifically, were the villains (aside from President Bush)? At best, speakers only called up generalities (e.g., oil and gas companies) rather than naming and denouncing specific enemies (e.g., ExxonMobil). What specific actions could audience members perform that could confront those enemies and disrupt their activities? Changing lightbulbs and petitioning congressional leaders with postcards or letters do not address that question. Suitable answers to those questions would likely arouse the audience's anger as well as evoke a meaningful sense of their own agency through performing actions that demand, rather than mildly request, changes in business as usual. An audience so disposed is more likely to respond in proportion with the magnitude of climate crisis.

The urgency of the climate situation was further defused when speakers and participants talked and acted in ways that minimized the situation's gravity. For example, one speaker said, in so many words, that the rally helped him learn that saving the planet could be great fun. Symbolic actions, both discursive and nondiscursive, were enacted throughout the day that conveyed "having fun" as a theme of the rally. As a result, one could experience the rally more as a fun-filled Saturday afternoon at the park than as a protest in response to an urgent crisis.

A festival-like atmosphere permeated the drama at the Common, minimizing any sense of the climate situation's threatening qualities.[15] One factor generating that atmosphere was the wide range of groups that turned out to proselytize their causes at the rally: The July 26th Coalition of Boston praised Cuba as leading the world in sustainable development; the Massachusetts Animal Rights Coalition (whose representatives were costumed as a pig, a cow, and a chicken) explained how vegetarianism relates to the fight against global warming; the advocacy group Environment Massachusetts provided information about its statewide actions; the World Can't Wait group distributed flyers advocating removal of the "Bush Regime" from office; members of the New American Communist Party presided over an extensive book table and talked about what made the party new; adherents of Falun Gong had several information tables with materials denouncing Chinese Communist treatment of its practitioners; and three young men costumed as Queen Elizabeth, Lord Bertrand Russell, and Sir Winston Churchill marched through the crowd performing a burlesque, apparently, of twentieth-century British colonial policy. Obviously, some groups were more relevant to the rally than others, but the mix contributed to the festival-like atmosphere at the Common.

Participants' efforts to be "creative" and "have fun" further contributed to the festival-like atmosphere. The organizers planned to form a human chain marking off the presumed future shoreline across the Common should global warming continue unabated. That plan was not realized, at least according to several mildly disappointed people who came equipped with goggles to participate. People displayed their protest art. Children and their parents held "recycling rope," adorned with recyclable objects, as they marched in a parade led by a Chinese-style "sea dragon," accompanied by a rhythmic drumbeat. Children drew and proudly displayed colorful fish posters, presumably to further symbolize what the future has in store for the Common. Indeed, children had a large presence throughout the day's events. Children stood on the bandstand along with the speakers. Children helped lead the recital of SIU's "final pledge." And, after the pledge, when the formal events concluded, the crowd was told to "make some noise" with the Chad Hollister Band accompanied by the Park School Freedom Choir that, of course, consisted of children. All of this was entertaining, and the children often were endearing, but the gravity of the climate situation is obscured by so fun-filled and festive a scene.

And finally, the speakers themselves minimized the threat by introducing "clean" technology as something that, reassuringly, could stabilize the climate crisis while enabling people to thrive materially. "We can build jobs," said a Harvard professor. "We can make smart technologies so we create better batteries for cars, use nontoxic chemicals in houses—for paints and carpets—and biodegradable plastics." Another speaker detailed a vision of dramatic technological transformation, calling the Cape Wind Project in Massachusetts "a major step toward rewiring the whole world with clean energy."[16]

A few speakers mentioned that revolutionary changes are needed to turn the crisis situation around, but overall, the rally left the impression that the revolution did not really demand much from participants. The intervention of a technological deus ex machina in "green" or "clean" costume brought promise of climate stability without major sacrifices during what presumably must be a major restructuring of the oil-based economy. A happy, though unconvincing, ending to the climate drama was promised after all.[17]

Those interviewed indicated how the performance was received at the Boston Common. They felt positive and optimistic about the situation, with few exceptions. The fact that they attended the rally left them feeling "encouraged," "hopeful," "better informed," "like you're part of something," and "happy to participate." Participants had watched children draw posters, played with dogs, heard about clean technologies, "made some noise" with the rock band, and generally felt good. They also were left without clarity about what political actions to do next and how to do them, as discussed earlier. Again, these attitudes and feelings are not in proportion to the climate situation.

The weak performance of protest at the Common serves as an instructive guiding exemplar for analyzing news coverage of the SIU drama nationwide. If other rallies were at all similar, one can adduce reasons why the mediated drama failed to convey a sense of urgency commensurate with the climate situation. One reason, as already mentioned, is that the reportage of "movement" claims probably seemed exaggerated and, thus, did not adequately dramatize the event's political significance as a proof of power. Another reason is that reportage of the climate situation's gravity and SIU's appeal to Congress seemed incongruous with varied activities that were used to illustrate stories, whether verbally or visually: parades, chants, cheers, slogans, children's art, enactments of warming (e.g., "It's hot in here," cheerfully exclaimed with SIU sign in hand; melting ice blocks on pavement), and dramatizations of coastlines (e.g., lower Manhattan's "human shoreline" holding blue sheets for waves and wearing masks, flippers, and related ocean garb). These efforts were good-hearted and cheerful overall; they often seemed more entertaining than political, more funny than poignant. When readers were given information about specific actions that they could perform, those actions (recycling, reducing garbage) did not seem in proportion with the climate crisis. Overall, news coverage of SIU events comported with my experience of the Common rally in that it generally did not convey a sense of dramatic conflict and urgency.[18]

In sum, the Common rally did not enact an adequate performance of protest that evoked attitudes commensurate with the purportedly threatening situation. Nor did mediated coverage of SIU events convey much urgency about that situation. I shall return to the question of what does constitute an adequate performance of protest in the next section.

Implications

Demonstrative rhetoric is inventive; it shows, manifests, and exhibits something previously unseen or concealed. SIU rallies were not made visible or present as an emerging protest movement of political consequence to those not already committed. Planners and organizers might reconsider the relationship of Internet organizing to public showing, exhibiting, or manifesting political power. As matters now stand, only the already committed and considerably patient would go to the SIU Web site to watch the post-April 14 slideshow to gain some tangible sense of the nationwide reach of the rallies. Protest movements or persuasive campaigns must become publicly visible to raise the prospect of having political traction with the wide swath of people who are uncommitted or have never even heard of SIU.

The civil rights movement reached a wide audience through conventional news coverage of public marches and public actions that *embodied* its power, its conflict with resisting authorities, and its ideological message. Setting aside SIU's ideological reasons against holding a march on Washington,[19] the March on Washington for Jobs and Freedom offers instructive insight into the capacity of this form of demonstrative protest action to bring the movement into wide public existence. Not only did the march exhibit large numbers of engaged people that ensured that this was a cause of movement proportions that would not quickly pass away, but it also embodied the otherwise abstract but traditional idea of ordinary citizens going to the central locus of federal power to petition their government for redress of their grievances. Marches to politically significant sites can make a cause visibly present to those not already involved at the organizational or participatory levels. And we need not turn to King's march on Washington for illustration. Consider, instead, the Million Man March or the Million Mom March, both of which garnered media attention that included speculations about the actual size and strength of the turnout.[20] Other instructive examples are found in the primary and presidential campaigns for Barack Obama. The Obama campaigns were well known for their politically savvy use of the Internet to organize, network, and raise money. Yet, the magnitude of Obama's political support also was demonstrated by the embodied presence of large numbers of committed supporters at planned events. To take only one example, 75,000 people turned out in Portland, Oregon, to hear Obama speak at a critical juncture of the primary campaign. A campaign or movement is made most powerfully present through embodied actions that demonstrate large numbers of committed people will turn out at planned public events.

SIU planners might consider similar strategies that could embody the magnitude and depth of its support. For example, organizing marches on every state capitol could disclose SIU's power while putting governors and state legislators on notice that they are responsible for advancing the climate-cause agenda. The relatively localized actions will maintain ideological consistency by not requiring the large carbon expenditures of a nationwide march on Washington. Since a march is no day at the park, people will exhibit their level of commitment when they bother to participate. Through this sort of demonstrative protest rhetoric, SIU might become more visible before a wide audience through the media—illustrated, preferably, with photogenic images of the marches.

One problem with planned marches is that they can expose superficial as much as committed support for a cause; low turnouts demonstrate weakness, not strength. An important resource of demonstrative protest rhetoric is the perceived size of the turnout, whether manifested in a large arena filled to capacity or in a relatively small town-meeting hall that can accommodate only two-thirds of those committed to attend. Are SIUers committed enough to participate in more demanding kinds of protest actions? Will they do so in numbers sufficient to exhibit proof of power? Or will the turnouts prove altogether too modest, exposing superficial interest and weak commitment? Regardless, unless organizers can answer these questions with some assurance, they are well advised to avoid claims about the public strength of their campaign that cannot be demonstratively established.

How to create and maintain public perception of a protest movement and also evoke a perspective that challenges the practices that constitute the problem protested against are interrelated rhetorical problems for SIU. The core of any movement as a *protest* movement is to issue a challenge to received frames of reference and imply alternative ways of thinking, discoursing, and acting. Addressing the two interrelated problems requires reaching audiences through conventional news coverage. Scholarly commentators have long recognized that demonstrative protests use "body rhetoric" (Griffin 1964, 127) in staged spectacles to capture that reportage and challenge readers' and viewers' frames of reference (Berg 1972). That depiction of embodied protests is a good partial description of image events, as shown in other studies (Deluca 1999a; Johnson 2007). Both rhetorical problems, then, could plausibly be addressed through carefully staged image events.

Other chapters in this book examine image events as resources for SIU's rhetoric of movement building (chapter 9 and chapter 11). As those studies note, most of the action reports at the SIU Web site are not likely to reach those who do not already adhere to the cause. Even should undecided or curious people look at those reports, they are more likely to encounter ordinary photographic or video images than striking images that evoke thought. Getting striking image events before the public through the conventional news media still is an important way to both build public perceptions of a protest movement and disclose that movement's distinctive perspective.

This takes us to the quality of demonstrative protest rhetoric as proof of a cause's power. The Boston Common did not provide proof of power for the uncommitted, for political opponents, or, perhaps, even for participants at the rally who identified with the cause. It is doubtful that politicians opposed to climate-change legislation or the leadership of oil, coal, and gas companies are concerned about SIU's Boston or nationwide actions. Participants at the Common did not seem especially focused on their next actions in service to the cause; instead, many left the next political actions up to the politicians, or were confused about what they could do specifically, or viewed the rallies as an opportunity for some people to get educated or informed more than as the start of an active political movement. Apparently, given the purported movement's loss of momentum, the same point may apply to participants at events across the country. There was no logic of demonstrative protest rhetoric that compelled or constrained opponents' or participants' subsequent discourse and actions so that they advanced the cause's moral or political presumptions.

To enact that logic, a robust protest rhetoric that has confrontation at its core is needed. We have seen that SIU rallies lacked compelling confrontations. This is not due to the absence of options for enacting confrontational rhetoric. Consider a few examples: holding demonstrations at the local offices of opposing senators and representatives, showing up in sufficient numbers at SUV dealers to make potential customers uncomfortable, staging events that draw attention to local companies that profit from or contribute significantly to carbon emissions. Such actions are confrontational because they make the normal conduct of business-as-usual uncomfortable. Further, mild acts of disruption, following the Alinsky model, would in embodied form demonstrate SIUers' commitment to the cause, demarcate an in-group from the out-group, exemplify the urgency of the climate situation, and demand in no uncertain terms immediate action. Whether disruptive or not, protest actions against parties most responsible for the problem sharpen political and moral divisions and confront participants, onlookers, opponents, and elected leaders alike with the necessity of decision. Without confrontational protest actions, there can be no proof of the cause's power that is sufficient to compel change.

Based on evidence gleaned from the Common rally, those in attendance were largely confused or divided about what was next expected of them. True, speakers exhorted a range of specific actions and made abstract appeals to the growing movement's political significance, but SIUers did not provide specific political guidance about what the assembled and committed should do next to further the political cause or "movement" through their own embodied, public actions. Specifics about what the committed were expected to do next to keep the movement going actually were lost amid the welter of suggestions about how to alter one's own life practices and exaggerated portrayals of the movement's power.

This problem of organization can be addressed through using the full range of social networking tools available online. For example, the primary and presidential political campaigns for Barack Obama made use of the whole panoply of those tools. Supporters were encouraged to start up their own neighborhood groups, create blogs, start an Obama Web page, or receive frequent campaign updates by e-mail. Obama supporters even received e-mailed video clips of a leading strategist who addressed viewers directly as he detailed campaign strategy and invited their continued participation. Consider parallel applications for SIU: a video clip of Bill McKibben explaining SIU's strategy that appeals directly to his viewers for their continued participation in enacting tactics to come; local organizers contacting SIUers who turned out for a rally to provide context, time, and place for the next political action; neighborhood discussion groups forming to share climate concerns and, perhaps, coordinate actions for addressing the problem locally. As these examples also indicate, social-networking tools can help create a balance between directed campaign leadership and creative grassroots actions. Moreover, continued contacts and reminders from the national or local organizers, other SIUers, or family, friends, and neighbors through social networking can help exert some "soft" pressure on participants to keep involved. Altogether, use of online organization in its full capacity can help SIU planners give what so many participants lacked after the April 14 events: a sense of momentum, or moving forward.

Finally, I have argued that SIU did not enact a compelling performance of protest. The Boston rally and others like it, as reported through the news media, did not evoke attitudes that were in proportion with the purportedly planet-threatening climate situation. If human-induced global warming is in fact threatening life as we know it, one might anticipate finding ourselves in a struggle of epic proportions or, worse, as players in a great tragedy. Having fun is a strikingly incongruous response within either dramatic framing of the situation.[21] Indeed, one could easily doubt that those who have fun are themselves convinced of the danger they purport is near at hand. Even more practical audience responses, such as making lifestyle changes to reduce "carbon footprints" or writing congressional representatives about the need for climate legislation, seem altogether too mild to encompass or sum up the gravity of the climate situation. A different sort of drama that generates attitudes more commensurate with that situation is needed. That drama, I have contended, involves the compelling performance of protest.

A performance of protest needs the dramatic elements of confrontation, conflict, and division. Such a drama makes clear one's identifications: "we" embody people who serve indisputable moral and political presumptions, while "they" embody moral poverty, political irresponsibility, or indifference. A performance of protest enacted through conflict would exhibit and evoke powerful emotions: anxieties about the future, pity for innocents who will suffer, anger toward those who are indifferent or who actively oppose needed change, courage from acting upon moral and political convictions, confidence that change can occur through collaborative action, and inspiration from sharing with others the sense of moving the cause forward. In such a drama, participants know what they are expected to do and what roles they are called upon to play. And, perhaps above all, supporters, onlookers, potential adversaries, and other witnesses of the performance of protest are left anything but mildly disposed toward the situation; rather, the drama leaves them no middle grounds for decision but instead confronts them with a stark choice that puts their own commitments and values at issue.

McKibben claimed that the magnitude of change required to address the global-warming problem needs a "much larger movement" than the environmental movement, a movement that is "as morally urgent, as committed, as passionate as the Civil Rights Movement if we're going to have any chance of turning this around" (quoted in Donnelly 2007; also see NPR 2007). That sort of movement requires agents of change who are willing to act in ways that embody "moral urgency, commitment, and passion." Civil rights demonstrators did not sign postcards, play with dogs, listen to music in the park, and call it a fun day; they sat at segregated lunch counters and invited the consequences or, more generally, created conditions for conflict with those in authority by breaking unjust laws and disrupting conventional, everyday practices. Perhaps SIU should organize a day of public protest against leading institutions that either resist change or benefit lavishly from its absence. Conflict, after all, is the soul of drama. And actions that confront and convey anger toward those who resist efforts to save the earth or who benefit from the earth's degradation seem much more in proportion with the situation than mildly petitioning the powerful for action.[22]

Burke warned us that we must be prepared to discount the errors of our metaphors whenever they are confronted by the "objective recalcitrance" of our situation (1985, 28). If human-induced global warming and its impact on the earth as we know it are realities, we are confronting a recalcitrant situation that resists the dominant metaphors of our times. McKibben acknowledged the far-reaching economic and social changes needed to address the climate-change problem,[23] but at SIU rallies, the benefits of those changes seemed too good and the sacrifices too meager to be true. Not only were we assured that our quality of life would not diminish, but we were told that it would even improve. The errors or limits of the metaphors of technological progress were not criticized; rather, they were perfected as "green" or "clean." Industrial capitalism and oil-driven technologies generated the means of production and afforded the consumption-based style of life that presumably left us with the global-warming crisis as a massive "externality." Now, however, we are told that we have a new "green" market capitalism that not only will prove enormously profitable, but also will enhance our quality of life through use of perfected, clean technologies that can fix the situation its hoary industrial predecessor has left us. This is not to say that technology has no role to play in this drama, but that alternative metaphors also are needed to address the limitations and temper the excesses of an idealized instrumentalism. Those metaphors, for example, must include depictions of who we believe we are as human beings on the earth, what we should aim for in a hopeful future, and the quality of actions that can realize those aims. Without such metaphors, we are unlikely to gain a mature understanding of the climate situation in all of its complexity. Without that mature understanding, as Burke would put it (1973, 131), we may literally find ourselves incapable of coming to terms with our situation and, instead, find our efforts hopeless as they crash against the objective recalcitrance of the climate-crisis situation.

Endnotes

1. Based on related historical and contemporary literature, I have elsewhere articulated the qualities characteristic of demonstrative rhetoric and use them in this chapter to guide analysis of how the SIU protest actions manifest a specific form of what I have characterized as the demonstrative dimension of rhetorical display (Prelli 2006).
2. It is important to stress that this paper is guided primarily by McKeon's conception of demonstrative rhetoric and its inventional capacity as a field-invariant or "universal" art. Thus, applications of that conception are not restricted to protest rhetoric generally or "image events" in particular.
3. Materials examined in this study are transcriptions of all speeches and interviews with participants; written materials collected at the rally; on-site written observations; photographs of speakers, audience members, and the Common and adjacent venue; and specific bodily displays, such as a children's parade. I am indebted to political science graduate student David Anderson and natural resources graduate student Lindsey Fong for their assistance conducting and transcribing the on-site interviews.
4. Overall, news coverage was positive about the SIU events, seldom offering an opposing or skeptical opinion. One of the few exceptions that proves the rule is an energy-industries spokesman's call for more practical engagement of issues (quoted in Bauers 2007).
5. The answer is that the previous and only march was organized by McKibben in Vermont and drew about one thousand people.
6. A knowledgeable reader would recall that estimates of turnout on the first Earth Day on April 22, 1970, were as high as 20 million, with nearly 25,000 in New York City alone (Davis 1998, 1).
7. Rhetoricians will note that the SIU campaign effort to turn individual events into exemplars of a wider political movement is "prefigured" by synecdoche, or the effort to understand the whole based on examination of parts. McKibben himself thought according to that figure, since he believed that the electronic linkages of the Internet would enable a whole to emerge that would be much greater than the sum of its dispersed parts (McKibben 2007a). Even though news reportage was prefigured similarly, the transference from part to whole was not fully convincing. For a study of synecdoche in relation to demonstrative protest rhetoric, see Foss and Domenici (2001).

8. The successful action carried the implied threat of future, similar disruptive actions, an implication that Blitefield maintained is the function of protest demonstrations, as what he called "premonitory proof" (2006, 256).
9. The earliest studies of American protest rhetoric were especially attuned to its distinctive confrontational quality, a quality that remains peculiar to demonstrative protest rhetoric (Benson and Johnson 1968; Andrews 1969; Scott and Smith 1969; Haiman 1967, 1982; Gregg 1971).
10. For an analysis of climate-change legislation proposed in the 110th Congress, see Larson and Heilmayr (2008).
11. SIU planners themselves evidently have difficulty identifying congressional leaders influenced by the rally that could reduce that doubt (chapter 5). This does not mean that congressional leaders who were vulnerable in the next election did not attempt to use the climate-change issue as evidence that they were in tune with their constituencies on environmental issues. It is not surprising, to take two examples, that New Hampshire Republican Senator John Sununu and Maine Republican Senator Olympia Snow crossed party lines to vote against the Republican-led filibuster against the Leiberman-Warner Climate Security Act (S. 3036) (U.S. Senate 2008).
12. The same point was made when one speaker created and another extended a Newtonian metaphor to capture the crisis: One said, "There is an increasing awareness that our current relationship with nature is on a collision course." The other later added, "When an irresistible force like nature meets an immovable object like our federal government, something has to give...[W]hen history and nature collide, nature always wins, and history always has to heal its wounds and pick itself up and start over again."
13. Some speakers appealed to bipartisanship, and none singled out individual Republican politicians, other than the president, for special scorn. Occasionally, one got the impression that a Democratic politician was shadowboxing with unnamed Republican adversaries. For example, one said, "[T]here will be those...[who] are involved in a disinformation campaign. There will be those...[who] will try to stop us every step of the way."
14. The final pledge follows:

    I pledge to make saving our planet a priority,
    to waste less and share more of Nature's gifts,
    and to do everything in my power to stop the climate crisis.

    I pledge to cherish Mother Earth and to make a difference today.
    I pledge to recycle as much as I can and help others to recycle,

and to walk, bike, or take public transportation whenever I can.
I pledge to make a plan with my family to save energy,
and to talk with my neighbors and friends about our concerns and hopes.

I pledge to keep speaking out until our government takes action.
I pledge to start here—before I leave—by signing a postcard
asking Congress to cut carbon 80% by 2050.

From this day forward, I'm going to Step It Up!

15. Speakers could have amplified the urgency of citizen action in response to the climate situation by using Boston Common rhetorically as the symbolic site of great political struggles. Not only does the Common resonate with historic meaning as a site of civic participation and collective public action (Fischer 2000), but the Massachusetts State House can be seen from the rally site. Only one speaker drew upon the Common as a symbolic resource for setting the scene for the rally (unlike the SIU event at the Alamo described in chapter 1). The opening speaker did so obliquely when he started the day's events by telling his audience that they "stood on the shoulders of idealists": Revolutionary War soldiers, Boston Tea Party participants, abolitionists, suffragists, participants in the civil rights movement. Another speaker referred perfunctorily to the statehouse when he introduced a state senator. If this rhetorical omission was due to trust that the audience already appreciated the site's political significance, that trust was largely misplaced. More people interviewed interpreted the Common's significance in personal or individual terms (e.g., "I've seen some theater productions here, and I would certainly be sad if this place was flooded. I grew up in Boston. I love Boston, so I wouldn't like to see it change that much.") than in political terms (e.g., "Well, I mean, it's so apt. The Commons! I mean...if there were...any place that people would meet, I mean, it's here, because there's always so much different activity going on that it's kind of the place for a rally, I think."). A majority of respondents thought the Common was significant for these two reasons, but a sizeable minority did not see the Common's significance at all. An opportunity was missed to use the Common as a symbolic vehicle for setting the stage from which to dramatize the call for urgent political action.
16. This speaker went on to amplify his vision: "With solar panels and windmills and machines that make electricity from ocean waves and tides, it would create millions and millions of jobs all over the world...And it would jump-start the clean-energy industry into becoming a central, driving engine of growth in the global economy. And it would activate an army of

clean-energy engineers who would make us safer and much more admired around the world than all of our tanks and bombs combined."

17. One speaker evoked the prospect of technological transformation with clumsy irony: "The good news is...we can turn this around. And as [the opening speaker] was pointing out, our heroic foremothers and forefathers knew how to do social change. And they also knew how to do technological change—moving from the horse and buggy to the airplane, the steamship, trains, the automobile...ironically enough. They knew how to go from the Pony Express to the silicon chip and the Internet and the World Wide Web."
18. For television coverage, see NBC News (2007) and ABC News (2007a, 2007b). For newspaper coverage, see as examples Anonymous (2007), Fernandez (2007), Lindberg (2007), Levenson (2007), Curran (2007), and Abramson (2007).
19. The main reason was that carbon expenditures needed to travel to Washington would contradict SIU campaign aims of reducing emissions.
20. For useful studies of the rhetoric of these two marches, see Arthos (2001) and Hayden (2003).
21. This point should not obscure the fact that protest rhetorics can also be framed comically, as Christiansen and Hanson (1996) have shown. Yet, even comedic framings are irreducible to simply "having fun."
22. Not at all incidentally, Cathcart (1978) argued that confrontation distinguishes movement rhetoric from other rhetorical forms.
23. McKibben stressed that fossil fuel is "at the white-hot center of our economy" and that the nation must deal "with the most fundamental parts of our economy" to address the climate crisis (Democracy Now! 2007; also see NPR 2007). An indication of the social changes required is McKibben's contention that our society needs a new conception of "happiness" that is not based on expansion, growth, and size (Democracy Now! 2007; NPR 2007). These are formidable and clearly radical tasks that will incur vigorous opposition. Opponents already suspect that a serious effort to address global warming will have radical consequences. For example, Senate Minority Leader Mitch McConnell of Kentucky defended a Republican-led filibuster of the Lieberman-Warner Climate Security Act (S. 3036) by claiming the legislation—designed to reduce carbon emissions 71 percent by 2050—would require "the largest restructuring of the American economy since the New Deal" (quoted in CBS News 2008).

References

ABC News. 2007a. *Good Morning America*, April 14, 2007. Lexis-Nexis *Academic* database. http://www.lexisnexis.com (accessed July 3, 2008).

———. 2007b. *World News Saturday*, April 14, 2007. LexisNexis *Academic* database. http://www.lexisnexis.com (accessed July 3, 2008).

Abramson, Mark. 2007. Locals fight global warming: Palo Alto joins cities for national day of climate action. KRTBN *Knight-Ridder Tribune Business News—Palo Alto Daily News*, April 15, 2007. Lexis-Nexis *Academic* database. http://www.lexisnexis.com (accessed July 3, 2008).

Andrews, James R. 1969. Confrontation at Columbia: A case study in coercive rhetoric. *Quarterly Journal of Speech* 55:37–46.

Anon. 2007. Mainers rally to curb carbon emissions. *Associated Press State and Local Wire*, April 14, 2007. LexisNexis *Academic* database. http://www.lexisnexis.com (accessed July 3, 2008).

Arthos, John Jr. 2001. The shaman-trickster's art of misdirection: The rhetoric of Farrakhan and the Million Men. *Quarterly Journal of Speech* 87:41–60.

Bauers, Sandy. 2007. Rallying the masses in support of the earth; Big buzz builds for Step It Up; A nationwide global-warming event tomorrow. *The Philadelphia Inquirer*, April 13, 2007. LexisNexis *Academic* database. http://www.lexisnexis.com (accessed July 3, 2008).

Benson, Thomas W., and Bonnie Johnson. 1968. The rhetoric of resistance: Confrontation with the warmakers, Washington, D.C., October, 1967. *Today's Speech* 16:35–42.

Berg, David. 1972. Rhetoric, reality, and mass media. *Quarterly Journal of Speech* 58:255–263.

Blitefield, Jerry. 2006. It's showtime! Staging public demonstrations, Alinsky style. In *Rhetorics of Display*, ed. Lawrence J. Prelli, 255–272. Columbia: University of South Carolina Press.

Burke, Kenneth. 1973. *The philosophy of literary form: Studies in symbolic action*. 3rd ed. Berkeley: University of California Press.

———. 1985. Dramatism as ontology or epistemology: A symposium. With Bernard L. Brock, Parke G. Burgess, and Herbert W. Simons. *Communication Quarterly* 33:17–33.

Cathcart, Robert S. 1978. Movements: Confrontation as rhetorical form. *Southern Speech Communication Journal* 43:233–247.

CBS News. 2008. Republicans stall climate change bill. CBSNews.com, June 6, 2008. http://www.cbsnews.com/stories/06/06/national/printable 4158834.shtml (accessed June 24, 2008).

Christiansen, Adrienne E., and Jeremy J. Hanson. 1996. Comedy as cure for tragedy: Act Up and the rhetoric of AIDS. *Quarterly Journal of Speech* 82:157–170.

City of Boston. n.d. History of climate action in Boston. http://www.cityofboston.gov/ climate/history.asp (accessed January 7, 2009).

Curran, John. 2007. Global warming "Day of Action" rallies planned nationwide Saturday. *Associated Press State & Local Wire*, April 13, 2007. LexisNexis *Academic* database. http://www.lexisnexis.com (accessed July 3, 2008).

Daley, Beth. 2007. N.E. chorus against global warming. *Boston Globe*, April 12, 2007. LexisNexis *Academic* database. http://www.lexisnexis.com (accessed July 3, 2008).

Davis, David Howard. 1998. *American environmental politics*. Chicago: Nelson-Hall.

Delicath, John W., and Kevin Michael DeLuca. 2003. Image events, the public sphere, and argumentative practice: The case of radical environmental groups. *Argumentation* 17:315–333.

DeLuca, Kevin Michael. 1999a. *Image politics: The new rhetoric of environmental activism*. New York: Guilford.

———. 1999b. Unruly arguments: The body rhetoric of Earth First!, Act Up, and Queer Nation. *Argumentation and Advocacy* 36:9–21.

Democracy Now! 2007. Step It Up: Thousands gather this weekend for largest ever rally against global warming. Interview transcript, April 13, 2007. http://www.democracynow.org/2007/ 4/13/step_it_up_thousands_gather_this (accessed July 3, 2008).

Donnelly, John. 2007. Early critic of warming steps up activist role. *The Boston Globe*, March 19, 2007. LexisNexis *Academic* database. http://www.lexisnexis.com (accessed July 3, 2008).

Fernandez, Manny. 2007. For the environment, rallies great and small (and unusual attire). *New York Times*, April 15, 2007. LexisNexis *Academic* database. http://www.lexisnexis.com (accessed July 3, 2008).

Fischer, David Hackett. 2000. Boston Common. In *American places: Encounters with history*, ed. William E. Leuchtenburg, 125–143. New York: Oxford University Press.

Foss, Karen A., and Kathy L. Domenici. 2001. Haunting Argentina: Synecdoche in the protests of the mothers of the Plaza de Mayo. *Quarterly Journal of Speech* 87:237–258.

Gregg, Richard B. 1971. The ego-function of the rhetoric of protest. *Philosophy and Rhetoric* 4:71–91.

Griffin, Leland M. 1964. The rhetorical structure of the "New Left" movement: Part I. *Quarterly Journal of Speech* 50 (2): 113–135.

Haiman, Franklyn S. 1967. The rhetoric of the streets revisited: Some legal and ethical considerations. *Quarterly Journal of Speech* 53: 99–114.

———. 1982. Nonverbal communication and the First Amendment: The rhetoric of the streets revisited. *Quarterly Journal of Speech* 68:371–383.

Hauser, Gerard A. 2006. Demonstrative displays of dissident rhetoric: The case of prisoner 885/63. In *Rhetorics of Display*, ed. Lawrence J. Prelli, 229–254. Columbia: University of South Carolina Press.

Hayden, Sara. 2003. Family metaphors and the nation: Promoting a politics of care through the Million Mom March. *Quarterly Journal of Speech* 89:196–215.

Johnson, Davi. 2007. Martin Luther King Jr.'s 1963 Birmingham campaign as image event. *Rhetoric and Public Affairs* 10 (1): 1–26.

Larson, John, and Robert Heilmayr. 2008. Comparison of legislative climate change targets in the 110th Congress. *World Resources Institute*, June 18, 2008. http://www.wri.org/publications/usclimatetargets (accessed June 24, 2008).

Levenson, Michael. 2007. Young, old urge action on global warming—From Maine to Boston, hundreds rally on dangers of climate change. *Boston Globe*, April 15, 2007. LexisNexis *Academic* database. http://www.lexisnexis.com (accessed July 7, 2008).

Lindberg, Anne. 2007. Local rally one voice in chorus of protest. *St. Petersburg* [Florida] *Times*, April 11, 2007. LexisNexis *Academic* database. http://www.lexisnexis.com (accessed July 7, 2008).

McKeon, Richard. 1968. Discourse, demonstration, verification, justification. In *Demonstration, verification, justification*, ed. Philippe Devaux, 37–55. Paris: Nauwelaerts.

———. 1971. The uses of rhetoric in a technological age: Architectonic productive arts. In *The Province of Rhetoric*, ed. Lloyd F. Bitzer and Edwin Black, 44–63. Englewood Cliffs, NJ: Prentice-Hall.

McKibben, Bill. 2007a. Hi, tech: Who needs newspapers when the Web can do the job? *Grist*. http://www.grist.org/comments/dispatches/2007/01/08/mckibben/index7.html (accessed June 9, 2008).

———. 2007b. Step It Up: Introducing a brand-new, mass-protest climate movement for 2007. *Grist*, http://www.grist.org/comments/dispatches/2007/01/08/ mckibben/index.html (accessed June 9, 2008).

———. 2007c. We stepped it up! Day of action was a raging success, with 1,400 climate events around the U.S. *Grist*. http://www.grist.org/comments/dispatches/ 2007/01/08/mckibben/index14.html (accessed June 9, 2008).

Menino, Thomas M. 2007. Remarks to the Greater Boston Chamber of Commerce on December 11, 2007. City of Boston. http://www.cityofboston.gov/mayor/ pdfs/2007Chamber.pdf (accessed July 3, 2008).

NBC News. 2007. *NBC nightly news*, April 24, 2007. LexisNexis *Academic* database. http://www.lexisnexis.com (accessed July 3, 2008).

NPR (National Public Radio). 2007. Rallies encourage U.S. to cut carbon emissions. *Talk of the nation*, April 13, 2007. LexisNexis *Academic* database. http://www.lexisnexis.com (accessed July 3, 2008).

Peters, Stephanie M. 2007. Global warming opponents demand cut in carbon emissions, creation of 5m "green jobs" urged. *Boston Globe*, November 4, 2007. LexisNexis *Academic* database. http://www.lexisnexis.com (accessed July 3, 2008).

Pew Reports. 2008. An even more partisan agenda for 2008. The Pew Research Center for the People and the Press, January 24, 2008. http://people-press.org/report/388/an-even-more-partisan-agenda-for-2008 (accessed July 3, 2008).

Prelli, Lawrence J. 2006. Rhetorics of display: An introduction. In *Rhetorics of display*, ed. Lawrence J. Prelli, 1–38. Columbia: University of South Carolina Press.

Sanders, Bernie. 2007. We must act now on global warming. *States News Service*, April 16, 2007. LexisNexis *Academic* database. http://www.lexisnexis.com (accessed July 3, 2008).

Scott, Robert L., and Donald K. Smith. 1969. The rhetoric of confrontation. *Quarterly Journal of Speech* 55:1–8.

Step It Up. 2007. The big one: One earth, one climate rally. Boston, MA. http://events.stepitup2007.org/events/show/573 (accessed April 15, 2007).

U.S. Senate. 2008. U.S. Senate roll call votes 110th Cong., 2nd sess., June 6, 2008. http://www.senate.gov/legislative/LIS/roll_call_lists/roll_call_vote_cfm.cfm?congress=110&session=2&vote=00145 (accessed June 24, 2008).

Chapter 4

Step What Up?

Rhetorical Framing and Dialectical Tensions in Salt Lake City's Step It Up Events

Danielle Endres, Deborah Cox Callister, Autumn Garrison, Samantha Senda-Cook, and Julie Kalil Schutten[1]

> While melting polar ice caps and rising sea levels seem to be far-away concerns, Utah's precious winter snow pack, our primary water source, is at risk due to climate change and will suffer more and more in the future unless we can work together as citizens and governments to dramatically cut carbon emissions. Please join us as we honor our life sustaining water supply by participating in a gathering-of-the-waters celebration at the Sugarhouse Park pond.
>
> —Salt Lake Spring Runoff Celebration (2007) organizers

> Gather to put motion behind our intentions for positive change! Protect the future by telling Congress to "Step It Up" in the face of our current climate crises and be part of the largest rally in history against Global Warming! Join in on 108 Sun Salutations at your local Salt Lake City Action, come meditate under our banners for change, or just come and watch to show your support!
>
> —People Protecting the Planet (2007) organizers

> Salt Lake is planning the grandest Step It Up event in the nation! The day will be packed full of music by Los Lobos, the Salt Lake Alternative Jazz Orchestra, Brenn Hill, Motherlode Canyon Band, Salty Rootz, and Blue Haiku. On the plaza grounds, there will be food, exhibits, presentations, vendors, informational booths, and people to help you get messages of concern to members of Congress...Our goal is to send thousands of messages to Congress asking them to pass legislation to cut carbon emissions by 80% by 2050. With local events like this happening across the country, Congress and the presidential candidates will have to listen and take action!
>
> —Step It Up! Salt Lake City (2007) organizers

The three epigraphs above demonstrate the variety of actions planned for the Step It Up 2007 (SIU) campaign to cut carbon dioxide (CO_2) emissions and address global warming. In addition to these Salt Lake City (SLC) events, nationwide there were over 1,400 SIU actions that ranged from rallies and marches at centers of government to image events (DeLuca 1999) underwater in coastal reefs and at sites of diminishing glaciers to group hikes in iconic national parks. The national SIU organizing team called for one message, "Step It Up, Congress: cut carbon 80 percent by 2050," to unite these varied actions.

In this chapter, we grapple with some of the communicative challenges faced by a national campaign that strives to loosely choreograph over 1,400 individual, locally organized actions. Using the SLC actions as an illustrative case study, we identify some barriers to movement building that may arise when there are disconnects, or ruptures, in communication between organizers and participants in environmental campaigns. We argue that one of the most important challenges of environmental campaigns is to rhetorically frame the campaign in a way that reaches multiple audiences and motivates them to act. Yet, campaigns can easily be crippled by communicative disconnects between organizers and the public. These disconnects have implications not only for the success of a single campaign but may also have implications for the larger movement.

We use two concepts from communication theory to identify sites of potential disconnect between organizers and participants: rhetorical framing and dialectical tensions. First, we draw from the concept of rhetorical framing (Burke 1966; Entman 1993; Goffman 1974; Lakoff 2004; Ott and Aoki 2002). To explain the concept of framing, Kenneth Burke stated, "Even if any given terminology is a *reflection* of reality, by its very nature as a terminology it must be a *selection* of reality; and to this extent it must also function as a *deflection* of reality" (1966, 45). Any message is, consciously or not, rhetorically constructed to emphasize certain things and de-emphasize other things, like wearing a pair of colored glasses. Rose-colored glasses, for instance, tend to emphasize positive aspects of a situation while the negative aspects of the situation fade out of view. Framing is a rhetorical device that can be employed by both individuals and groups or organizations. In the case of the SIU actions in SLC, we identify several tensions in the ways that national organizers, local organizers, and participants at the SIU actions rhetorically framed the action, the participant, the problem, and the solution.

In order to further our understanding of these tensions in framing, we also draw on the concept of dialectical tensions (Bakhtin 1981; Baxter 1988; Frey and Barge 1998; Johnson and Long 2002; Kramer 2004; Mabry 1999). A dialectical tension exists when two ideas are valid on their own but become contradictory when paired together (see chapter 5). However, dialectical tensions should not be seen as mutually exclusive options from which one must choose. Rather, dialectical tensions are conflicts, or opposing pulls, that need to be communicatively managed (Kramer 2004). We cannot solve a dialectical tension, but we can work to use communication strategies and tactics to manage the situation.

In this chapter, we identify four dialectical tensions within the rhetorical framing of the SLC SIU actions: protest-oriented/awareness-oriented, collective affiliation/individual activity, unified/fragmented, and political/symbolic. Management of these tensions through precise rhetorical framing can be beneficial to most campaigns or movements in widening the appeal of their message and reaching multiple contradictory but not mutually exclusive audiences.

Our chapter begins with a short description of the SLC SIU actions as the context for our case study. The next section analyzes the three SLC SIU actions, focusing specifically on the four dialectical tensions that emerged from our data. The chapter concludes with implications and suggestions for environmental movements and campaign practitioners to manage dialectical tensions through rhetorical framing.

SALT LAKE CITY AS A SITE

Utah is known for many things—skiing the "greatest snow on earth," the Church of Jesus Christ of Latter-day Saints (LDS), and magnificent national parks—but protesting and activism are usually not included in the list. The majority of Utahns (though not the majority in SLC) are politically conservative, registered Republican, and members of the LDS religion (Canham 2007). Yet, SLC, the state capital and largest city in Utah, was the site of three SIU actions to call for Congress to cut CO_2 emissions.

The first action took place at Washington Square Park in the downtown area; the park surrounds the Salt Lake City and County Building. The largest of the three SLC actions (with an estimated five thousand participants) was billed as a free concert and featured Los Lobos, several local bands, over thirty informational and commercial booths, food and beer stands, and a bank of computers for sending e-mail messages to Congressional representatives. An interview with one of the organizers after the action illustrated the organizers' intent to create a "family-friendly" venue that would reach multiple audiences, such as "people in the middle class with normal jobs, families, members of the LDS church, everyday citizens that watch the news [who] are too busy to get involved...the Wal-Mart crowd, [and] the Costco crowd," as opposed to the progressive Democrat, traditional protester, or the "Prius-driving Wild Oats crowd."

The other two actions in SLC were much smaller—drawing between thirty and fifty people each—and took place in the morning, before the Washington Square action. The Liberty Park action invited participants to do yoga as a form of collective intention setting to cut CO_2 emissions. About two-thirds of the participants did 108 sun salutation yoga formations while the other third watched or meditated. According to an interview with the organizer at the action, "The sun salutation is usually done once a year. It is a cleansing ritual. It is how you put out intentions...We're setting an intention to gather awareness and ask Congress to step it up." At another popular SLC park, Sugarhouse Park, graduate students from a local university arranged the spring runoff or "gathering-of-the-waters" celebration that drew about fifty participants. According to one of the organizers, "We felt that to best illustrate our community's stake in the climate-change issue, we needed to focus on our dependence upon snowfall and snowpack for our water supply." The climax of the action was a symbolic act in which participants poured the water they brought (from seven local watersheds and beyond) into one common vessel. This action engaged participants in human interconnection with the water cycle, and the dousing of a symbolic fire of global warming that burned in an adjacent vessel. An organizer reflected about the participants at this action:

> We wanted to attract our friends and families and our neighbors to our event, and for the most part, we were able to do that...We felt that our action was primarily attractive to an audience [that was] already concerned about the potential impacts of climate change.

Unlike the Washington Square and Liberty Park actions, the organizers of the Sugarhouse Park action intended to reach and motivate a sympathetic audience of people already concerned about global warming.

RHETORICAL FRAMING AND DIALECTICAL TENSIONS IN SIU

Using participant observation, our five-member research team gathered texts from all three SLC actions, including field notes, photos, face-to-face interviews, one-page questionnaires, and recordings of speeches. We also quantitatively analyzed seventy-four short questionnaires completed by participants. From these data, we uncovered dialectical tensions within the rhetorical framing of (1) the action, (2) the participant, (3) the problem, and (4) the solution.

Framing the Action

Our analysis of the three SLC actions reveals that organizers chose to frame their actions in very different ways. The national organizers' call for local organizers to create their own "actions" is broad and nebulous. This resulted in actions ranging from free concerts to rallies, festivals, image events, and ceremonies. Examination of the framing of the SLC SIU actions reveals a dialectical tension between framing actions as *protest-oriented* and framing them as *awareness-oriented.* Protest-oriented actions are designed to challenge the status quo and promote change through confrontation and dissent (Gamson 1975; Stewart, Smith, and Denton 2007). Participants in a protest-oriented action generally already accept the message of the campaign or movement and work together in a collective action targeted to an outside audience of decision makers. A "protest" calls forth images of people marching and chanting in dissent. At the same time, it deflects alternate ways of engaging in collective action and potentially limits participation by people who do not consider themselves "protesters" or do not want to engage in certain actions that may be associated with "protesting." Awareness-oriented actions are designed to educate the public and potential movement members about the problem that the social movement or campaign is addressing and the need for individual and collective action to address the problem (e.g., feminist consciousness raising; Campbell 1973). Participants in these actions often include a broad range of people, from those who already accept the message to those who are undecided or know very little about the message. In awareness-oriented actions, the goal is to educate and potentially gather more supporters who may be willing to be involved in collective action, such as protesting, in the future.

The SLC actions were all framed more as awareness-oriented. Even though at one time there was a plan for an additional SLC action in which participants would march to the Utah capitol and demand action on climate change, SLC had no protest-oriented actions on April 14, 2007. To demonstrate how choices in rhetorical framing of an action have implications for who shows up to an action, what publics or participants expect from an action, and how the campaign's message is received, we turn our focus to the Washington Square action.

Participants and organizers used many ways to describe the Washington Square action, including "concert," "festival," "event," "beer garden," and "rally." However, the action was most prominently framed as a "free concert." In addition to being advertised in local media as a concert (Salt Lake City 2007; Gadette 2007; 570 KNRS 2007), the first thing we saw when we walked toward the action was a sign that proclaimed in bold red letters, "Free Concert Tonight! 3–10 PM" (see figure 6). Framing this action as a free concert situates it as an awareness-oriented action. One of the organizers of this action noted that it was the rock bands that attracted people who might not otherwise have been at a climate-change rally, but that the informational booths, speeches, and bank of computers for sending messages to Congress were intended to educate and motivate people to become involved in the campaign.

Our interviews at the downtown action indicate that many of the participants were there for the concert, specifically to see Los Lobos perform, but knew neither that it was an action about climate change nor that the main message of the action was to cut CO_2 emissions. While the framing of the action as a concert did draw a larger, more diverse crowd than the other actions in SLC, our interview data also show a potential disadvantage of this strategy. Several participants were unaware that the concert was related to climate change or were unable to respond to the question,

FIGURE 6. Sign at the downtown Salt Lake City SIU action.

"What is Step It Up?" This reveals a communicative disconnect between the local organizers and the people who showed up at the action. This disconnect could be communicatively managed by local organizers realizing the necessity of cultivating understanding and educating the large group of new people they attracted with the free concert about climate change. However, the main messages of the speeches given between the musical groups mismatched the awareness-raising framing of the Washington Square action. Although the speeches did emphasize and support the SIU message about climate change, they were designed not to educate or raise awareness, but to motivate people who already believed in the need to cut CO_2 emissions and in the SIU campaign. For example, the emcee repeatedly called for the audience to "step it up" without explaining what SIU was. And, a speech by a local university professor gave attendees a list of things to do, such as changing lightbulbs and riding bikes, but did not educate people about the problem of climate change and how these activities related to climate change. The potential disadvantage of awareness-raising actions was compounded at Washington Square with messages that did not align with the awareness-raising goal of the event.

Another risk in broadly framing the action as a way to raise awareness and educate new audiences is that it may alienate participants who already understand and accept the premise that we need to take action to prevent further human-caused climate change. Our interview data include participants who thought the action was "lame" because it was "not a protest" and "nobody [at the action] care[d]" about the issue. Participants ready for a protest-oriented action may, therefore, express frustration when the action frame is pulled too much toward an awareness-oriented action. As researchers who expected and wanted a protest-oriented action, our field notes attest to this sense of frustration. A more protest-oriented frame would have reached this particular audience better. In a campaign to cut CO_2 emissions, the SIU organizers would do well to reach a diverse set of audiences, including those who already believe in the need to protest or take political action to prevent global warming, undecided or unaware publics, and even those who do not believe in climate change or are disinclined to take action.

The framing of SIU actions has important implications for the larger campaign with regard to audiences, participant expectations, and perceptions of success. In essence, participants are interpellated or hailed (Althusser 1984) in certain ways through the framing of actions (e.g., attend a free concert or come do yoga). Hailing is the idea that messages are addressed to particular kinds of people and that if we respond to the message, then we accept this identity. Although each of the SLC actions entailed unique framing and attracted different audiences, they all fell into the category of awareness-oriented actions. As such, local organizers hailed participants as concert-goers or yoga enthusiasts, which deflected attention away from the SIU campaign's goal for actions focused on congressional action to cut CO_2 emissions. These actions missed their opportunity to reach both the not-a-protestor and the protestor audiences.

FRAMING PARTICIPATION

As we interviewed people at the three actions, we noticed that attendees offered information about themselves while explaining why they came to the action. While some participants talked about collective action and identification (e.g., affiliation with Moveon.org or the climate-change movement), others identified their individual activities (e.g., taking public transportation and recycling). Our data suggest a second tension, which concerns how participants framed their identities in relation to SIU (or not).[2] The *collective action/individual activity* tension concerns whether participants framed themselves as acting in coordination with others toward collective social change or whether they specified individual activities that do not necessarily imply identification with a movement or campaign. Even though the SIU actions were collective to the extent that there were groups of people, we argue that success for social-movement campaigns depends on participants specifically identifying as part of collective action for social change (Gamson 1995; Melucci 1995). Individual activities like recycling can be a part of collective action if the individual collectively identifies with the movement. Collective action and individual activities stand in tension as opposing approaches and can create a disconnect between organizers and participants. Campaign organizers must not simply take showing up at an action as a sign of participation in the campaign/movement and must therefore find ways to negotiate the tension between participants who frame their participation as collective action versus individual activity.

Campaigns and movements need to draw on collective action and individual activities to be successful; however, these are often perceived to be in tension by both participants and organizers. Indeed, Bill McKibben's (2007) call for people to "do something" other than change out lightbulbs and be part of "the first nationwide do-it-yourself mass protest" indicates a perception that individual activity is not as effective as collective action. As Michael Specter has argued, in the case of climate change, "personal choices, no matter how virtuous, cannot do enough. It will also take laws and money" (cited in Pollan 2008). Tension arises when individual activity precludes collective political solutions; individual lifestyle choices can satisfy people's personal environmental values or desire to do something about climate change in the absence of identifying as a part of collective social change. Nonetheless, collective action is often fortified by individual activities, such as climate-change movement members who reduce their carbon footprints, invest in alternative-energy sources, and talk to neighbors about climate change. These individual activities contribute to the broad goal of curbing global warming while simultaneously cultivating movement members who embody movement values by demonstrating their commitment to political representatives. Therefore, organizers must manage the potential tension between collective and individual by constructing how collective action and individual activities align with the campaign goal and each other.

Collective action toward social change is the basic goal of any social movement or campaign (Gamson 1995; Melucci 1995; Stewart, Smith, and Denton 2001). Collective action can take many forms, from protest rallies to letter-writing campaigns to boycotts. While collective action can be defined by numerous examples, Alberto Melucci (1995) argued that development and negotiation of a sense of collective identity is crucial to effective collective action. One way to foster collective identity is through identification with organizations or movements. C. B. Bhattacharya and Kimberly Elsbach (2002) contended that people who identify with an organization are likely to take action and get involved in related campaigns or movements. Collective action, therefore, requires collective identification with the movement and also entails taking actions that are geared toward social change as part of a collective. Individual activities are also important to social movements; people often have to make individual changes in their behavior in order to engage in collective action. Moreover, Kent Granzin and Janeen Olson's research (1991) indicated that people who see friends and family members engaging in environmentally conscientious behaviors are more likely to adopt these behaviors themselves. Therefore, individual activities can impact people nearby. However, individual activities are not inherently linked to collective action for social change. Melucci (1995) suggested that collective identity is key to determining if people are acting individually or acting together with other individuals for collective change. Examination of how participants framed themselves in relation to SIU can help to tease out whether they are part of the movement.

In our interviews, participants evoked their identities to justify why they had attended an action. Interestingly, most participants did not identify as part of the SIU campaign, but rather, either identified with related organizations and movements or listed the individual activities in which they engaged. When asked, "Why are you here?" or other interview questions, some of our participant interviewees noted their affiliation with existing organizations. For example, a participant at the Sugarhouse Park action said she decided to participate in the SIU campaign because she is a member of "Al Gore's Listserv" (i.e., the Climate Project), and a participant at the Washington Square action indicated that he was there because "I've been an e2 citizen for about a month."[3] While some participants listed their affiliation with particular organizations as their reason for participating in the SIU action, others mentioned collective identification with the broader climate-change or environmental movement. Several participants stated being "an environmentalist" as their reason for participating in the action. Our survey of 74 participants indicates that 63 percent of our survey participants perceived themselves to be part of a movement to prevent human-caused climate change, 22 percent were not sure, and 15 percent did not see themselves as part of a movement. Those who collectively identify with organizations or a movement are likely to engage in collective action toward social change (Melucci 1995). Because these participants already framed themselves as part of collective action, they participated in SIU as part of a movement for collective social action on climate change.

Although many participants cited collective identification and collective action as their reasons for participation, other participants cited personal behaviors—individual activity—as an explanation for their participation in the SIU event. For example, a Liberty Park participant said that he was participating in the SIU action because "I bike everywhere [motioning toward the bike he was leaning on]." In response to the question "What is Step It Up?" a participant at the Liberty Park action talked about taking individual actions and gave examples of her efforts to ride her bike, be a vegetarian, and use compact florescent lightbulbs. Often participants answered our questions about their participation in SIU by offering examples of activities from their own lives that they believed supported environmental or climate-change efforts. However, people who use individual activities to define the campaign (e.g., bike riding in answer to "what is Step It Up?") may not have collective identification with the campaign and hence may not understand the need for collective action for social change. These responses indicate that some participants framed their participation in SIU as individuals taking individual actions. If the person is not identified with the collective, he or she may see individual actions as doing enough. Moreover, when individuals frame themselves in relation to their individual activities as opposed to collective identification with the movement, the campaign's central collective goal can be lost. There is no clear connection between bike riding and SIU's goal of congressional legislation to cut CO_2 emissions. These comments suggest instances where individual activities are standing in for collective goals. It weakens the campaign when participants are not able to express collective goals to the media or when politicians hear mixed messages about what the campaign wants. Moreover, this supports the idea that solving climate change through individual activities is possible, which is not true in the case of climate change (McKibben et al. 2007).

Organizers should be cautious not to conflate showing up at an event with participation and collective identification with a movement. Our analysis has shown that while many participants did frame themselves as part of collective action for social change, there were also many people whose references to individual activities indicated that they may not have viewed themselves as part of collective action. It would be ridiculous to suggest that organizers should spend all of their efforts trying to find out if participants identify with the movement or not; it is important for them to recognize that there will likely be people from both ends of the dialectic at their actions. In order to manage the collective-action/individual-activity tension, organizers could focus some of their energies on establishing how individual activities can be a means of supporting the collective action. In other words, participants on the individual-activity end of the dialectic may need to be persuaded to identify as a part of collective action for social change. Campaign and movement organizers should explicitly frame participation for the participants in an attempt to move participants into the collective action rather than assuming that participation automatically implies or creates collective identification.

Framing the Problem

The national SIU organizers clearly framed human-caused climate change as the problem the campaign was created to address. Phil, one of the organizers of SIU, stated on the Web site, "If we don't start to curb carbon emissions now, we're in for a very different future for us and our children, even beyond the devastation that Hurricane Katrina wreaked on the gulf states" (Step It Up 2007). Additionally, an interview with two of the national organizers confirms the primacy of framing climate change as the problem. Despite the clarity of the national SIU problem frame, our analysis of local actions and interviews with participants reveals that this was not the only frame adopted by the participants at actions. Some participants linked climate change with the problem of local snowpack. The presence of more than thirty local organization booths with divergent agendas persuaded people that animal rights or workers' rights were the real problem. In our interviews with participants at the SLC actions, we found confusion among participants about the campaign-defined problem. In this section, we discuss the dialectical tension between the *unified* and *fragmented* framing of the problem.

There was some unity in the framing of the problem across the SLC actions. The national SIU Web site included marketing materials such as printable posters, flyers, and personal checklists presenting climate change as the major problem. We observed these at each of the actions. At the Sugarhouse action, the SIU materials were a central focus and were prominently displayed. A poster about global warming from the national SIU Web site, a banner with the "Step It Up Congress: Cut Carbon Emissions" message, and an SIU checklist for what people can do in their everyday lives to reduce their impact on global warming were also prominently displayed (see figure 7). At the Washington Square action, a large "Step It Up, Congress: Cut Carbon 80% by 2050" banner reminded people of the campaign and the problem. Our interview data across all three actions reveal that the local organizers consistently framed the problem, at least in part, as "global warming," "climate change," and "carbon emissions."

Despite this unification of the national and local messages, local actions also displayed other ideas about the problem that fragmented the problem frame. Instead of framing climate change as the main problem, local

FIGURE 7. SIU materials displayed at the Sugarhouse action.

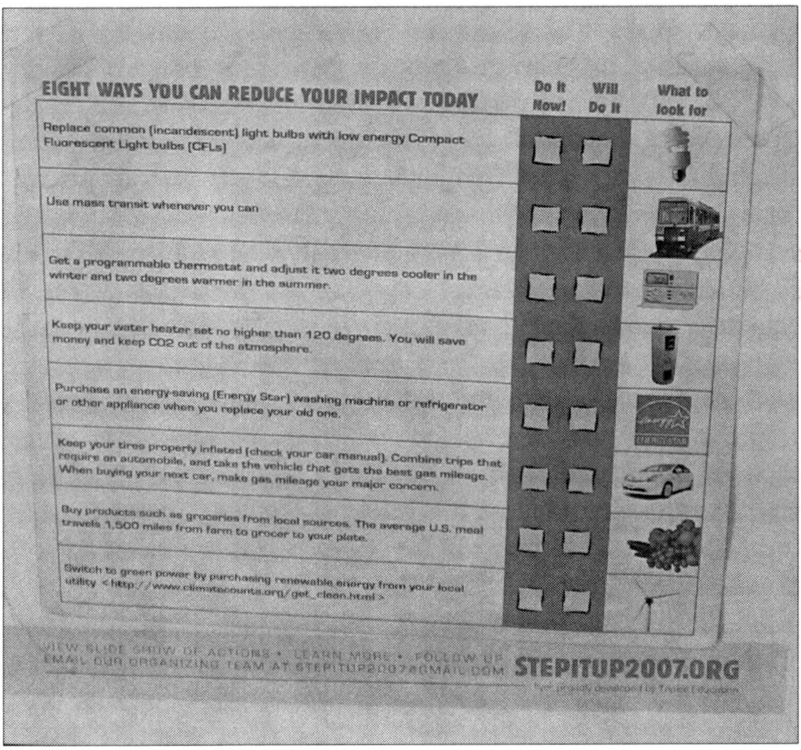

organizers and participants often framed the main problem as something other than climate change. At the Washington Square action, this fragmentation was due in part to the more than thirty "advocacy-type" organizations with disparate messages that were invited to set up informational booths. Each of these groups had its own mission, name, and frame of the problem. Although three out of the thirty groups (e.g., Post Carbon Salt Lake) addressed climate change specifically, most of the groups identified a different problem, not specifically related to climate change but generally related to the environment. For instance, the Utah Animal Rights Coalition (UARC) advocated that eating meat and animal testing were the main problems to be addressed. In this case, the local organizers, knowingly or not, invited competing organizations with different campaigns and problem frames that not only resulted in fragmentation between the local actions and the SIU national campaign, but also risked confusing participants about the goal of the action and the SIU campaign. This opened discursive space for participants to articulate their own ideas about the main problem to be addressed. If everyone has a different idea of what the problem is, then it may be difficult for the campaign to advance a consistent message about the solution. This point was made by a Washington Square participant who claimed that health effects were the main concern; he stated, "The message is askew. They seem totally uninterested in the health effects that automobiles are having on people's health." Similarly, another Washington Square participant claimed that a large part of the problem is automobiles: "People use cars for convenience...It comes down to ego problems, convenience, and power. People feel God-like driving fast." These participant perceptions of the problem, while certainly related, serve as important evidence of the lack of a unified frame of the problem as climate change at SLC SIU actions.

Alternate problem frames are not completely detrimental to a movement or evidence that SIU has "failed" *per se*. There can be value in having "something for everyone" at such actions to spark interest, identification, and mobilization in the movement (see also chapter 11). However, this dialectical tension highlights the importance of communicative management to help link these possible alternate problem frames together to develop realistic solutions that individuals can work toward. For example, some organizers and participants attempted to link the problem of global warming with local issues. The Sugarhouse Park action framed the problem as the risks to local water supply but related those risks to climate change. Similarly, the HEAL Utah booth at the Washington Square action linked climate change with the local controversy over nuclear power. It is important for local SIU organizers to recognize and understand local concerns so that messaging can be used to link these local concerns to the central target. When done well, strategic linking of local issues to the campaign's main problem frame can not only garner support for a general national campaign, but can also raise awareness about the interconnections between local water or air quality and climate change.

The dialectical tension between unification and fragmentation of the problem frame helps us to understand a site of disconnect between the common national SIU message, the diverse messages about the problem at local actions, and SIU participants' perceptions. While it is important for local SIU organizers to be consistent in the message to cut carbon emissions, they also must be prepared to speak to other associated and immediate problems with which audience members are concerned. While some people may realize that climate change exists, they may be more likely to identify with and therefore move to action as a result of a concern closer to home (e.g., nuclear power in Utah or a freeway being built nearby). In other words, a general message about global warming may fall on deaf ears for those who are concerned with other problems. However, it is also important for practitioners to be able to manage the tension between the need for unification while still attending to other, fragmented messages. Organizers must also be able to educate and make people aware of how the problems and concerns they have do, in fact, relate to the bigger problem of global warming due to human-related CO_2 emissions. This will inherently impact how interested parties devise a solution.

Framing the Solution

The overall solution presented by the SIU campaign is clear: "Step It Up, Congress: cut carbon 80 percent by 2050." Our analysis reveals that while the loosely organized national campaign clearly targeted congressional legislation as the solution, it did not specify for local actions *how* to achieve this solution. Consequently, local organizers interpreted various ways to affect climate change that were influenced by local customs and cultures. This disconnect resulted in a dialectical tension between *political* and *symbolic solutions* in the various ways that local SIU organizers and their respective participants internalized the national SIU campaign solution. By political solutions, we mean messages specifically crafted to influence policy and governmental decision makers. Symbolic solutions, on the other hand, target nonlegislative changes within the body politic through enactments, promises, and shifts in language. Both political and symbolic actions can be performed on individual and collective levels, so we are not making a distinction between individual and collective action in this dialectical tension. This political-/symbolic-solution tension relates to ruptures between the congressional solution framed by the SIU national political campaign and the (near-) absence of this frame in local organizer solutions. Only one of the three SLC actions directed political messages toward a congressional audience.

In anticipation of national climate-change day, SIU founder Bill McKibben wrote, "Many more [actions] will still be organized by people who aren't official activists at all, just so concerned about climate change that they're ready to *do something*" (2007; original emphasis). Interestingly, this action phrase, "do something," is prominent throughout our data. Doing something to effect change (through congressional action or otherwise) is obviously essential and critical to dealing with global warming, but this vague rhetorical framing at the national level might have created disconnects regarding the intended audience(s) at the local level.

The local SLC SIU actions called for solutions to climate change that pulled toward the symbolic-solution end of the dialectic. Consequently, the national campaign organizers' vision for congressional action to cut carbon emissions largely faded from view. For example, the Liberty Park action framed symbolic solutions through sun salutations that simultaneously set individual intentions about climate change. The Washington Square organizers framed the solution through speeches (e.g., a list of twenty nonlegislative things that individuals can do to lower their carbon footprint) and diverse messages found in the advocacy booths on the perimeter of the space (e.g., what individuals can buy to reduce carbon emissions).

Unfortunately, the foregrounding of free music relegated these solution messages to background noise. However, this was the only SIU SLC action that did offer e-mail options targeting congressional decision makers (an advocacy option on par with the national organizers' congressional solution frame). The Sugarhouse Park organizers spanned the political/symbolic- solution tension by organizing a collective symbolic-response ritual (the gathering of the waters) and creating a collection of individual political messages on a banner that they had planned to deliver to Utah's governor (see figure 8).[4] While this action framed solutions as both symbolic and political, the symbolic activity entailed individuals collectively promising to effect change, and the political activity involved a collection of individual calls for policy changes at a gubernatorial, not a congressional, level. This solution frame acknowledges that both the body politic and the governmental representatives are needed to mitigate the effects of exponential increases of carbon dioxide in the earth's atmosphere.

Stanley Deetz (1992) inferred that resistance to collective political action is not unique to SLC. In discussing resistance to collective action embedded in U.S.-American discourses, he wrote,

> Clearly there has been a fear of participation equal to or greater than the fear of tyranny, whether tyranny by either autocrats or the majority…The emphasis has been on the freedom *from* the decisions of others rather than the freedom *to* participate in collective decisions. (155; original emphasis)

The absence of political protests as a solution to global warming at the SLC SIU actions seems to reflect this tacit fear of collective political action; that is, the actions tended to offer a means for simultaneous individual actions short of traditional political protesting. One participant explained

FIGURE 8. Individual political action.

that protesting, from her perspective, meant inaction. She said, "I don't think it does a lot of good. It means people are not doing things, not taking action." In short, the absence of collective political action from the SLC SIU solution frames could be construed as local conservative influence, but it also could reflect broader, national political-action discourses.

The focus of the SLC actions toward individual symbolic activities created some tension for those who value collective political action as a viable solution to effect change. Feeling marginalized, a group of participants we interviewed expressed a cynicism toward their apolitical peers. They pointed to signs they were carrying that read: "Stop Global Warming" and "Cut emissions by 80% by 2050." These participants explicitly called for more political solutions, such as what California is doing to outlaw specific problematic elements, at the federal level. In short, the absence of a specific call for collective political action entailing traditional political protesting at the national organizers' level resulted in a missed opportunity in SLC (and perhaps in many other locations) to unite voices in a collective body politic calling for congressional action to cut carbon emissions. Instead, the vague "do something" frame manifested in this dialectical political-/symbolic-solution tension. This offers us a site where we can see how solution disconnects occurred between the national and local organizer solution frames. Unfortunately, the creative license granted to local organizers to "do something" thwarted Congress as the target audience. With the exception of the e-mail advocacy option, local organizers reframed the climate-change solution site away from congressional decision makers and toward an array of personal (yoga and prayer flags), local (gathering of the local waters, listening to free music), or state (gubernatorial advocacy) solutions.

In summary, the consistent call by national SIU organizers for local political action as a solution to cut carbon emissions, coupled with the loosely organized national campaign structure, allowed for local creativity and empowerment but resulted in an array of solutions that were primarily incongruent with the overall national campaign message targeting Congress as the audience (or locus for change). The local political action messages also fell short of reaching protest audiences. Thus, ruptures in communication occurred between the national SIU message and the solutions developed by local actions. Only one aspect of the Washington Square action targeted a congressional audience. Otherwise, the symbolic solutions acted in tension with and potentially undermined the national campaign's call for a political solution. Political and symbolic solutions, as is the case with all dialectical tensions, are not mutually exclusive. Most campaigns/movements will call for both types of solution. However, they come in tension when they are not complementary. In the context of a loosely organized, open-source campaign like SIU, local organizers' focus on symbolic solutions could undermine the national organizers' call for political solutions. This political-/symbolic-solution dialectic requires careful management by national organizers to ensure that their emphasis on a political/symbolic solution is not undermined by the other.

Conclusion and Practical Suggestions

Bill McKibben touted SIU as "the first nationwide do-it-yourself mass protest," a coordinated set of actions with a shared message that would be more effective than a traditional march on Washington (2007). Our chapter points to the benefits and challenges of this new type of environmental campaign. Although a loosely organized structure allows for personalization of actions for local audiences and has the potential to reach more people with the message, it also allows for interpretation and framing of the message that could lessen the effectiveness of the campaign's message (see chapter 11). The dialectical tensions we identified represent the seemingly contradictory pulls faced by the national organizers and local organizers that we believe are typical of many environmental movements and campaigns today. The awareness and careful management of these tensions by national and local organizers can help to take advantage of the benefits as well as handle the challenges that arise from a loosely organized national campaign. Therefore, we will conclude our chapter with a series of practical yet theoretically and empirically informed guidelines for managing these dialectical tensions for the purpose of minimizing communicative disconnects and maximizing the campaign or movement's goals.[5]

Be aware of tensions. Although obvious, our first suggestion is that national organizers and local organizers should be aware of the dialectical tensions we discussed and decide how these tensions can more effectively be managed to achieve the particular goals of the campaign or movement.

Coordinate to reach multiple audiences. If the campaign's goal is to reach a wide audience of participants, national campaigns should make efforts to coordinate actions that are both designed to reach participants who favor protest-oriented events and to raise awareness with participants who would not be likely to attend a traditional protest. Along with the do-it-yourself cookbook of materials for potential local organizers (see chapter 9 for more discussion of the cookbook), national campaign organizers should also provide explicit guidance about managing the dialectical tension between protest-oriented and awareness-oriented actions in particular locations. If there will be only one action in a particular location, an action that begins with an awareness-raising component (e.g., a festival-like rally with information booths) and is followed by a more protest-oriented collective political action (e.g., a march to the local government center with a mandate to take political action) could be one way to manage this tension. The action could be framed as an "information fair and march," and publicity could emphasize different elements of the action in different venues (e.g., focus on the information fair in broad media venues and focus on the march and political action in specialized Listservs and media related to the campaign's topic). If there will be multiple actions in one location, encourage the organizers to coordinate the actions so that some are more oriented toward raising awareness and some are more oriented toward political protest. Moreover, the coordinated local organizers could cross-promote each other's actions. The organizers of each action could be well informed about the nature of the other actions and could let people at their actions know if the other actions may be of interest to them.[6]

Frame what counts as participation. Movements and campaigns should attend to how the movement/campaign frames participant identification with and participation in the movement. If a person's attendance at an action is framed as participation in the movement, participants may not take any next steps after the action, even if this is the goal of the campaign. National and local organizers should be clear about the goals of the campaign (e.g., for Congress to cut carbon emissions) and clearly frame participation by defining how actions relate to the campaign goal (i.e., continuing to ride your bike is a good thing, but pressuring Congress to cut greenhouse-gas emissions will require that you contact your senators on a regular basis). Specifically, campaigns should design and disseminate specific action plans that are tailored for specific audiences/people. These action plans should create relationships between individual and collective actions.

Address fragmentation through interconnection. Part of the success of a campaign/movement comes from the development of a shared understanding of the problem in the status quo being addressed (Stewart, Smith, and Denton 2007). Some fragmentation of the framing of the problem is inevitable and can benefit the movement/campaign. However, fragmentation can also make it difficult to convince participants that the campaign's problem is the most important one, which can lead to fragmentation of solutions or lack of adherence to the campaign. By anticipating the alternate problems/concerns of the audience(s), local organizers can frame them as being interconnected with the campaign's definition of the problem. For example, the organizers of the Washington Square action could have framed the problem of human-caused climate change as an umbrella problem that included the problems of asthma, nuclear power, and so forth in a way that was consistent with the movement/campaign's goals.

Decide whether to emphasize political or symbolic solutions. It is important to recognize the value of the symbolic and the political in a movement/campaign's solution. Symbolic actions and political actions are not entirely separable concepts, yet a tension exists for campaigns seeking both symbolic (constitutive) change and political change (Buechler 2000). We recommend that campaign organizers decide what balance of symbolic and political actions is important for the campaign and strive for a complementary blend of both in the framing of the solution. Considering the unique context in which a campaign is initiated, organizers could ask themselves whether the presence of many people at a protest as a symbol of resistance to the status quo is more important than advocating for a specific legislative change, or vice versa (DeLuca 1999; Schutten 2008). In an awareness-raising phase of a campaign, for example, it might be more appropriate to emphasize symbolic solutions that can lead to the desire to advocate for political change. Or, if the campaign has a sufficient base of supporters who are already engaged in nonpolitical actions (like lowering their carbon footprints), it would be most appropriate to emphasize political solutions. Otherwise, more emphasis on symbolic solutions may just reinforce behaviors instead of pushing for change.

Use push and pull methods of message dissemination. We noticed that the Washington Square action was planned and spatially constructed to be a *pull* medium, not a *push* medium. A pull medium, such as the Internet, relies on people taking initiative to find information, whereas a push medium, such as television advertising, disseminates information to anyone watching. At the Washington Square action, having informational booths was very helpful, but their placement around the perimeter of the action, away from the music stage, meant people had to choose to look at the information at the booths. These booths might have only pulled in people already interested in the movement. Combined with a majority of the focus of the stage on the musical acts and just a few speeches or articulations of the SIU message, the people who showed up for the free concert may not have felt the pull of the booths and hence did not leave with a clear understanding of the action. Designing this action to have more aggressive push techniques for getting the message across might have more effectively spread the message. This could include more focus on the SIU message from the stage. Or, drawing on a word-of-mouth approach, the organizers could have pushed the message by having organizers walk through the crowd to "spread the word" about the message by asking simple questions such as "Have you learned today how you can step it up?" A combination of push and pull techniques can be helpful in managing communicative disconnects at the action itself.

These suggestions are not the only ways to manage the dialectical tensions we have developed in this chapter, but they provide a good starting point for examination of how rhetorical framing can be used by campaigns and movements to further their goals and minimize communicative disconnects. Furthermore, these are not the only dialectical tensions that exist for movements and campaigns. For example, in addition to the tensions we discussed in this paper, we also saw tensions in vagueness/specificity and long-term/short-term articulations of the campaign's solution. We believe that further research into the common dialectical tensions that affect campaigns and movements, particularly ones that follow the SIU model of movement building, is extremely important for understanding and utilizing new social movements as a proactive method for creating positive change in society.

ENDNOTES

1. Danielle Endres (assistant professor of communication at the University of Utah) is the first author of this chapter. Deborah Cox Callister, Autumn Garrison, and Samantha Senda-Cook (in alphabetical order, PhD students at the University of Utah) are co-second authors of this chapter. Julie Kalil Schutten (lecturer, Northern Arizona University) is the last author. The authors would also like to thank Craig O. Rich, Rulon Wood, and Wayne Davis for their help in gathering materials for analysis.
2. We treat identity as a rhetorical construction as opposed to an innate quality, which means that people make (un)conscious decisions about who they are and how they express that identity. We do not think that an essential, natural identity exists that simply shines through behaviors and attitudes.
3. e2 Citizen stands for the Salt Lake City Environmentally and Economically Sustainable Program. Started by the mayor's office, the program is "designed to educate and support citizens who take steps in their own lives to address climate change and the further degradation of the planet" (http://www.slcgreen.com/e2citizen/default.htm).
4. In our "member-check" meeting with the Sugarhouse organizers, they told us that they were not able to deliver the sign to the governor. They tried multiple times to contact the governor to set up a time to give him the banner, but they were unable to set up a meeting. The governor signed the Western Regional Climate Initiative in May 2007 (Bauman 2007), and it is unclear if he was aware of the Sugarhouse Park action banner at that time.
5. We call these guidelines instead of rules because the context, goals, and audiences are different for each movement or campaign; there is no foolproof list of steps to take.
6. We do recognize the challenges to coordination of local actions. Indeed, our interviews with the organizers of the Sugarhouse Park action reveal that there was some tension among the organizers of the three actions in SLC. However, we do not take the existence of tension between the SLC organizers as a sign that coordination between the actions is impossible. On the contrary, if each action is primarily designed for a different audience, the organizers would not be in competition for the same audiences. Similarly, if actions are coordinated to take place at different times, it could also reduce competition over audiences.

References

570 KNRS. 2007. Community calendar, April 14, 2007. http://www.570knews.com/slc-common/community_calendar/ (accessed May 14, 2007).

Althusser, Louis. 1984. *Essays on ideology*. London: Verso.

Bakhtin, Mikhail. 1981. *The dialogic imagination: Four essays by M. M. Bakhtin*. Austin: University of Texas Press.

Bauman, Joe. 2007. Guv commits Utah to climate accord. *Deseret Morning News*, May 22, 2007. http://deseretnews.com/dn/view/0,1249,660222802,00.html (accessed November 27, 2007).

Baxter, Leslie A. 1988. A dialectical perspective on communication strategies in relationship development. In *Handbook of personal relationships*, ed. Steve Duck, 257–274. London: Wiley.

Bhattacharya, C. B., and Kimberly D. Elsbach. 2002. Us versus them: The roles of organizational identification and disidentification in social marketing initiatives. *Journal of Public Policy & Marketing* 21:26–36.

Buechler, Steven. M. 2000. *Social movements in advanced capitalism: The political economy and cultural construction of social activism*. New York: Oxford University Press.

Burke, Kenneth. 1966. *Language as symbolic action*. Berkeley: University of California Press.

Campbell, Karlyn K. 1973. The rhetoric of women's liberation: An oxymoron. *Quarterly Journal of Speech* 59:74–86.

Canham, Matt. 2007. "Utah less Mormon than ever." *The Salt Lake Tribune*, November 18, 2007.

Deetz, Stanley, A. 1992. *Democracy in an age of corporate colonization*. Albany: State University of New York Press.

DeLuca, Kevin M. 1999. *Image politics: The new rhetoric of environmental activism*. New York: Guilford Press.

Entman, Robert M. 1993. Framing: Toward clarification of a fractured paradigm. *Journal of Communication* 43:51–59.

Frey, Lawrence, and Kevin Barge. 1998. *Managing group life: Communication in decision-making groups*. Boston: Houghton Mifflin Co.

Gadette, Jamie. 2007. Music picks. *Salt Lake City Weekly*, April 12, 2007.

Gamson, William. 1975. *Strategy of social protest*. Homewood, IL: Dorsey Press.

———. 1995. Constructing social protest. In *Social Movements and Culture*, ed. Hank Johnston and Bert Klandermans, 85–106. Minneapolis: University of Minnesota Press.

Goffman, Erving. 1974. *Frame analysis: An essay on the organization of experience*. Cambridge, MA: Harvard University Press.

Granzin, Kent L., and Janeen E. Olson. 1991. Characterizing participants in activities protecting the environment: A focus on donating, recycling, and conservation behaviors. *Journal of Public Policy & Marketing* 10:1–27.

Johnson, Scott D., and L. Long. 2002. "Being a part and being apart": A dialectical perspective on group communication. In *New directions in group communication research*, ed. Larry Frey, 25–42. Thousand Oaks, CA: Sage.

Kramer, Michael. 2004. Toward a communication theory of group dynamics: An ethnographic study of a community theater group. *Communication Monographs* 71:311–322.

Lakoff, George. 2004. *Don't think of an elephant: Know your values and frame the debate*. New York: Chelsea Green Publishers.

Mabry, Edward A. 1999. The systems metaphor in group communication. In *The handbook of group communication theory & research*, ed. Larry R. Frey, Dennis S. Gouran, and Marshall S. Poole, 71–91. Thousand Oaks, CA: Sage.

McKibben, Bill. 2007. Step It Up: Introducing a brand-new, mass-protest climate movement for 2007. *Grist: Environmental News and Commentary*. http://www.grist.org/comments/ dispatches/2007/01/08/mckibben/index.html (accessed February 2, 2007).

McKibben, Bill, Phil Aroneanu, Will Bates, May Boeve, Jamie Henn, Jeremy Osborn, and Jon Warnow. 2007. *Fight global warming now: The handbook for taking action in your community*. New York: Holt.

Melucci, Alberto. 1995. The process of collective identity. In *Social Movements and Culture*, ed. Hank Johnston and Bert Klandermans, 41–63. Minneapolis: University of Minnesota Press.

Ott, Brian L., and Eric Aoki. 2002. The politics of negotiating public tragedy: Media framing of the Matthew Shepard murder. *Rhetoric & Public Affairs* 5 (3): 483–505.

People Protecting the Planet. 2007. http://events.stepitup2007.org/april/events/show/389 (accessed August 25, 2009).

Pollan, Michael. 2008. Why Bother? *New York Times Magazine*, April 20, 2008. http://www.nytimes.com/indexes/2008/04/19/magazine/index.html (accessed June 1, 2008).

Reese, Stephen D. 2001. Prologue—framing public life: A bridging model for media research. In *Framing public life: Perspectives on media and our understanding of the social world*, ed. Stephen D. Reese, Oscar H. Gandy Jr., and August E. Grant, 7–32. Mahwah, NJ: Lawrence Erlbaum.

Salt Lake City. 2007. http://www.ci.slc.ut.us/ (accessed April 14, 2007).

Salt Lake Spring Runoff Celebration. 2007. http://events.stepitup2007.org/april/events/show/607 (accessed August 25, 2009).

Schutten, Julie K. 2008. *Coming out of the coven: Contemporary witches, hidden populations, and new social movements*. Saarbrücken, Germany: VDM Verlag Dr. Muller.

Step It Up. 2007. The Step It Up team. http://april.stepitup2007.org/ article.php?id=79 (accessed June 10, 2008).

Step It Up Salt Lake City. 2007. http://events.stepitup2007.org/april/ events/show/1583 (accessed August 25, 2009).

Stewart, Charles J., Craig A. Smith, and Robert E. Denton. 2007. *Persuasion and social movements.* 5th ed. Long Grove, IL: Waveland Press.

Chapter Interlude

Interview with Adele Bealer, Co-Organizer of Salt Lake City Gathering of the Waters SIU Event

Samantha Senda-Cook

Background note: Adele Bealer was one of several primary organizers for Step It Up (SUI) Salt Lake City, Sugarhouse. At the time, she was a graduate student in the environmental humanities program at the University of Utah. As part of a class project, the students had to organize an SIU event. What follows is an edited transcript of an e-mail interview, returned on May 8, 2007. Senda-Cook and Bealer met in a class they had together in the spring of 2007.

SENDA-COOK: **What was the vision for the water celebration? (What were motivating factors? Do you think you achieved that? If so, how? In other words, what contributed to the completion of that vision?)**

BEALER: We [the students who organized the SIU action] felt that to best illustrate our community's stake in the climate-change issue, we needed to focus on our dependence upon snowfall and snowpack for our water supply. Inspired by Thomas Berry's call for a new ritual in *Evening Thoughts* and by other authors whose works we've shared this year [in the environmental humanities program] (Gregory Bateson, Elizabeth Kolbert, E.O. Wilson), we wanted to create a celebration of our local watersheds that would encourage participants to feel connected to each other through that most vital element, and to also feel a sense of the fragility of this precious and threatened resource. Participants were invited to bring water from their home watershed to commingle with the waters from each of our seven primary watersheds here in the Salt Lake Valley, and then to take small quantities of that mixture of waters to pour over a wood fire, symbolizing the threat posed by global warming. There was a moment of silent reverence during that symbolic extinguishment that seemed to unite every participant. Just as our voices united in the refrain, we read together following the acknowledgement of each watershed.

SENDA-COOK: **What were your standards when organizing this event? (How did you decide to do what you did? Did you have a template? Where did it come from? How would you judge if the event was successful?)**

BEALER: We contacted Rob Thayer,[1] author of *LifePlace*, and asked for a copy of the watershed ritual he developed for his home watershed of Putah Creek (California) and that he describes in his book. He directed us to a Gary Snyder poem as the inspiration for his ritual, and urged us to create our own. There are 10 of us in this graduate class, and several are very talented writers. We just let them go and trusted their instincts. The result was beautiful, moving, and meaningful on many levels. For all of us, the event was successful just in the creation of it, having the 50 or so participants who joined us in this celebration was just [a] lagniappe.

SENDA-COOK: **How did you learn about SIU?**

BEALER: Terry Tempest Williams, our class discussion leader, mentioned Bill's plans in one of our evening classes, and several of us had seen the early buzz on *Grist*.

SENDA-COOK: **Why did you choose the place that you did? (How did that contribute to challenges or successes you faced? Was it appropriate for this event? Why? What elements were you looking for/ did you get in the selection of this place? Did you feel anything different in this place on that day than you normally did there?)**

BEALER: Sugarhouse Park represents a multicultural slice of Salt Lake City life that seemed utterly appropriate for an action designed to bring awareness on a personal level to friends, families, and neighbors. With one of our celebrated creeks running right through the middle of the park, the natural outdoor setting was the perfect location for what we had in mind. During the event, I was so focused on the language of the ritual that I honestly have NO idea what else was going on in the park at that moment. It was a very special experience.

SENDA-COOK: **What was your role in the planning process and how did it change over time? (I remember you telling me a little about how your event changed and the leadership changed. What were those changes specifically? How did these changes affect the overall feel of the event? How did they affect the group dynamics?)**

BEALER: We did all of our organizing by email, with weekly meetings to keep everyone on the same page. We delegated general categories of activity (ritual, PR, day-of-details) and relied on each other to keep the enthusiasm going. My role was as the primary liaison for the national SIU organizers, so I concentrated on keeping those emails forwarded to all the group and on summarizing group meeting results for those who missed a meeting. I think our event changed most in our perception of its size. We started out thinking of doing something on a much larger scale but quickly realized that there were other events that were going for that kind of audience. We really wanted our ritual to have a more personal and connected feel, and I think that choosing to go that route really involved each of us in a more personal way. I think that our ritual became far more intimate and meaningful because of that change.

SENDA-COOK: **What do you think was the most important message of that day? (What did you think about the rally? What was the purpose? Who was the intended audience? Did you draw attention for those not involved?)**

BEALER: Certainly our event focused on the fact that we are all implicated in this ecological community, and that we must each do what we can to make a difference. We wanted to attract our friends and families and our neighbors to our event, and for the most part we were able to do that. A few of our participants were strangers to us, drawn by local press coverage, but most were persons who are already concerned about climate change issues. While we were especially gratified at the participation of those who were not familiar to us, we felt that our action was primarily attractive to an audience who were already concerned about the potential impacts of climate change.

SENDA-COOK: **What actions should happen next? (What would your ideal world look like with regard to climate change and politics? What will happen if we don't take any action?)**

BEALER: We'd like to see our governor, our senators and representatives, and our state legislators demonstrate very real awareness that the well-being of their constituents depends on their "stepping up" to this issue. We are still trying to make an appointment to deliver the banner that all our participants signed to Governor Huntsman. Ideally, we'd like our elected officials to demand stricter emissions standards and to quit suggesting that there's any question about the scientific community's stance on fossil-fuel consumption and its link to climate change. If we don't take any action, then every day can be a red burn day here in the Salt Lake Valley. 365 days a year of particulate particles obscuring our blue skies. Now that's something we can all demand that we NOT look forward to in the future.

SENDA-COOK: **Thanks again!**

Endnote

1. Rob Thayer is professor emeritus of landscape architecture at the University of California, Davis.

Part II

Modes of Organizing

Chapter 5

Organizing Step It Up 2007

Social-Movement Organizations as Collective Resistance

Todd Norton and Travis Paveglio

Organizing is a fundamental process in the progression of social movements, yet too few practitioners and academics explicitly focus on the *organizational nature* of movements. This is an unfortunate oversight, given the increasing role internal-organizing structures (bureaucratic versus collectivist) and processes (decision making, power, and so forth) play on the public accomplishment of a social movement.

Given the increasingly sophisticated demands on social movements (e.g., chapter 9), we need more nuanced yet practically oriented knowledge about the organizational dynamics that sustain and facilitate them.

As Fitzgerald and Rodgers (2000) clearly indicated, the realm of social movement is no longer about a temporarily organized citizenry targeting a specific collective action such as marching on a political center. Social movements are increasingly driven and structured by organizational entities—institutions established for the explicit and primary purpose of organizing citizen action. Though not a legally established not-for-profit organization, Step It Up (SIU) is a social-movement organization, an entity created for the primary purpose of organizing action among networks of organizations and interested publics (see chapter 7).

In this chapter, we examine SIU to highlight techniques and strategies used to advocate for climate change, specifically the role resistance plays in bringing about change in broader social and political systems. As calls for change, social movements function as unequivocal challenges to the status quo to change policy objectives or processes. This is clear in SIU's primary message demanding that Congress "cut carbon emissions in America 80 percent by 2050."

There is thus a fundamental tension in SIU, and in many other similar situations, between the social movement and the organizing entity that orchestrates the movement. On one hand, the movement is couched in terms of a bottom-up or grassroots challenge to the dominant structures of control, such as corporations or government. On the other hand, social-movement organizers provide centralized control and decision processes for the movement—thus undercutting the bottom-up nature of the movement. In fact, when we interviewed SIU national organizers, they noted,

> We definitely recognized that tension. We got a lot of e-mails and a lot of calls from people that were telling us, "You know, 80 percent is too much or 80 percent is too little," and a lot of people who were talking about completely different things like becoming a vegetarian or not idling in your car.

As a campaign organized by a central decision-making unit, SIU provides us with an opportunity to learn how resistance is successfully orchestrated and expressed. The question "What constitutes meaningful resistance and change?" is more complex than one might expect, with organizational perspectives largely divided over the issue. Researchers debate whether individual and largely symbolic forms of resistance are adequate and meaningful (Mumby 1997, 2005) or if resistance needs to be collective and change the material conditions of a situation (Ganesh, Zoller, and Cheney 2005). These different perspectives are surely recognizable to social-movement organizers, as the debates parallel each other and raise important questions regarding how to successfully organize a movement for change.

Our objective is to understand whether movement organizations such as SIU are (1) primary beginnings of collective action to genuinely shift larger social systems, including their material conditions, or (2) reproducing a constantly evolving narrative of control/resistance between organizations of environmentalists and business interests—an ongoing and largely unproductive conflict. We thus hope to offer a theoretically informed and practically oriented account of this debate that may inform future movement-organizing processes. We begin by discussing the ever-increasing role of social-movement organizations in public issues such as climate-change policies.

Social-Movement Organizations

Social-movement organizations are becoming increasingly prevalent in our political landscape and warrant additional consideration. They are significant because the networks among them constitute a semi-invisible but well-organized and often highly orchestrated space in society. Take, for example, two significant social-movement organizations related to land management and environmental issues: the Blue Ribbon Coalition champions motorized recreation and conservation of public or private lands, while the Sierra Club strives to conserve and protect wild places from human-induced degradation. An exploration of either organization's Web site uncovers prominent "related links," pages connecting users to countless other groups advocating similar positions and commensurate ideologies on the uses of public lands or opportunities for activism surrounding different environmental issues.

These coalitions may enhance the movement by diversifying the resources available to it, or they may hinder it by further establishing insular cultures of activism that fail to carry the organizational message to a larger audience. In fact, as they suggested in an interview, the national organizers of SIU took advantage of these pre-existing networks with groups such as Youth Climate Movement, Sierra Club, and National Resource Defense Council and the e-mail lists of MoveOn.org. Simply put, the movement spread quickly because of existing interorganizational structures.

Though these organizations and groups differ considerably, they form a temporary, strategic alliance based upon similar norms, opinions, and ideologies regarding the direction of the movement and how things *ought to be*. Because SIU explicitly challenged Congress, our analysis highlights the role of resistance in movement organizing.

SOCIAL-MOVEMENT ORGANIZATIONS AS RESISTANCE

As forces for change, social-movement organizations generally utilize resistance and oppositional strategies to foster change in the status quo. Given that opposition is a central goal, a means of assessing how to foster effective resistance for change is a useful consideration. The difficulty is that there are multiple ways to assess oppositional politics such as those that played out in SIU: a dialectical view and a transformative perspective.

Resistance Organizing as Dialectical

Dialectical approaches to organization studies problematize the socially and historically contingent dimensions of organizations. A dialectic generally refers to the tension experienced by people over two equally possible but opposing tensions. So, in a relationship context, a desire for stability in a partner and a desire for change and excitement are dialectic inasmuch as having more of one may mean having less of the other. In the context of organizing social movements, dialectical tensions arise when stakeholders simultaneously oppose one another and sustain one another's existence (see chapter 4).

Dialectics encompasses notions of conflict and emphasizes the ways organizational actors create and renegotiate meanings for organizational processes through interaction and shared meaning (Grossberg 2003; Collinson 1992; Clegg 1981). It is based upon the premise that organizations and larger social structures are composed of competing discourses (Fairhurst and Putnam 2004; Ezzamel, Willmott, and Worthington 2001).

In a dialectic of control/resistance, opposing organizations struggle to control a situation. As Mumby argued, "resistance does not simply occur in opposition to dominance, but is often implicated with it in mutually defining ways" (1997, 363). As such, two groups vigorously opposing one another are equally implicated in sustaining one another's efforts. Norton (2008), for example, demonstrated how the environmentally favorable policies of then-president Bill Clinton led to and strengthened the emergence of property-rights advocate groups in the Western states. Such is the case with SIU and local environmental groups being caught in a dialectical tension of whether to join a larger movement. While joining the larger movement will bring additional attention, this additional attention comes at the cost of having to sacrifice some of their specific focus.

Understanding dialectics helps us consider the role resistance plays in a social-movement organization like SIU and the institutions it opposes, such as corporations and government agencies. As Ashcraft clarified, "resistance and control are both moves of power" (2005, 70). The ability to influence the meanings underpinning organization is thus one of the most important forms of power contestation, as such processes legitimize certain interests exercising power over others (Deetz and Mumby 1990).

Approaching the problem traditionally, SIU represents the politically marginalized segment of society in contemporary environmental politics. As the "underdog," its resistant practices are not only justified but expected. Approaching the problem dialectically, however, SIU could be viewed both as a means of resistance and as an attempt at control. From this dialectical vantage point, we can begin to explore what it means to practice meaningful resistance: whether resistance is located in symbolic performances of protest (see chapter 8), if it changes the system it opposes, or if it simply reproduces that system (Deetz 1985).

By encouraging us to question our assumptions about the function of social-movement discourse, a dialectical view helps us analyze opposition more systematically (Peterson and Norton 2007). For example, at the same time as organizing or attending an SIU event may give participants a sense of their own political power, it may also validate the political system SIU organizers seek to change. A dialectical analysis demonstrates not only the way resistance and control are locked in a systemic and continual struggle, but also that resistance discourse may function to undermine its oppositional goal. In short, advocates of a dialectical perspective see everyday practices as a meaningful site of organized resistance because social actors may choose to deviate from the system of domination (see chapter 10). The dialectical reading of organized resistance thus explores how people's everyday actions both contest and reproduce an established system of dominance (Mumby 2005).

A Transformative View of Collective Resistance

Some calls for collective action question whether we place too much emphasis on the relationship between opposing forces in social movements (Ganesh, Zoller, and Cheney 2005). While the complexities involved in mutually opposing and sustaining forces are interesting, what matters for social-movement organizations is that they actually transform the system through changes in material conditions such as new policy. From this perspective, symbolic and individual-level performances of resistance are inadequate precisely because they do not make any practical change or transform policy.

Collective-action scholars argue that meaningful resistance is limited to those practices that have some ability to "effect large-scale, collective changes in the domains of state policy, corporate practice, social structure, cultural norms and daily lived experience" (Ganesh, Zoller, and Cheney 2005, 177). This is particularly important for environmental social-movement organizations because the natural-resources issues are both impacted by and cut across political boundaries (Peterson and Norton 2007).

Viewing resistance as collective action borrows, in part, from the long tradition of research on social activism, including social- and new-social-movement theories (Fitzgerald and Rodgers 2000). This classic view of organized resistance includes informal organizing and mobilization based upon shared beliefs regarding a social or political injustice of some kind (Della Porta and Diani 1999), and can either complement or contrast established organizational models (Stohl and Stohl 2005). Hubbard simplified this shift in global resistance by building off a familiar phrase: "think globally; act locally; for global change" (2006, 321). Resistance organizations create networks of groups that are linked by larger purposes but operate in different locations and on various issues.

Calls for collective action, then, add a transformative criterion to our assessment of social-movement resistance. From this perspective, symbolic acts and individual gestures of opposition are interesting but largely impotent because the basic structure of a situation may be unaltered—for example, the SIU attempt to persuade Congress to reduce emissions in the United States may or may not influence policy. In this way, SIU challenges us to think about not just the symbolism of resistance but also the possibilities of changing material conditions.

Both of these approaches to organized resistance contribute to our ability to assess effective resistance in a social-movement context: the dialectical perspective, suggesting that we consider the way opposed forces are mutually sustaining, and the collective-action approach, suggesting that movements must somehow transform the structure of a situation (e.g., emissions policy) to be considered successful. With these distinct ideas in mind, we engage SIU as collective resistance.

Engaging SIU 2007

Our analysis of the SIU campaign employs a discourse-analysis approach (Titscher et al. 2000) through which we uncover the meanings in the written and spoken communication of SIU participants. We use this approach to explore how SIU participants view the symbolic and material dimensions of their campaign and compare it to their ability to transform the material conditions of global warming and emissions policy. Spoken texts used in this analysis include site descriptions and interviews with SIU participants at events across the country, including individual events in Washington state, Utah, North Carolina, Michigan, and Texas. Written texts such as the SIU Web site, promotional materials, and individual event presentations are also included as important sites to explore the interplay of intended goals, specific ideologies, and the diversity of participants engaging the movement.

We used discourse analysis because it acknowledges the power of language in creating or modifying social relationships and the way these interactions help individuals construct their reality (Titscher, Meyer, and Vetter 2000). Power and control are two major themes underlying any organization striving for social change, because their efforts imply a perceived imbalance of power and suggest that discourse may influence the relevant material conditions. For the purposes of this study, we focused on the motivation of SIU as a control and a resistance dialectic between similar "green" organizations and big business/government interests or as a transformative event intended to break free of this cycle. Hence, we also borrow from critical discourse analysis (Fairclough 1993) to identify the power of language in creating the shared agreement necessary for the creation and exercise of power on the topic of carbon emissions. This is important to SIU, as its organizers challenged existing discourse about climate change by introducing contradictory messages related to current power articulations.

Our analysis of the SIU event reveals a social-movement campaign fueled by both dialectical and collective-action frameworks. Our discussion focuses on the lessons SIU offers emerging forms of environmental activism and social change. Of particular importance is how the collectively and individually motivated nature of campaigns such as SIU modify participants' attachment to an organizational cause and the viability of their actions in achieving those goals.

Interpreting SIU 2007

We found that participants made sense of SIU in the following ways: (1) as a highly symbolic interplay of control and resistance tensions between their movement and government/big business, and (2) as a catalyst for collective actions leading to new policy on carbon emissions. However, we also found that participants' contradictory views of SIU undermined the above tendencies. The following sections explain these dynamics.

SIU as Dialectical Opposition

Participants' use of symbolic language and frequent mention of competing interests were a clear indication of tensions in SIU. Frequently reported goals included issues such as "raising of awareness" or "making voices heard," implying a symbolic struggle of competing discourses or attempts to gather momentum toward collective or individual action.

Many participants across the nation framed their understanding of SIU and its goals as purely symbolic and never mentioned any specific steps toward collective action. This is evident in the following statement from a Utah participant about the purpose of SIU:

> In effect, [the goal is] to raise awareness of the state of affairs that life faces in the future. I'm a component of life on the planet. I'm a species that can use words and communicate. It's best to think of all life versus the greasy money in wallets.

Other examples included the organizers of the Sugarhouse Park SIU event in Utah, who reframed their day of climate action as a "ritual" because of its spiritual potential. As a participant in Michigan said, "Sometimes, even though we are a democracy, we become voiceless, and this is a chance for us to have our voices heard." Here, the goal is recognition and symbolic worth—a chance to participate in governance—but only through representation.

These discourses point to the symbolic nature of SIU resistance without specific action items to transform the problem of emission levels. In this way, SIU effectively functions as a symbolic acknowledgment that a problem exists, and attention is not directed at transformative potential of solutions. As one participant in eastern Washington said,

> I mean, people are starting to realize that this isn't just a slight change in the weather, like Grandpa saying, "It was much worse when I was a kid..." So I think we are more at awareness; we aren't anywhere near figuring out what can be done about that.

Other participants conceded that while awareness of and education about carbon emissions are vital steps in the progression of the climate movement, some sort of collective action is needed to achieve the goals of reduced carbon emissions.

SIU Opposes the Status Quo

Perhaps the most telling indication of dialectic tensions underlying SIU was the frequent and explicit opposition of SIU to competing against status quo interests such as big business or government. These sectors (important institutions in the management of emissions and global-warming policy) were treated as the collective enemy of SIU and environmental movements generally. As one Seattle resident said, "The climate crisis is a failure of business. It's an economic and ecological failure." A speaker at an event in Texas suggested a similar point: "You know, we're a great democracy...some might call it 'economacry.' That's a democracy driven by big climate money, whose [stakeholders] have [a] special interest in keeping their big, private money keeping-on coming! At your and [my] expense[s]."

In this way, SIU participants blamed business as the cause of global warming, yet failed to talk about the consumer's role in emissions in the private sector. More nuanced conversations regarding the mutual role consumers and businesses play was evident in a lengthy interview with a small group of Spokane residents. In that conversation, a local resident who worked for Eco Depot spoke at length about the energy (and fiscal) savings available through energy-efficient products. In the bulk of data available from the numerous research sites, however, these details were left out of the attention-grabbing discourse of speeches. At the Spokane site, the media-focused aspects, such as taking a picture and somehow incorporating the central message regarding the reduction of emissions that national-level organizers demanded be included in each local event, did little if anything to explore these practical challenges and possibilities for emissions reduction (see chapter 8). Media events surrounding the central message simply positioned business and especially government in opposition to SIU goals and objectives. In the words of one North Carolina resident, SIU is attempting to "deliver a message to legislators on all levels who just do not appear to be listening to us." Similarly, a Seattle resident said,

> What we are trying to do is get Congress to realize that there is a huge group of people in this country who are waiting for them to lead us where we need to go…We need them to take up the gauntlet and go for it, and they need to force us to go there.

In summary, SIU used an oppositional tactic to symbolically but abstractly represent the potential of the climate movement and to justify its resistance of the status quo and what it perceived to be the dominant voices in emissions policy. The strength of this tactic, based upon the significant attendance at SIU events across the United States, is that it provides a jarring distinction between the perceived status quo and the opposition infused by the social movement. The limitation of it, however, is that it is not collaborative or productive; by holding just one or two sectors responsible for the entire problem, such a strategy further polarizes discussion surrounding climate-change politics. While this oppositional tactic is useful in the short-term as a rallying point, in the long-term, it *constrains* movement participants from collaborating with organizations with contrasting views. A slogan challenging Congress and event discourse blaming business not only fails to invite members from those organizations, it places them in an oppositional and perhaps defensive position.

The oppositional stance of SIU also limits the possible participants by including ideologically supportive organizations (e.g., Sierra Club, Youth Climate Movement, and a host of locally based environmental groups) and excluding alternative but potentially influential perspectives (e.g., local chambers of commerce, industry experts) that could undoubtedly offer considerable insight into the practical challenges and possibilities involved in reducing carbon emissions by 80 percent. In short, this sort of simple opposition seems successful because it unifies large numbers of participants. Such rallying cries do little, however, to analyze or transform the practical complexities of climate change.

SIU as Collective Action

The movement-organizing tensions in SIU include the move toward collective action with transformative potential. These actions included defined legislative goals and propositions for regulation of carbon emissions, including strategies for integrating business interests. Such practices attempted to affect large-scale change in the systems underlying the issue and the symbolic arguments behind them. This is apparent in the message SIU sent to Congress about carbon emissions and in its ideas for tangible actions each region could mandate among business interests. Many SIU events included information booths advocating "green building," energy-saving products, or alternative-fuel vehicle manufacturers. Local politicians and event organizers also mentioned regional legislation attempting to foster community-wide change. As the mayor of East Lansing, Michigan, explained to participants at an SIU event,

> I learned what an RPS is: renewable portfolio standard. Which means we're looking at our energy use and making it renewable. And we've established a goal of 10 by '10, 15 by '15, and 20 by '20. The 20 percent is more aggressive than a lot of communities. And it will require the city to purchase energy-efficient products and use biofuels, and we're working on all that.

From a collective-action perspective, participants framed SIU not in terms of working collaboratively with those controlling the political system but, rather, in terms of taking over that political system. As one Michigan participant stated, "It's a grassroots initiative. It is time for *us to take control* because the federal government is not acting. We need to start at the local level." The emphasis here is a clear distinction between active and passive resistance; collective action entails people actually *performing* change through *organizing* rather than *requesting* change to be made. Active resistance would have proactively targeted a discernible change in the problem, such as the possibility of reducing energy usage (a key component in the production of energy, hence emissions). SIU was largely passive inasmuch as it did not target actions to transform the system producing emissions; it simply demanded that some other entity—that is, Congress—cut emissions.

Through calls for collective action in SIU, participants reclaimed their power to effect change by recognizing the power their shared collective can have. As one North Carolina participant said,

> You would expect that the leaders would be ahead of the people. That is the way it used to be: the leaders were ahead of the people; they knew more, and they led the people. This is not the case now. The people are ahead of the leaders.

The important difference here is that participants viewed the campaign in terms of shared action rather than individual measures, including ideology and practices.

Collective Versus Individual Action

The difficulty of managing collective action is the maintenance or balance of both individual efficacy and the interest of the broader movement. While the passages above clearly point to the possibilities of a social collective wresting control from the status quo, these ideas also minimized the important role of individuals. In short, collective action must be managed in concert with individual responsibility and calls for "everyone to do their part" in reducing carbon emissions.

In contrast to the rallying function of symbolic opposition reviewed above, a focus on individual action effectively reduced the ideological strength of the movement by undermining the need to gather in the first place. This was evident in people both participating in and distancing from the movement. As one Michigan resident said, "I ride my bike and walk. I don't own a car. I'm not an eco-Nazi or anything, but I do my part." A participant in Utah voiced a similar sentiment: "If the change is one at a time, so be it. We need to take responsibility for our own actions. We only can take responsibility for our own actions. It can't be just other people." Yet another Utah participant said, "These kinds of concerns will only be solved through a populist movement," while a participant from Seattle synthesized the entire event as such: "The most important message is that if people work together, then we can bring change."

Focusing on individual responsibility (clearly an important dimension in many environmental social movements) inevitably draws attention away from the broader collective. Collective-action approaches to social movement and change are thus caught in an inevitable tension: emphasizing the collective level of action underplays the important role of individual choice; focusing on the individual level of choice reduces the eminence of the movement generally. The conception of individual responsibility, whether in concert with larger collective actions or as the ultimate solution to reducing carbon emissions, draws power away from the group and its shared message by assuming that collective concern cannot motivate others to change. It also negates the call for regulation and policy in *mandating* carbon emissions, including the primary goal of SIU, by presupposing that only individual choice can make a difference.

LEARNING FROM SIU

Our analysis shows that organized resistance contains dialectical tensions of control/resistance and elements of collective action intended to produce change in broader social systems. In the following section, we discuss what can be learned from SIU, with particular attention paid to the success of various resistance tactics and what these mean for organizers attempting to mobilize for social change. The goal here is to explain why additional consideration of both dialectical approaches and collective action are useful to future social-movement organizations.

We conclude by outlining the role of regional organizing in SIU and how it may influence the movement in the future.

1. *Both symbolic and material considerations are necessary to produce meaningful social change.* Symbolic advances of awareness, education, and support for the ideological basis of the movement are necessary before participants can achieve material gains such as new policy, regulation, or mandates regarding conservation. Surely, activist organizations understand the power of symbolism in invoking people's willingness to participate. Likewise, activists must ensure that advances in the symbolic strength of the movement continue in order to sustain the interest necessary to push for broader social or material change.

We found evidence of symbolic, dialectical tensions and collective, material actions working in concert to foster social change during the SIU event. Organizers of many events invited local or national businesses producing "green products" to have informational booths at their events, integrating symbolic messages with broader advances in social or economical systems. This bridged the tensions many participants experienced when asked to vilify business interests, by transforming their views of production. Likewise, many individual events discussed local or statewide policy related to the reduction of carbon emissions and the primary goal of the event, cutting carbon emissions in America 80 percent by 2050, encompassing a symbolic stance toward the issue and a plausible, material change to achieve it. Yet, as we discuss below, there is a delicate balance between these perspectives that organizers must keep in mind during any attempt at organized resistance.

2. *Social-movement organizations must balance issues of power and accountability.* Organized resistance should be careful of undermining its own efforts. By holding a public institution accountable, it simultaneously repositions that organization in a dominant role. Many participants of SIU were limited by the central message because it conceptualized Congress as the only body capable of producing change, thus reducing other institutions' empowerment or ownership of the problem or solution. Note, we are not arguing that Congress is, in fact, the only institution responsible for managing climate-change policies but, rather,

that SIU discourse placed Congress in this position. Alternatively, for instance, the SIU campaign could have empowered its members by including localized social contracts or formal legislation among communities involved, providing messages challenging individuals to reduce consumption, or rallying around messages to save the environment for future generations. Such efforts may have created a shared sense of accomplishment among members while the long process of national legislative change takes place.

The national organizing unit for SIU made a strategic decision to reduce the complexity of climate change and focus solely on Congress. The strength of this strategy, based upon available data from several research sites, is that it provided an easy target for critique—all climate issues are the responsibility of Congress. The limitation of this tactic is that it paradoxically undermined the empowerment of other potential change agents. By positioning Congress at the center of climate policies, SIU obscured the role of other actors, including individual consumption choices, the entrepreneurial possibilities of doing business in environmentally sustainable ways, and the potential for local governmental bodies to development eco-friendly policies. The central point is that these possibilities are not implicated in the central message of SIU: "Step It Up, Congress! Reduce carbon emissions by 80 percent by 2050."

An additional limitation of the 80 percent goal, the central mantra of the SIU organization, is that the resistance and challenge embedded within the message are at odds with the grassroots appearance of the movement. The SIU events across the nation were intended to demonstrate a groundswell of citizen demands that Congress "step it up." At the same time, the message and objective of the campaign were not collaboratively generated among participants. Rather, they were selected by a small group of national-level organizers, thus contradicting the grassroots image. When asked where the central message and objective came from, national organizers explained,

> Well, we were all, except for Will, we were all still students at Middlebury College at the point where we decided on that slogan.

> And basically we sat down with Bill McKibben, and we decided that we really wanted to set the bar; we wanted to make sure that people knew that we were talking about real carbon cuts. You know, there's been a lot of talk about cap and trade, there's been a lot of talk about specific policy. But nothing that really, we thought, could galvanize a lot of people—a goal. And so, Bill is constantly in contact with climate scientists around the country, it's pretty much his job as an author, as a writer, and an expert on that issue. And so he spoke with James Hanson, the NASA head climate scientist, and that's basically—Jim Hanson told Bill that 80 percent by 2050 is what we need to get at the very least. And that's how we pretty much decided on that goal.

The contradiction here arises between the hierarchical decision making and how that centralized message actually fits with local events or the grassroots enactment of the campaign. SIU organizers said they stipulated that all local events would include the same message, because

> having everybody put that slogan on their banner and take a picture of it was (1) an easy ask, and (2) to make sure that everybody was saying the same thing, even if they were saying other things, they were definitely saying our slogan as well.

The problem is that the message did not resonate equally in all sites, which led to some events being separated into two processes: an image-event to comply with the national organizers' requirement to emphasize the central message, followed by a very different process involving speeches and informal discussion that rarely touched on the 80-percent objective (see, for example, the Spokane event described in chapter 8).

Conversely, organized resistance can also undermine the possibilities of critiquing powerful institutions by making individuals accountable for collective problems. We saw participants' solutions to global warming conforming to the old axiom, "if everyone does their part." While this mindset contributes to the solution of any social problem, it also limits possibilities for larger social change by placing all responsibility for change at the individual level.

As a means to balance this tension, organizers need to focus on creating a collective consciousness that looks beyond the everyday actions of citizens and encompasses the complex political, social, and organizational tasks needed to achieve broader change. Individual actions are indeed crucial to the success of any change; however, organizers should carefully evaluate the way their members conceptualize individual actions performed in pursuit of a solution. Individual action can be conceptualized as a community- or society-wide responsibility when it is presented as a social norm rather than an individual choice. Social networks create new meanings for the movement goals, or allow support and insight into practical, everyday solutions regulated by other citizens. This could include citywide initiatives, county legislation, or neighborhood programs designed to reduce carbon emissions or develop alternative transportation.

3. *Social-movement organizations must avoid simply "preaching to the choir."* At the beginning of this chapter, we discussed social-movement organizations as one example of the interconnected webs of influence undergirding society. Groups like the Sierra Club, Greenpeace, and countless numbers of smaller regional environmental and recreational groups constitute an organizational structure within community and regional affairs. These organizations often share members or collaborate on projects, which is precisely how SIU grew so quickly. National organizers of the SIU events knew of these linkages and drew from this network of relationships to gather support quickly for their movement. Participants of the SIU events often mentioned their memberships in national organizations, such as the Audubon Society, or regional organizations, such as the Bike Collective in Utah or the Earth Ministry in Seattle, as central to their participation in the event.

This method of organizing has both positive and negative implications for the success of social movements. On one hand, the shared resources and quick mobilization of support were essential to the strength of the message SIU organizers wanted to send to national policymakers. However, the reliance on established networks whose members already identified as environmentalists led SIU events (both nationally and locally) to reproduce a message among those already supportive of that message.

Simply put, the use of well-known networks and familiar relationships produced rapid growth in SIU events across the country in the short-term, but likewise undermined the long-term change potential because they failed to reach new organizational segments of society (e.g., local chambers of commerce, recreation groups, builders' associations, automotive or transportation sectors). Event participants noticed this unfortunate but predictable consequence of limiting networking efforts. As one participant at an Eastern Washington event suggested, "I think we are preaching to the choir. I think it is tougher, you know, with the hunting crowd in Idaho. It is tough to get them to change their habits. But it needs to be done, I think."

We are not arguing that established environmental networks should not be used to mobilize support for new movements. Organizers should use these contacts in order to learn from their experiences and draw strength from those with similar views. But efforts must be made to incorporate people from outside existing networks and to create an identity specific to the new movement. In the case of climate change, this requires accessing social networks not previously associated with environmentalism and directing promotion toward new target audiences. Likewise, campaigns such as SIU should clearly establish their uniqueness to carve out a niche within the increasingly crowded world of social-movement organizations.

Several SIU events made efforts to incorporate groups outside of the established environmentalism networks, such as companies that produce eco-friendly products. The visible and nonconfrontational nature of most events also encouraged broad participation. Ultimately, though, most people involved were already engaged in some sort of environmental activism and seemed to identify the events as outgrowths of their existing social-movement networks. The failure to achieve strong identification among participants is evident in a statement from one North Carolina participant: "I am not really sure what the SIU group is, but I guess it is kind of like other environmental or conservation groups."

To make a discernible change in global-warming policy, SIU must shed its image as just another environmental group. Similarly, promotion of the climate-change movement could transcend stereotypical associations of environmentalism by focusing on different aspects of the issue such as health concerns (e.g., asthma), economics (e.g., projected costs of global warming), and development (e.g., green building). For example, the mayor of Lansing, Michigan, told listeners that "sustainable economic development is where it's at...It's the growth model for the future." And local green-builders' associations orchestrated the events in Austin and San Antonio, Texas.

4. *The SIU mode of organizing is both innovative and troubling.* In many ways, SIU marks a departure from the traditional organization of the environmental movement because its primary goals and conceptions originated among key individuals conceiving of the problem at a national level. At the same time, this means that SIU organizers maintained a hierarchical decision-making model which excluded other participants from formative decision processes such as the central message and the choice to target Congress. Organizers then passed the message on to local stakeholders, who were asked to organize local events to give the campaign a grassroots flavor. Most of today's well-known environmental organizations started as small regional groups that eventually grew through disparate membership and gained significant measures of social power in public policy making. As part of the environmental movement, groups like the Sierra Club, the Audubon Society, and the Defenders of Wildlife exemplify this pattern.

Contemporary social movements require that members develop a collective identity that extends beyond individual events but may also appear to outsiders to represent more of a disparate collectivity (see chapters 9 and 11). The form of organization employed by SIU was clearly successful in developing a single, rallying event, but the constraints it placed on local organizers may contribute to its downfall. On one hand, initial proclamation of movement goals from a central source of authority may fast-track a campaign, facilitating national awareness much more quickly than otherwise comparable social-movement organizations. By starting with established goals and ideology in mind, however, organizers of SIU may have reduced the sense of ownership local participants developed for the issue of global warming or its proposed solutions. Less ownership in the formulation of the organization could translate into decreased participation in later phases of the movement. While we cannot predict what effects this form of organizing will have, we do anticipate that the SIU model of simultaneously organizing from the top-down and bottom-up may not account sufficiently for variance arising at each level—differences that may become apparent when the two processes eventually meet. Our analysis suggests that because movement leaders and organizers must work at multiple scales, they should reflect on the appropriateness of various ways of organizing and develop a broad repertoire of different but complementary strategies, depending on their goals and objectives. An organizational strategy that is effective for executing a single event (for example, April 14, 2007) may be less effective for mounting a campaign or for sustaining a social movement.

Concluding Thoughts

In this chapter, we explored the degree to which new forms of organizing continue to replicate the same dialectical, symbolic tensions that spawned them or to what degree members are able to incorporate the capacity to transform those tensions in a productive way. Our analysis of SIU indicates the possibility of the latter condition, given careful attention to the symbolic and material functions of social-movement organizing and its initial framing by participants. Our analysis also indicates that simply providing a new form of organizing is not the ultimate solution to political gridlock, for this too can create possible conflicts among members with no clear understanding of the movement.

The *organization* of social movements in our postbureaucratic age will continue to yield interesting insights into the motivations and pathways for societal change. Other movements will likely adopt new strategies of organizing similar to SIU's by capitalizing on or renegotiating the interconnected webs of influence that uphold our society. Future work should track the progression of these new forms of organizing to uncover what effects they will have in the social sphere and how they may influence their members' performance of organized resistance.

REFERENCES

Ashcraft, Karen Lee. 2005. Resistance through consent? *Management Communication Quarterly* 19 (1): 67–90.

Clegg, Stewart. 1981. Organization and control. *Administrative Science Quarterly* 26:545–562.

Collinson, David. 1992. *Managing the shopfloor: Subjectivity, masculinity and workplace culture*. New York: Walter de Gruyter.

Deetz, Stanley. 1985. Critical cultural research: New sensibilities and new realities. *Journal of Management* 11 (2): 121–136.

Deetz, Stanley, and Dennis K. Mumby. 1990. Power, discourse, and the workplace: Reclaiming the critical tradition. In *Communication Yearbook*, vol. 13, ed. James Anderson, 18–47. Newbury Park, CA: Sage.

Della Porta, Donatella, and Mario Diani. 1999. *Social movements: An introduction.* Oxford: Blackwell.

Ezzamel, Mahmoud, Hugh Willmott, and Frank Worthington. 2001. Power, control and resistance in the factory that time forgot. *Journal of Management Studies* 38:1053–1078.

Fairclough, Norman. 1993. Critical discourse analysis and the marketization of public discourse: The universities. *Discourse and Society* 4:133–168.

Fairhurst, Gail, and Linda Putnam. 2004. Organizations as discursive constructions. *Communication Theory* 14:5–26.

Fitzgerald, Kathleen, and Diane Rodgers. 2000. Radical social movement organizations: A theoretical model. *Sociological Quarterly* 41:573–592.

Ganesh, Shiv, Heather Zoller, and George Cheney. 2005. Transforming resistance, broadening our boundaries: Critical organizational communication meets globalization from below. *Communication Monographs* 72 (2): 169–191.

Grossberg, Larry. 2003. History, politics and postmodernism: Stuart Hall and cultural studies. In *Critical Dialogues in Cultural Studies*, ed. David Morley and Kuan-Hsing Chen, 150–173. New York: Routledge.

Hubbard, Dean. 2006. What kind of globalization? Organizing for workers' human rights. *The Journal of Labor and Society* 9:315–335.

Mumby, Dennis. 1997. The problem of hegemony: Rereading Gramsci for organizational communication studies. *Western Journal of Communication* 61 (4): 343–375.

———. 2005. Theorizing resistance in organizational studies: A dialectical approach. *Management Communication Quarterly* 19 (1): 19–44.

Norton, Todd. 2008. Property as interorganizational discourse: Rights in the politics of public spaces. *Communication Theory* 18 (2): 210–239.

Peterson, Tarla Rai, and Todd Norton. 2007. Discourses of sustainability in today's public sphere. In *The debate over corporate social responsibility*, ed. Steven May, George Cheney, and Juliet Roper, 786–819. Oxford: Oxford University Press.

Stohl, Michael, and Cynthia Stohl. 2005. Human rights, nation states and NGOs: Structural holes and the emergence of global regimes. *Communication Monographs* 72 (4): 442–467.

Titscher, Stefan, Michael Meyer, Ruth Wodak, and Eva Vetter. 2000. *Methods of text and discourse analysis*. London: Sage.

Chapter 6

Toward Just Climate-Change Coalitions

Challenges and Possibilities in the Step It Up 2007 Campaign

Danielle Endres, Tracylee Clarke, Autumn Garrison, and Tarla Rai Peterson

The environmental justice (EJ) movement is premised on the rights of all people to benefit from a healthy environment and to be meaningfully involved in and treated fairly during environmental decision making. When these rights are violated, environmental injustice occurs, often in the form of toxic waste, pollution, and environmental degradation, disproportionately causing harm to underresourced communities (i.e., communities of color, women, and the poor). According to the International Climate Justice Network (2002), "The biggest injustice of climate change is that the hardest hit are the least responsible for contributing to the problem." To make things worse, the 2007 Intergovernmental Panel on Climate Change (IPCC) indicated that there is a growing climate divide between rich and poor communities regarding both where the effects of climate change will be the most acute and who is best prepared to address these effects. The report states,

> Poor communities can be especially vulnerable [to the effects of climate change], in particular those concentrated in high-risk areas. They tend to have more limited adaptive capacities, and are more dependent on climate-sensitive resources such as local water and food supplies. (2007, 8)

The Gulf Coast and Desert Southwest are two examples of vulnerable regions in the United States given in the IPCC report. Put simply, those in underresourced communities may not have the luxury of leaving an area that is experiencing the effects of global warming. Robert D. Bullard, Glenn S. Johnson, and Angel O. Torres argued that

> as we search for ways to rectify global climate change, we desperately need the input of the populations most likely to be negatively affected: people of color and other poor people in the North and in the developing countries of the South. (2005, 292)

The Bali Principles of Climate Justice sought to accomplish this goal by "broadening the constituency that provides leadership on climate change" (International Climate Justice Network 2002).

Campaigns such as Step It Up 2007 (SIU) and Al Gore's *An Inconvenient Truth* represent the emergence of climate change as perhaps the most important issue for the mainstream environmental (ME) movement. Yet, these campaigns often neglect to highlight the EJ implications of climate change. Even if mainstream climate-change campaigns acknowledge the disproportionate effects of global warming on traditionally underresourced and marginalized peoples, the principles of EJ, such as equal representation, are not central to campaign development. We argue that a coalition between the EJ and ME movements is essential to address climate change, and we suggest ways to encourage development of such a coalition.

This type of collaboration is crucial for several reasons. First, environmental injustices are inextricably embedded in the climate crisis. Second, the maintenance of separate movements/campaigns on climate change can perpetuate injustice; the ME movement and its campaigns may receive more resources than the EJ movement/campaigns to address climate change, a common grievance of members of the EJ movement. Third, a united campaign for climate justice will be a stronger force for social change. Specifically, building a coalition between EJ groups and ME groups would reach broader constituencies, which is crucial for a problem as big as climate change. Finally, the people who are most directly affected by climate change tend to be already marginalized groups (IPCC 2007), so they may not be considered without the prodding of the EJ movement. A coalition campaign would embody the Principles of Environmental Justice, particularly principle seven, which recognizes "the right [of marginalized voices] to participate as equal partners at every level of decision-making" (First National People of Color Environmental Leadership Summit 1991).

In this chapter, SIU acts as our case study to examine the possibilities and challenges of building a coalition campaign between the ME movement and the EJ movement to address climate change. We call for the two movements to strategically work toward a jointly developed *just climate-change campaign*. We begin by reviewing the possibility of coalition building between the two movements. Then, we set the scene for our analysis by presenting data on the (lack of) presence of EJ organizations and messages and of marginalized people at the events we attended. Next, we examine SIU as a case of two challenges to building environmental coalitions—accessibility of Internet technologies and green consumerism. Then, we offer some possibilities for coalitions, including the use of faith-based groups as bridges between the two movements. The final section of the chapter lays out practical suggestions for ME movement practitioners interested in cooperating with the EJ movement.

THE POSSIBILITY OF COALITIONS

Our chapter assumes that the ME movement (specifically, the climate-change movement) and the EJ movement (specifically, the climate-justice movement) can indeed be compatible despite a long history of division and contention. Phaedra Pezzullo and Ronald Sandler suggested, "Since at least the early 1990s, activists from the EJ movement consistently have criticized what they consider the ME movement's racism, classism, and limited activist agenda" (2007a, 2). One Seattle interviewee echoed this sentiment when she shared her frustration with the lack of minorities involved in SIU events and the white, middle-class agenda of the SIU campaign:

> [I would] like to see more minorities at these events. Once again, it's about the middle-class white people. This is what I have been struggling with since college. How do I go to minority communities and connect those people to these issues? Yet, we need to figure out how to talk to more than Fremont and Ballard [two primarily white, middle-class neighborhoods in Seattle]. That's too easy.

This comment reflects that even if there is desire within mainstream environmentalists to add diversity to the movement, they may not know how to encourage involvement from marginalized peoples. Yet, this type of reaching out may meet resistance from members of the EJ movement who believe the whiteness of the ME movement provides a barrier to collaboration. Beverly A Wright, Pat Bryant, and Robert D. Bullard (1994) argued, "these [ME] groups cannot reach out to African Americans and people of color as long as they are nearly all white" (1994, 121).

Despite the history of animosity between the EJ and ME movements, Pezzullo and Sander argued that further institutionalization and receptivity of the EJ movement and the development of active EJ campaigns in ME organizations like Sierra Club and Greenpeace are signs of the potential for discussing collaboration and coalitions. "Ongoing conversations within both movements regarding the viability and direction of their futures" may be the catalyst for exploring coalition building and cooperation between the movements (2007a, 12). Other scholars also argue for the potential for and necessity of collaboration between the EJ movement and the ME movement (Agyeman 2005; Agyeman, Bullard, and Evans 2002; Agyeman and Evans 2003; Faber 2007; Jamieson 2007).

There are at least two competing perspectives on the possibilities for collaboration between the EJ and ME movements: movement fusion and strategic coalitions. Movement fusion is the "coming together of two (or more) social movements by developing a common agenda" (Cole and Foster 2001, 164). The concept of *just sustainability*, or the "nexus of theoretical compatibility between sustainability and EJ" (Agyeman and Evans 2003, 48), exemplifies movement fusion. Through providing examples of successful cooperation between sustainability/environmental organizations and EJ organizations, proponents of just sustainability show that it is possible and desirable. Rather than fusing the two movements together, however, Pezzullo and Sandler argue for developing strategic coalitions between the movements. They suggest that certain fundamental differences limit "the prospects or even the desirability of...merging the two movements into one single, unified movement" (2007b, 311). Indeed, the conflict between the two movements can be productive for both movements.

Although we do not ascribe fully to either of these two models of coalition building, our chapter builds from Pezzullo and Sandler's (2007b) advocacy of the value of "campaign or issue-specific collaborations or alliances" between the EJ movement and the ME movement. We believe that the climate-change crisis is ripe for collaboration between the two movements and that specific campaigns can and should be developed to address climate change. Yet, we are not ready to give up the concept of movement fusion. While we do not advocate for the complete fusion of the EJ and ME movements because of their differences and the value of conflict between the two, we believe we should strive for fusion in a variety of strategic campaigns on climate change. In the next few sections, we will use SIU as a case study in the challenges and potential for collaboration on specific climate-change campaigns.

Setting the Scene

When we interviewed the national SIU organizers following the completion of the first SIU campaign, they claimed they had thought about the EJ implications of climate change when planning the campaign. Despite this awareness, however, our analysis reveals that there was little presence from the EJ movement or marginalized peoples in the SIU campaign. First, we observed a lack of EJ groups at the SIU actions. At actions that included tables and booths for other organizations, we did not observe representation from local EJ groups. In some places, this may be due to a lack of such groups. However, Oakland and San Antonio are home to several prominent EJ organizations that were not present at the SIU actions. Awareness of the lack of participation by EJ groups led the SIU national team to attempt to work with and learn from EJ leaders and groups, such Van Jones and the Ella Baker Center for Human Rights in Oakland, California, to increase their participation at the SIU2 events that took place in November 2007.

Second, our analysis of speeches and messages of the events indicates limited inclusion of EJ principles. Instead, we observed that the SIU event messages focused more on what behaviors people could perform to slow global warming (i.e., buying lightbulbs and pressuring Congress) than on the consequences of global warming that disproportionately affect marginalized people and on the injustices of climate change. One of the national SIU organizers stated, "My sense is that that issue [environmental justice] was not dealt with...openly in most of these events."

There were, however, speakers at the Raleigh and Seattle events who highlighted EJ principles and issues. At the Raleigh action, Carlos Velazquez, Otomi Indian and director of natural resources and environmental affairs for the Southern Cherokee Nation, spoke about the effects climate change would have on indigenous people, especially the Inuit in the Arctic. Moreover, at the Seattle action, JiJi Jally also spoke about the EJ implications of climate change:

> As a Marshallese, I worry about the social injustice of the poor and developing people in the world suffering much more of the

> hardship linked to climate change…My country is a tiny country; it just has 70 square miles of land. Our entire social structure is linked to the land. If we lose our land to the rising seas, we cannot simply pick up and move to another location just as people in America can if their land is damaged or disappears.

At this event, Jally was the first speaker, and her comments were explicitly referenced by the speakers who followed her. A Presbyterian minister also talked about the religious community's commitment to environmental justice. Although EJ was not the theme of the entire event, Jally's comments were presented first and seemed to set the tone for the rest of the event. These two examples articulate the opportunities for weaving EJ implications of climate change into ME-movement campaigns on climate change.

Finally, our examination of field notes from researcher attendees and author observations reveal that a majority of participants and speakers at the SIU actions were white and middle class. An interview with the national SIU organizers confirms this finding across the national actions:

> We did do a really good job of engaging, you know, middle-class, white America—well, the middle-class, liberal white America—in our actions, but one thing that I think we should recognize that we failed at is something that the environmental movement in general has failed at, is really engaging minorities.

Although this result may not be surprising in places with large majorities of white people, such as Salt Lake City (79.2 percent White), Seattle (70.1 percent White), Raleigh (63.3 percent White), East Lansing (80.9 percent White), and Spokane (89.5 percent White), it should be surprising for the actions in Oakland (35.7 percent African American, 15.2 percent Asian American, and 21.9 percent Hispanic/Latina/Latino), Boston (25.3 percent African American, 7.5 percent Asian American, and 14.4 percent Hispanic/Latina/Latino), and San Antonio, (6.8 percent African American, 1.6 percent Asian American, and 58.7 percent Hispanic/Latina/Latino).[1] One participant in Oakland, California, noted her disappointment in the organization of the SIU action: "I looked on the Web page and I saw the pictures coming in, and it was all white people. There was no diversity. Look around, everybody in this room is white." Moreover, the San Antonio event was organized by a young (thirty-something), white-collar professional woman, and the speakers, other than one Latino state representative, were all non-Latino white-collar professionals. The participants mirrored this demographic. That is not insignificant in San Antonio, where a majority of the population is Latino. Interviews and field notes from the other SIU locations reveal a lack of people of color among action organizers, speakers, and participants.[2]

Our data also address participants' economic class. At the San Antonio action, one researcher spoke briefly with twenty-five to thirty people attending the rally. With two exceptions, all of the interviewees were either white-collar professionals or students at an expensive private university. Of the two exceptions, neither lacked privilege. One has an advanced degree and explained that he "works full-time on environmental issues." The second said she has a small inheritance that enables her to "live a simple life and be involved in environmental and social justice issues" in San Antonio. The Salt Lake City Liberty Park action focused on using yoga, an activity that is primarily associated with middle- to upper-class people in the United States, as a way to address climate change. Our observations confirm a majority of middle-class to upper-class participants at actions across the nation.[3]

Despite the lack of participation from marginalized and underresourced people at the SIU actions, we did observe at least one action that succeeded in attracting a diverse set of participants: an action at Washington Square Park in downtown Salt Lake City, which drew up to five thousand people, had a noticeable presence of people of color. Interview data from the action reveal that the organizers specifically tried to raise awareness and educate a diverse audience, particularly the Latino population in Salt Lake City and the "Wal-Mart crowd." Interview and field note data from this event confirm the success of the attempt.

A just climate-change campaign would not only be based in a coalition between EJ and ME groups but would also make present the voices and perspectives of marginalized and underresourced people—people of color, indigenous people, women, and poor people. Our data indicate that SIU did not successfully integrate EJ issues, members of the movement, and marginalized people into its campaign. One reason for this paucity may be the extant divide between the EJ movement and the ME movement, which we discussed above. Although the past divide poses a challenge, it does not need to be an insurmountable barrier to contemporary issue-based coalitions between the two movements. Certainly, an important part of making a just climate-change campaign is for the ME movement to *meaningfully* reach out, work with, and share its resources with EJ groups. Moreover, a just climate-change campaign must also make efforts to include marginalized and underresourced people, whether or not they are involved in the EJ movement. Such attempts will not be easy but are necessary.

Challenges to Collaboration

In this section, we will discuss two challenges to collaboration that we observed in the data from local SIU events: (1) the relationship between access and the digital divide, and (2) the role of green consumption in the campaign. Analysis of each reveals limitations in how the SIU campaign reached out to marginalized and underresourced peoples and the EJ movement. Our analysis also positions these shortcomings in the SIU campaign as challenges that must be addressed for future just climate-change campaigns.

Access and the Digital Divide

One reason for the lack of participation by marginalized and underresourced people may be related to the SIU campaign's reliance on the Internet as its main form of outreach and organizing. The organic, open-source structure of SIU promoted the potential for marginalized people and the EJ movement to sign up to host an event with their own take on SIU, learn about local SIU events, and generally participate in the campaign (see chapter 7, chapter 8, and chapter 11 for more discussion of SIU as an open-source campaign). However, we argue that using the Internet as the primary means for disseminating information about the campaign may have widened the divide between Internet users and non-users. This digital divide extends beyond race and income to include geographical location, such as rural areas, where rates of Internet use are often lower (Whitacre and Mills 2007). Understanding the digital divide can help to explain why heavily relying on the use of the Internet for movement participation (i.e., online sign-ups to host events, dissemination of information about the campaign through the SIU Web site, blogs, and *Grist*, etc.) can result in a lack of participation of marginalized people, who may suffer the most from climate change.

The digital divide refers to the disparity of use of the Internet and unequal access to information communication technology (ICT) across classifications of race, gender, age, income, education, and geography (NTIA 2000a, 2000b, 2002; Rogers 2001; Mossberger, Tolbert, and Gilbert 2006). General concerns about information inequalities have come to the forefront in public and political debate over the last decade (Selwyn 2004; Van Dijk and Hacker 2000). However, much of the discussion about new technologies and social equality has presented an oversimplified notion of a digital divide between "haves and have-nots." This approach provides a limited view of the complex relationship between technological development and social justice (Rogers 2001). Selwyn argued that "there needs to be a political recognition that the crucial issues of the digital divide are not just technological—they are social, economic, cultural, and political" (2004, 357).

ICT, specifically the Internet, has tremendous potential to overcome existing social division, achieve greater social equity and empowerment, and improve everyday life for those on the margins of society (Mehra, Merkel, and Bishop 2004; Selwyn 2004). One critical way ICT can empower a community is by providing its members the capacity to more fully participate in society, creating what Mossberger et al. called "digital citizenship" (2006, 585). Among other things, ICT provides the opportunity to engage marginalized people in planning, policy implementation, and community-specific activities such as educational campaigns. This helps to reengage underresourced and marginalized people with each other and the process of public-policy development, and may contribute to a more active and participative democracy (Bekkers, Duivenboden, and Thaens 2006; Mehra, Merkel, and Bishop 2004; chapter 8).

Grassroots environmental organizations have increasingly relied on the Internet to mobilize their members and organize protests (Kellogg 1999). Environmental advocates use a variety of electronic tools such as Geographic Information Systems (GIS) and the Internet to advance their cause and address environmental problems. Use of such technologies, however, requires a set of technical prerequisites. Environmentally burdened communities tend to be economically and technologically disadvantaged, with fewer resources to access scientific and technical information. Lack of meaningful access to computer technology, as well as a lack of computer skills, makes hope for an Internet-technology-enhanced just climate-change campaign contingent. To be more in line with the principles of EJ,[4] campaigns relying on ICT must address basic equity and enfranchisement issues (Schuler 2000). Although SIU's use of the Internet as the primary means for signing up event hosts, disseminating information about the event, and monitoring initiative progress had the potential to engage marginalized people in public-policy development, the lack of representation indicates that that potential was not achieved.

Researchers and practitioners (Schuler 2000; Tesh 2000; Warschauer 2003) have identified several ways to use ICTs to reach disadvantaged communities and provide opportunities for previously disenfranchised groups to have an equal voice in public-policy debates. These include: (1) leveraging existing local and regional social resources and networks, (2) extending the social capital of marginalized people, and (3) combining technological approaches with a broad range of traditional face-to-face advocacy approaches. When combined, these three strategies use ICTs *and* work with local social networks to minimize the effects of the digital divide and empower marginalized people to be digital citizens. In the remainder of this section, we discuss each of these strategies in the context of the SIU campaign. The national SIU campaign organizers did not use these strategies. Through more systematic use of these approaches in relation to ICTs, specifically the Internet, environmental movements and campaigns can more fully engage traditionally marginalized people and create a more just climate-change movement.

Leveraging Local Networks

David Schlosberg (1999) has shown that the EJ movement typically relies on an organizational model that gathers strength through networks of diverse and geographically scattered local struggles. Online EJ resources that do not tap into local communities and social networks will not be able to mobilize environmentally burdened groups on the ground. Internet communication technology is most likely to increase the visibility and success of EJ in instances where existing people-based networks are well established and used in policy initiatives (Schuler 2000; Tesh 2000). By working directly with community-based organizations, local leaders, and educators, ME and EJ environmental practitioners will be more able to reach members of these marginalized and underresourced communities (Warschauer 2003).

Despite their desire to include underresourced people and address EJ issues, our data show that the national SIU organizers did not successfully access networks of underresourced people or reach out to established EJ organizations (by their own admission). One solution is for the central organizers to work with established EJ groups in future campaigns, which the SIU team did to a small extent for SIU2. But this is not enough. It is also crucially important for movements that are increasingly relying on ICTs to understand their audiences' relationships with ICTs. In order for the Internet to play a greater role in social and personal empowerment, we need to understand the practices, systems of relations, and context of particular minority and marginalized users of ICTs in order to identify local goals and how they use—or do not use—the Internet for meeting their objectives. This approach is crucial for several reasons. First, if we rely on the Internet as the primary means of campaign organization and participation, campaign success requires reaching diverse populations, which is only possible when we understand how marginalized and underresourced groups use (or don't use) the Internet. The SIU campaign was designed so everyday people could sign up to be part of the campaign by organizing their own events; however, this design did not factor in the digital divide and the social and technological practices of marginalized people, making it less likely that some audiences would find out about the campaign, sign up to organize an event, or participate at a rally. Second, when used well, ICTs can empower marginalized people and help foster digital citizenship.

By penetrating existing social networks, working with established EJ groups, and better understanding the social practices of marginalized people, environmental practitioners can more fully empower these communities and engage them in critical policy initiatives. So, while we do not advocate for SIU and other campaigns to devote all of their resources to learning about how marginalized people use ICTs, they cannot rely solely on working with EJ groups to increase access and participation by marginalized people.

Extending Social Capital

To more fully engage marginalized people in movements and campaigns, practitioners must recognize the mediating role of economic, cultural, and social forms of capital in shaping individuals' engagements with ICT (Selwyn 2004). To be successful in their endeavors, practitioners must extend the social capital of these communities through the use of Internet technologies (Shulman et al. 2005, 504).[5] One way to accomplish this is by directly working with local organizations to educate community members on issues that impact them and train community members in Internet technologies so they can manage information for their own purposes, thereby increasing their social capital (Kellogg and Mathur 2003, 578). Environmentally burdened communities, if given access to and training in Internet technologies, would be able to better address environmental issues with more comprehensive information and greater political clout.

Our analysis of the SIU national Web page (including its recipe book for local organizers) and an interview with two of the organizers show that the national SIU organizers neither provided training in Internet technologies to potential participants nor encouraged local-action organizers to provide such training. The national format of the SIU campaign assumed that access to the Internet was a given. This assumption further marginalized those communities that do not use the Internet or have the necessary technological skills or social capital to address climate change through involvement in SIU. Moreover, our data from the local events we attended show that even in events that used Internet technologies (e.g., the Salt Lake City Washington Square event's computer bank for e-mailing representatives), the local organizers did not assist or train people to use ICTs for things such as sending e-mail messages to Congress. Nor did local SIU events address technological possibilities for continued long-term involvement with climate-change issues, which would increase the social capital of marginalized groups. Given the diverse people in attendance, the Salt Lake City action suggests the potential for providing access to ICTs, helping people learn how to send an e-mail and use other technologies, and demonstrating the value of ICTs as a long-term tool for citizen participation in decision making. Armed with information from the national SIU organizers about the Internet-usage practices of marginalized people and the value of providing training in Internet and other ICTs, local events could engage with, instead of marginalize, groups who are less likely to understand and use the Internet.

Using Internet Technologies to Augment Traditional Approaches

In order for a technologically enhanced movement to be successful and fully engage marginalized people, the Internet must be used in connection with other, more traditional means of social activism (Shulman et al. 2005). Internet resources can often augment local movement efforts and provide support for initiatives but cannot replace face-to face social activism. Personal interaction among advocates is critical. Internet-based communication systems can be used effectively for community development *only* after its users have developed, through other means, trust in each other and a shared community identity or vision (Resnick 2002).

> Strategies that take into account the social nature of access, recognize the interaction between face-to-face and online communications, and combine Internet use with a broad range of other new and old media provide the best opportunity for promoting social inclusion through use of ICT. (Warschauer 2003, 197)

SIU used ICTs to organize face-to-face gatherings at the actions on April 14. This is the opposite of drawing on ICTs to supplement existing relationships and activist communities. Future campaigns might aim to develop a reciprocal relationship between ICTs and social networks by encouraging local meetings before and after the day of action. Connecting with existing social networks can also help reach this goal.

Internet technologies can be tools for community empowerment. Used creatively, they have the capability to engage traditionally marginalized people in discussions about the environment and strengthen the role of marginalized people in political decision-making processes. A combination of leveraging local networks, extending social capital though ICT skills, and not letting ICTs replace traditional networking can be effective strategies for engaging marginalized people. If a just climate-change campaign is going to rely on a open-source, Web-based model, as SIU did, the campaign must make sure that the campaign is accessible to marginalized and underresourced people who will be affected the most by climate change, not only through building coalitions with established EJ groups but also through understanding technology use and facilitating engagement.

Green Consumerism

A second challenge to integrating EJ and ME campaigns from our analysis concerns how green consumerism was often (perhaps inadvertently) presented as a way to solve global warming—a message that detracted from the justice implications of climate change. Human-caused climate change is closely linked to global patterns of consumption, economic injustice, and social injustice. Underresourced and marginalized people in the United States and the world are poised to suffer the most from global climate change, and in some cases, rich and privileged people will actually benefit from it (IPCC 2007; Bullard, Johnson, and Torres 2005; Jamieson 2007). According to J. Timmons Roberts,

> The cry for "climate justice"—that is, EJ on the issue of climate change—is growing louder as impacts are being increasingly felt in poor nations threatened by the changes. These nations are at the same time tragically unable to cope with and respond to climate disasters, such as spreading drought and agricultural instability (most notably affecting sub-Saharan Africa), sea level rise (Pacific island atolls and Bangladesh being the most vulnerable), and hurricane risk. (2007, 293–294)

From an EJ perspective, climate change will affect the basic survival of underresourced people living in vulnerable regions. Both the Principles of EJ and the Bali Principles of Climate Justice emphasize the need to prevent the disproportionate effects of environmental problems, pollution, and climate change on marginalized peoples as primary. Despite both documents' recognition of the need for consumption practices to change, the EJ movement is emphatically not focused on hybrid cars or voluntary simplicity.

The SIU campaign, along with many other ME approaches to climate change, displaces EJ concerns with climate change by emphasizing changes in individual consumption patterns. At each SIU event, we saw strong advocacy for changes in individual consumption patterns (i.e., compact florescent lightbulbs [CFLs], hybrid cars, and other green products) that was sometimes framed as a complement to congressional action on climate change but was also often framed as a means to take individual action on climate change to the exclusion of congressional action.

Green consumerism has been identified by some as a prosocial, socially conscious form of consumer behavior (Moisander 2007) with the ability to raise awareness of climate change (Grønhøj 2006) and empower people to feel as though they can make change through a few easy changes in their individual actions (i.e., walking instead of driving when possible, changing lightbulbs, and greening one's home). We recognize the importance of individual actions in a society with a strong individualistic bias; audiences like to learn of ways to solve problems in their own lives, and it is important to appeal to this aspect of audiences. However, we question the value of advocating green consumption when it may further marginalize and alienate underresourced people and exclude them from being a part of the solution to climate change. Many of the green products and services may be less accessible to underresourced and marginalized people in the United States. Moreover, focusing attention on individual consumer solutions can trade off with raising awareness about climate justice and the effects of continued human-caused climate change on underresourced and marginalized people. The high profile of consumer products at local SIU events, often overwhelming the calls for congressional action (e.g., chapter 4), was a sign to participants of what the SIU campaign was about. If the movement is primarily about changing consumption patterns, and maybe also about pressuring Congress to cut CO_2 emissions, it may rebuff EJ groups who are concerned with justice and effects on marginalized peoples.

At local SIU events, we observed a plethora of consumer solutions to climate change, and no critical examination of how consumerist solutions may detract from attention to environmental justice. Several of the events specifically catered to consumerism; booths and activities highlighted at the Austin, Bloomington, Lansing, Moscow, Raleigh, San Antonio, Salt Lake City, and Seattle actions encouraged attendees to make purchases to green their homes and make their next car purchase a hybrid to reduce their carbon emissions. While these individual consumer purchases do help mitigate the effects of global warming, they also encourage green consumers to assuage their own guilt about consumption as if it occurred in a political vacuum. Many of the vendors present at SIU actions took advantage of guilt by offering attendees opportunities to combat global warming by consuming. For example, at the Salt Lake City action, a booth offered ReDirect Guides (i.e., yellow pages for green products) with flyers that touted "10 easy things you can do to spend with intention" and introduced "the planet's first card that helps prevent climate change with every purchase" (ReDirect Guide n.d.). This type of advertising clearly taps into the green consumer's implicit desire to do the right thing (e.g., fight global warming) by consuming "with intention," which can shortchange political action. The green consumer is able to feel that by making responsible consumer choices, they have done their part to address climate change. These actions prioritize consumer changes over understanding and addressing the social-justice implications of climate change, not to mention that underresourced people with less discretionary income may be excluded from these individual actions.

Many of the proposed solutions to global warming at the SIU events (i.e., hybrid cars, solar power, and green homes) are far too expensive and therefore unrealistic options for a large portion of society. While some affluent sections of the population are willing to pay up to 25 percent more for green products (Prothero 1990), many people cannot afford such prices. Johanna Moisander (2007) explained that the ability to engage in green behaviors is a function of personal resources (i.e., money) and opportunity (i.e., availability of products and services) for the consumer to perform the behavior. McCarty and Shrum (2001) added that it is difficult for those who are impeded by the cost of green items to see the long-term benefits of such items. These higher price tags can be discouraging for families that struggle to make ends meet. Emphasizing green consumerism as a primary part of the solution to the problem of global warming effectively excludes a large portion of the population from being a part of the "solution."

Green consumerism does not have to be just for the rich, and can be an important part of the solution to climate change. We must start thinking differently about consumerist solutions; this thinking must include marginalized voices and the EJ movement. For example, CFLs could be appealing to underresourced people concerned about the environment and the high price of electricity. Providing information about the long-term cost-saving benefits of CFLs could encourage even those with little discretionary income to start using them. Instead of providing free CFLs to middle- and upper-class white people who often attend ME rallies on climate change and can afford to buy their own lightbulbs, ME groups could work with EJ groups to start distributing CFLs to underresourced people (modeling the urban gardens that many EJ groups have started). Changing consumer practices does not have to be mutually exclusive with pursuing EJ issues. Indeed, the seventeenth principle of the Principles of Environmental Justice states, "Environmental Justice requires that we, as individuals, make personal and consumer choices to consume as little of Mother Earth's resources and to produce as little waste as possible" (First National People of Color Environmental Leadership Summit 1991). What the EJ perspective does is highlight the injustices associated with consumerism and environmental problems such as climate change. The Bali Principles of Climate Justice state that "unsustainable production and consumption practices are at the root of this [climate change] and other global environmental problem[s]" and "combating climate change must entail profound shifts from unsustainable production, consumption and lifestyles, with industrialized countries taking the lead" (International Climate Justice Network 2002). Moreover, the Bali Principles argue that climate justice means that all people have access to necessities such as sustainable energy.

The design, implementation, and vision of the national SIU campaign did not sufficiently include EJ organizations and EJ principles into their campaigns. This insufficiency in the national campaign trickled down to the local actions with few exceptions, resulting in a primary emphasis in many events on consumerist solutions to climate change that exclude marginalized and underresourced people and detract from justice issues.

Possibilities for Collaboration

While our analysis of the SIU campaign reveals several challenges to a collaborative just climate-change campaign, we also observed possibilities for collaboration. An effective coalition between the two movements would not only have to increase access and participation but also address the institutional social capital, economic, racist, and classist assumptions in the proposed solutions to climate change. One way to do this is for the ME movement to use existing networks. Since the EJ movement is already strongly linked to faith-based organizations (Agyeman and Evans 2004; Gottlieb 2006; Allen, Daro, and Holland 2007), one way to build a coalition between the ME and EJ movements is to work with faith-based organizations and groups that have social justice and EJ campaigns. Just as churches were co-opted by the civil rights movement (McAdam 1982) and various antiwar movements, faith-based organizations can play a strong role in a just climate movement.

Faith-based organizations—representing Buddhism, Hinduism, Christianity, Islam, and Judaism—are addressing environmental problems, including climate change. According to Roger Gottlieb, "religious environmentalism is truly a global phenomenon, involving members of virtually every religious group, race, and culture on the planet" (2006, 471). Several interfaith groups highlight that climate change will disproportionately affect poor people and countries (Gottlieb 2006). Environmental organizers such as SIU could take advantage of the strong, albeit sometimes factional and contentious, eco-justice movement rooted in Christian churches in the United States, where "those who take the eco-justice perspective seek to pursue the struggle for justice and for a healthy environment together" (Bakken, Engel, and Engel 1995, 5). As Gottlieb reminded us, the EJ movement was "spearheaded initially not by Greenpeace, the Sierra Club, or the World Wildlife Federation, but by the United Church of Christ" (2006, 491). The strong role that churches played in the civil rights movement, the antinukes movement, and various antiwar movements reveals that faith-based organizations can play strong and important roles in organized resistance.

Faith-based communities were not absent from the SIU campaign. The national campaign was allied with five faith-based organizations that focus on climate change or other environmental issues.[6] At the Raleigh, North Carolina, SIU event, at least one faith-based group, Interfaith Power and Light, was present with a table; in Seattle and Boston, religious leaders gave speeches that drew on environmental-justice principles; participants across locations self-identified as members of churches or faith-based groups. A Presbyterian minister spoke at the Seattle main rally:

> People from our faith communities, are you out there? Can I hear a witness? [*shouting*] All right. I heard that we had twenty-seven churches, seven synagogues, three faith organizations, and lots of clergy today. That that's just the ones we counted as we were walking, so we had a terrific turnout. All these people wearing red are representing local faith communities. And they are here to witness to the fact that caring for the earth is an essential component of faithful living.

She went on to talk about how a commitment to environmental justice motivates religious groups to action, demonstrating that coalitions with religious organizations can effectively expand the reach of a campaign (e.g., members of thirty-seven religious organizations at the Seattle action) and frame environmental justice as a central motivation for action on climate change. An interview with a member of the First Baptist Church at the Kalamazoo, Michigan, event revealed that "as Christians, we need to practice stewardship of the earth." The church member had heard of the event through an article in the *Christian Century* magazine that called for Christians to participate in the climate-change movement. These instances suggest the potential organizational and moral resources for more fully co-opting religious networks toward building a just climate campaign.

Although we recognize the difficulty of creating alliances between the ME movement and the EJ movement, we argue that faith-based environmental groups may provide an interesting and fruitful bridge between the two movements. Because of the focus on justice and the linkages between the EJ movement and churches, incorporation of faith-based groups into SIU and future campaigns can both bring justice issues to the forefront and add new social networks of the movement. As Kim Allen, Vinci Daro, and Dorothy Holland suggested, "Alliances provide at least a temporary community of practice where different environmentalisms can be learned and where new forms can be created. This is the most likely positive path for the future" (2007, 130).

CONCLUSION

The response to the global crisis of climate change has produced patterns of advantage and disadvantage that plague the global community (Agarwal, Narain, and Sharma 2002). These injustices must be addressed in our pursuit of a response to global climate change. "The need to insure that public policy—environmental or otherwise—does not disproportionately disadvantage any particular social group, and affords opportunity for all, must be a precondition for the move toward just and sustainable societies" (Agyeman and Evans 2004, 163). In our analysis of the SIU events, we found that environmental justice principles were not sufficiently integrated into the campaign and message. Addressing climate change requires a coalition built between the ME movement and the EJ movement, not only for the power of collective action but also to address the justice implications of climate change. As an outgrowth of the ME movement, the SIU campaign could have made better efforts to draw in members of the EJ movement into its campaign. Pezzullo and Sandler suggested that "when groups from both movements come to work together with respect for the other within the context of a well-defined, campaign-specific effort, domestically or internationally, a just partnership is not only possible, but desirable" (2007b, 317). In retrospect, the SIU campaign could have been an ideal opportunity to create such an alliance. But what could have been done in the SIU campaign to create alliances, invite new social networks to participate, and include the principles of environmental justice into the campaign? In the remainder of this paper, we lay out some practical suggestions for future movement builders and campaign organizers to reach out to the EJ movement and marginalized peoples, build coalitions through specific campaigns, and propose solutions that work to overcome (and at the least not further entrench) environmental injustices.

First, it is crucial to *infuse the Principles of EJ into a just climate-change campaign*. We suggest that ME practitioners can begin to do this by creating alliances with established EJ organizations, ensuring participation venues for the people who will be affected the most from the effects of climate change, and developing solutions that address both climate change and its concomitant injustices. Because of the history of contention between the two movements, it is important for the ME movement to reach out to EJ organizations early in the development of the campaign. Rather than developing a campaign and then inviting EJ groups to be involved, a coalition campaign would be codeveloped by ME and EJ leaders or organizations and would draw on the principles of EJ in the development and implementation of the campaign. In such a campaign, it would be possible, for example, to reenvision the role of green consumerism and individual action, along with advocating policy changes, in ways that do not further marginalize underresourced people and entrench injustices. As we discussed above, changing consumer practices and sustainable living is articulated in the seventeenth principle of the Principles of Environmental Justice. Although an alliance with an established EJ organization is the ideal, it may not be possible for some campaigns to do so. In these cases, the Principles of EJ (see appendix A) can still be incorporated in the development of the campaign.

Second, in order for a just climate-change campaign to create opportunities for input from the marginalized and underresourced people who will be affected the most by the effects of climate change, the campaign organizers must *reach out to new social networks.* Overall, the SIU campaign was not successful in reaching new social networks (outside of the ME movement) of marginalized and underresourced people. In addition to working with established EJ organizations, we argue that tapping into existing networks of churches and faith-based groups can be an excellent way to reach not only marginalized and underresourced people but also networks of people who are already concerned about social and environmental-justice issues. On both national and local levels, organizers should reach out to faith-based groups that can rally their members to be part of the cause. We have already seen the success of using the social networks of churches in the civil rights movement.

Third, since marginalized and underresourced people may be less likely to have access to ICTs, *do not rely solely on the Internet for organizing*. In the case of the SIU campaign, the organizers relied on the Internet as the main means of participation in the campaign—local organizers signed up to host events on the SIU Web site; people found out about the SIU campaign on *Grist*, blogs, or through e-mail messages; and many actions called on participants to use e-mail to contact congressional decision makers. While we recognize the value of the Internet and other ICTs for social movements in the twenty-first century, we cannot assume that everyone is "plugged in." Organizers should combine ICTs with traditional organizing tactics such as face-to-face meetings. Moreover, making efforts to access new social networks cannot take place solely through the Internet. Campaign organizers can reach out to established EJ groups, faith-based groups, and community leaders through traditional means.

Fourth, when using the Internet and other ICTs, it is crucial that campaigns *understand how their potential audiences use ICTs* and *offer training in digital citizenship*. As we discussed, inequities in access to and use of ICTs make it so that we cannot assume that everyone is able to participate in a campaign digitally. Although it may require some shifting of resources within a campaign, learning about the usage practices of marginalized and underresourced peoples is crucial to a campaign that upholds the principles of environmental justice and creates opportunities for their input. Even if it is not realistic to perform a study of the use of ICTs for all audiences, it is safe to assume that a campaign that relies mostly on ICTs is not going to reach all audiences. In addition to learning about ICT usage, campaigns can offer training in using the Internet. On the national level, the SIU organizers set up a user-friendly Web-based system for local organizers to sign up for and design their events. The organizers were also available to talk over the phone. However, more effort could have been made to train new organizers in using the Internet. On the local level, events and actions could make available access to computers and offer on-the-spot training on how to use them to participate in the campaign. While the Salt Lake City event did provide computers for sending e-mails to Congress, the setup seemed to assume that people were already knowledgeable in computer usage.

Finally, when developing a national just climate-change movement, national organizers must *teach local organizers how to engage the principles of environmental justice.* As we discovered in our analysis, the events that did a reasonable job of incorporating EJ into their events had organizers who were accustomed to working with EJ groups, as indicated in an interview with some of the organizers of the Seattle events. For a national campaign to be serious about the principles of EJ, it must think about its audience of local organizers with varied levels of understanding of EJ. The SIU campaign did an excellent job of providing resources for the local organizers to learn how to work with the media, for example. The campaign could have also produced guidelines for working with local EJ and faith-based organizations, incorporating the principles of EJ, and reaching out to new social networks.

ENDNOTES

1. All percentages came from the 2000 census (http://quickfacts.census.gov/qfd/).
2. Although we were able to gather rough estimates of the presence of people of color at the actions because of our participant observation at the events, it is important to note that visual markers of racial identity (i.e., skin color) are not the only way to measure racial and ethnic diversity.
3. Although our data are limited to events attended by the research team, they are fully consistent with other research on public participation in environmental campaigns (e.g., Pezzullo and Sandler 2007a).
4. The Principles of Environmental Justice were adopted in 1991 by the delegates to the First National People of Color Environmental Leadership Summit in Washington, DC. A copy of the principles is available in appendix A.
5. Although extending social capital does not necessarily guarantee ecological protection and better democratic practice (Peterson et al. 2006), in this case, social capital is important for an understanding of the relationship between access and the digital divide.
6. These groups were Eco-Justice Ministries, FUSE (Faiths United for Sustainable Energy), Presbyterians for Restoring Creation, the Academy of Evangelical Scientists and Ethicists, and the Regeneration Project.

References

Agarwal, A., S. Narain, and A. Sharma. 2002. The global commons and environmental justice—climate change. In *Environmental justice: Discourses in international political economy*, ed. J. Byrne, L. Glover, and Cecilia Martinez, 171–199. New Brunswick, NJ: Transition Publishers.

Agyeman, Julian. 2005. *Sustainable communities and the challenge of environmental justice*. New York: NYU Press.

Agyeman, Julian, R. Bullard, and T. Evans. 2002. Exploring the nexus: Bringing together sustainability, environmental justice and equity. *Space and Polity* 6:70–90

Agyeman, Julian, and T. Evans. 2003. Towards just sustainability in urban communities: Building equity rights with sustainable solutions. *Annals of American Academy of Political and Social Science* 590:35–53.

Agyeman, Julian, and B. Evans. 2004. "Just sustainability": The emerging discourse of environmental justice in Britain? *Geographical Journal* 170:155–164.

Allen, Kim, Vinci Daro, and Dorothy C. Holland. 2007. Becoming an environmental justice activist. In *Environmental justice and environmentalism: The social justice challenge to the environmental movement*, ed. Ronald Sandler and Phaedra Pezzullo, 1–24. Cambridge, MA: MIT Press.

Bakken, Peter W., Joan Gigg Engel, and Ronald J. Engel. 1995. *Ecology, justice and Christian faith: A critical guide to the literature*. Westport, CT: Greenwood Press.

Bekkers, V., H. V. Duivenboden, and M. Thaens. 2006. *Information and communication technology and public innovation: Assessing the ICT-driven modernization of public administration*. Fairfax, VA: IOS Press.

Bullard, Robert D., G. S. Johnson, and A. O. Torres. 2005. Addressing global poverty, pollution, and human rights. In *The quest for environmental justice: Human rights and the politics of pollution*, ed. R. D. Bullard, 279–297. San Francisco: Sierra Club Books.

Cole, Luke, and S. Foster. 2001. *From the ground up: Environmental racism and the rise of the environmental justice movement*. New York: NYU Press.

Faber, Daniel. 2007. A more "productive" environmental justice politics: Movement alliances in Massachusetts for clean production and regional equity. In *Environmental justice and environmentalism: The social justice challenge to the environmental movement*, ed. Ronald Sandler and Phaedra Pezzullo, 309–320. Cambridge, MA: MIT Press.

First National People of Color Environmental Leadership Summit. 1991. Principles of environmental justice. Web resources for environmental justice activists. http://www.ejnet.org/ej/principles.html (accessed September 23, 2008).

Gottlieb, Roger S. 2006. Religious environmentalism in action. In *The Oxford handbook of religion and ecology*, ed. R. S. Gottlieb, 467–509. Cambridge: Oxford University Press.

Grønhøj, Alice. 2006. Communication about consumption: A family process perspective on "green" consumer practices. *Journal of Consumer Behavior* 5:491–503.

Intergovernmental Panel on Climate Change (IPCC). 2007. *Climate change 2007: Impacts, adaptation and vulnerability; working group II contribution to the Intergovernmental Panel on Climate Change Fourth Assessment Report; summary for policymakers*. http://www.ipcc.ch/SPM13apr07.pdf (accessed August 1, 2007).

International Climate Justice Network. 2002. Bali principles of climate justice. Web resources for environmental justice activists. www.ejnet.org/ej/bali.pdf (accessed April 17, 2008).

Jamieson, Dale. 2007. Justice: The heart of environmentalism. In *Environmental justice and environmentalism: The social justice challenge to the environmental movement*, ed. Ronald Sandler and Phaedra Pezzullo, 85–101. Cambridge, MA: MIT Press.

Kellogg, W. 1999. Community-based organizations and neighborhood environmental problem solving: A framework for adoption of information technologies. *Journal of Environmental Planning and Management* 42 (4): 445–469.

Kellogg, W., and A. Mathur. 2003. Environmental justice and information technologies: Overcoming the information-access paradox in urban communities. *Public Administration Review* 63 (5): 573–585.

McAdam, Doug. 1982. *Political process and the development of black insurgency*. Chicago: University of Chicago Press.

McCarty, John A., and L. J. Shrum. 2001. The influence of individualism, collectivism, and locus of control on environmental beliefs and behavior. *Journal of Public Policy and Marketing* 20:93–104.

Mehra, B., C. Merkel, and A. P. Bishop. 2004. The Internet for empowerment of minority and marginalized users. *New Media Society* 6:718–802.

Moisander, Johanna. 2007. Motivational complexity of green consumerism. *International Journal of Consumer Studies* 31:404–409.

Mossberger, K., C. J. Tolbert, and M. Gilbert. 2006. Race, place and information technology. *Urban Affairs Review* 41:583–620.

National Telecommunications and Information Administration (NTIA). U.S. Department of Commerce. *Falling through the net, toward digital inclusion*. 2000a. http://www.ntia.doc.gov (accessed September 26, 2008).

———. *National telecommunications and information administration annual report 2000*. 2000b. http://www.ntia.doc.gov (accessed September 26, 2008).

———. *A nation online: How Americans are expanding their use of the Internet*. 2002. http://www.ntia.doc.gov (accessed September 26, 2008).

Peterson, Tarla Rai, M. Nils Peterson, Markus J. Peterson, Stacy A. Allison, and David Gore. 2006. To play the fool: Can environmental conservation and democracy survive social capital? *Communication and Critical Cultural Studies* 3:116–140.

Pezzullo, Phaedra, and Ronald Sandler. 2007a. Revisiting the environmental justice challenge to environmentalism. In *Environmental justice and environmentalism: The social justice challenge to the environmental movement*, ed. Ronald Sandler and Phaedra Pezzullo, 1–24. Cambridge, MA: MIT Press.

———. 2007b. Working together and working apart. In *Environmental justice and environmentalism: The social justice challenge to the environmental movement*, ed. Ronald Sandler and Phaedra Pezzullo, 309–320. Cambridge, MA: MIT Press.

Prothero, A. 1990. Green consumerism and the societal marketing concept: marketing strategies for the 1990s. *Journal of Marketing Management* 6:87–103.

ReDirect Guide. n.d. *Make change with every dollar: 10 easy things you can do to spend with intention.* Handout.

Resnick, P. 2002. Beyond bowling together: Sociotechnical capital. In *Human computer interaction in the new millennium*, ed. J. Carroll, 247–272. New York: Addison-Wesley.

Roberts, J. Timmons. 2007. Globalizing environmental justice. In *Environmental justice and environmentalism: The social justice challenge to the environmental movement*, ed. Ronald Sandler and Phaedra Pezzullo, 85–101. Cambridge, MA: MIT Press.

Rogers, E. M. 2001. The digital divide. *Convergence: The International Journal of Research into New Media Technologies* 7:96.

Schlosberg, David. 1999. *Environmental justice and the new pluralism*. New York: Oxford University Press.

Schuler, D. 2000. New communities and new community networks. In *Community informatics: Enabling communications with information and communications technologies*, ed. M. Gurstein, 174–189. Hershey, PA: Idea Group Publishing.

Selwyn, N. 2004. Reconsidering political and popular understandings of the digital divide. *New Media Society* 6:341.

Shulman, S., J. Katz, C. Quinn, and P. Srivastava. 2005. Empowering environmental burdened communities in the US: A primer on the emerging role for information technology. *Local Environment* 10:501–512.

Tesh, S. N. 2000. *Uncertain hazards: Environmental activists and scientific proof*. Ithaca, NY: Cornell University Press.

U.S. Bureau of the Census. 2000. Quick Facts. http://quickfacts.census.gov/qfd/ (accessed September 26, 2008).

Van Dijk, J., and K. Hacker. 2000. The digital divide as a complex and dynamic phenomenon. Paper presented at the 50th Annual Conference of the International Communication Association, Acapulco.

Warschauer, M. 2003. *Technology and social inclusion: Rethinking the digital divide*. Cambridge, MA: MIT Press.

Whitacre, B. E., and B. F. Mills. 2007. Infrastructure and the rural-urban divide in high-speed residential Internet access. *International Regional Science Review* 30:249–273.

Wright, B. H., P. Bryant, and R. D. Bullard. 1994. Coping with poisons in Cancer Alley. In *Unequal protection: Environmental justice and communities of color*, ed. R. D. Bullard, 110–129. San Francisco: Sierra Club Books.

CHAPTER 7

A SOCIAL MOVEMENT SUCCESS STORY?

ASSESSING A SELF-IDENTIFIED MOVEMENT FOR CLIMATE ACTION

William J. Kinsella, Nick Temple, and Jim Shields

Look what's happening out in the streets—Got a revolution
  got to revolution
Hey I'm dancing down the streets—Got a revolution got to
  revolution
Ain't it amazing all the people I meet—Got a revolution got to
  revolution
One generation got old—One generation got soul
This generation got no destination to hold—Pick up the cry!
Hey now it's time for you and me—Got a revolution got to
revolution

> Come on now we're marching to the sea...
> We are volunteers of America...
>
> —Jefferson Airplane, "Volunteers"

From its inception, Bill McKibben and his associates have envisioned the Step It Up (SIU) project as a new social movement.[1] McKibben (2008) recounted that upon returning from a trip to India in 2006,

> I came back home to Vermont from that trip with Gandhi on the brain. I'd kept crossing his tracks time and again in village India, and his example kept crossing my mind as I sank into a kind of despair about American inaction on global warming, the great question of our day...[F]or some reason, the idea of [a] movement had an almost visceral appeal. I'd always loved the story of Gandhi's march to the sea to make salt. It seemed to me that the time had come to march here too, odd as it seemed in an Internet age. I emailed a few friends, and they emailed a bunch more, and we had some meetings, and three weeks later we started out, leaving from Robert Frost's old summer writing cabin in the Green Mountains...And by the time we got to the outskirts of Burlington, the state's one small city, there were a thousand of us marching...

Linking a tool of the Internet age with Gandhi's time-tested strategy for engagement and recruitment, McKibben and his friends achieved a statewide impact. By the end of their march, McKibben (2008) wrote,

> All the candidates for federal office in Vermont agreed to come meet with us for our final rally...liberal Democrat and conservative Republican, they filed up on stage and signed the pledge that they would support our demand: the quite radical pledge to cut carbon emissions by...80 percent...The mere act of marching—of walking side by side, of making ourselves visible and a little vulnerable—seemed to have set something in motion. We wanted to see if that something could translate beyond Vermont, after all the second-smallest state in the union. And so, a few months later in January, we launched a website: Stepitup2007.org.

A look at that Web site, as it appeared nine months later, makes the organizers' sense of excitement and shared identity evident (Step It Up

2007b). By then, they had achieved a national impact, using their Web-based message to recruit participants who organized rallies held throughout the United States on April 14, 2007. At the Web site's top-right corner (the place of priority in traditional print media) is a box with the heading "Join the Movement," inviting the viewer to enter a name, address, and zip code to become part of the action—specifically, a second "National Day of Climate Action" on November 3, 2007. The accompanying text counts down the days, updating automatically overnight. A message from "the Step It Up Team" tells us that its members "enjoy…ringing a small bell every time someone registers a new action for Step It Up 2007." Closing with a rousing call, "Onward with the movement," the message hails viewers to pick up the cry. Scrolling down the page, we see pictures and personal statements posted by the team:

- Will's longish hair, thick brown beard, and collarless shirt would not be out of place on a 1960s student protester, although the laptop on his lap would be. He describes himself as a "climate activist," but the quotation marks are his own; perhaps he finds the term a bit problematic. He frames the situation in urgent terms: there is a "climate crisis," and "time is of the essence in this movement."
- May sits in a tree; the camera views her from below, reading Plato's *Republic* (hardly a manual for grassroots democracy). Will has told us that he's "only twenty-three," and May looks to be of a similar age. She tells us that she has just had her first radio interview for SIU, on a Los Angeles program called *Uprisings*. She, too, describes the project as a "climate movement."
- Jon relaxes, cross-legged, on the hood of an orange VW camper. As SIU's Webmaster, he remarks that he's moved from visions of "occasional acts of civil disobedience" to "typing and clicking [his] way around a blossoming Internet-based social revolution."
- Phil, also only twenty-three, rejects the conventional wisdom that "people my age aren't engaged in social issues, that we're self-interested." From his work "with thousands of student activists trying to stop global warming," he knows that "young people are

creative and pack a serious punch when it comes to getting things done." He believes "that's because it's our future at stake"—perhaps there is some self-interest, after all.

- Jamie stresses the sense of community that has always meant so much to him, and that he feels in his work with SIU. Impressed by "the power of the Internet, blogs, and e-mails in organizing this movement," he marvels at "what a network we have built in only these few short months!" But the technology is only a tool: "it's really the power of all of you, the power of community, that is going to make this change happen."
- Kelly is dressed in a Lady Liberty costume—a motif we saw, as well, at our local SIU action in Raleigh, North Carolina. Having served as "logistics lackey" and "permit princess" for one of the April events, she's "ready to lace up [her] organizing boots once again." She's also "working on tapping into this on-line network of some, oh, 7.5 million progressive folks, called Care2, in order to spread the word about Step It Up." So far, she tells us, she has only three friends on the Care2 social networking site. She gives us a link so we can sign on as friends and as members of the SIU group (Step It Up 2007b).

McKibben's January 2007 remark that "we have all the parts of a movement except the movement itself" is paradoxical but insightful. The standard parts are here—youthful energy and optimism, an urgent and important issue, commitment, community, and comradeship—albeit modeled more on the 1960s social movements than on Gandhi's. A new element is present as well: the technologies that supported some of the 1960s social revolutions, mind-expanding chemicals and birth control, have now been replaced by online technologies for information distribution and networking. Still, the new movement's leaders seem to be having an equally good time.

But is SIU truly a new social movement? If so, is it a successful one? What counts as success, for social movements generally and for this self-proclaimed movement in particular? One notable aspect of SIU is its self-conscious self-identification: the organizers are setting out, deliberately and strategically, to build a movement. But are movements built, or do they grow organically (a fitting metaphor for a movement dedicated to environmental concerns)? Is it now possible to build a movement in months, rather than years or decades, thanks to "the power of the internet, blogs, and e-mails"?

This chapter explores these and related questions using resources from social-movement theory, rhetoric and narrative theory, environmental communication, media theory, and theories of democracy and political participation. We draw upon these resources to outline a framework for assessing SIU in terms of its organizers' own vision of an emerging social movement. Our assessment addresses the following concerns:

- What counts as a "true" social movement in terms of issue scope and definition, breadth and scale of participation, member identification and sense of community, goals, objectives, impacts, and methods of organizing and communicating? Is it valid and useful to view SIU as such a case?
- What counts as "success" for a social movement, in general terms, and what would count as success for SIU? How successful has SIU been to date, irrespective of its formal status as a social movement?
- What is distinctive about the case of SIU, and how can this case inform other efforts within and beyond the environmental public sphere?

To address these questions, we draw upon a range of data including interviews, field notes, and images archived by the eight communication-research teams that attended the April 2007 rallies,[2] and messages posted on the national and local SIU Web sites. A number of the topics explored here, such as the relationships between collective and individual member identity, organizing strategies, and uses of new communication technologies, are developed further in subsequent chapters. Our goal is to assess SIU in terms of its own vision as an emerging social movement, and we organize our analysis accordingly. Before proceeding with that analysis, however, we situate the concept of a "social movement" within a broader perspective on contemporary public discourse and rhetoric.

Global Warming, Rhetorical Exigence, and Communicative Action

In his classic essay on "the rhetorical situation," Lloyd Bitzer noted that rhetoric always responds to a particular *exigence*—a circumstance demanding attention and action. An exigence is "an imperfection marked by urgency; it is a defect, an obstacle, something waiting to be done, a thing which is other than it should be" (1968, 6). Using an ecological example, Bitzer observed that "the pollution of our air…is a rhetorical exigence because its positive modification—reduction of pollution—strongly invites the assistance of discourse producing public awareness, indignation, and action of the right kind" (7). His choice of this example, when the contemporary environmental movement was new, now seems prescient in light of the exigence we have come to call global warming. We differ from Bitzer in one regard, however. Consistent with rhetorical models prevailing at that time, he took a view that may now seem insufficiently "social constructionist" or "constitutive," cautioning his readers not to "assume that a rhetorical address gives existence to the situation; on the contrary, it is the situation which calls the discourse into existence" (2). We propose, instead, that a true social movement does more than *respond* to a preexisting exigence; it also helps to *produce* that exigence by putting a social problem on the agenda for attention and action (for more on this approach, see Kinsella and Mullen 2007).

Contemporary environmental politics reflect the conditions of the "risk society" described by Ulrich Beck (1992, 1995a, 1995b), a society that finds itself threatened by the products of its own modernization. Global warming (along with other risks such as nuclear extinction and general ecological collapse) is an example *par excellence* of those conditions. Accordingly, Niklas Luhmann began his analysis of "ecological communication" by observing that society has "become alarmed as never before" (1989, 1) as "the system registers the effects of its own behavior upon the environment" (23). For Luhmann, the communicative act of "registering" establishes a set of concerns that become objects of attention and foundations for further communication. Without denying the material reality of these self-induced effects, Luhmann highlighted communication about them as the focus of analytical concern.

We might expect that such ecological risks, which Beck described ironically as "democratic" because they threaten all members of society, would become topics of broad public debate. Nevertheless, as narrative theorist Walter Fisher argued, they are typically addressed through restricted, technical analyses by experts, reducing members of the public to "spectators whose choice becomes only a nonrational choice between experts on a stage" (1987, 72). Thus, for Fisher, issues that are fundamentally *public* and *moral* become topics of narrow, technocratic decision making. Such conditions undermine the prospects for a pluralistic and inclusive politics as envisioned by public-sphere theorists such as Jürgen Habermas (1974, 1984, 1989); instead, issues that affect the broad community are shunted from the "public" to the "technical" spheres of argument (Goodnight 1982). These conditions not only undermine the prospects for "strong democracy" (Barber 1984), they also undermine the legitimacy of political institutions as "system rationality increasingly loses its claim to be world-rationality" (Luhmann 1989, 138).

The rising alarm over global climate change, however, has the potential to breach these discursive boundaries. In classical rhetorical terms, global warming provides a moment of *kairos*, an exigence calling for timely intervention. Furthermore, the *globality* of global warming—a compelling case of Beck's "democratic" risk—marks it as among the most public of problems in Fisher's sense. Communicative actions such as those of SIU have the potential to expand the conversation about global warming to a wider public audience and to restore a sense of moral import to the issue. McKibben's initial "despair about American inaction on global warming, the great question of our day," and the "almost visceral appeal" of action in the tradition of Gandhi suggest the emergence of one crucial component of a social movement—a sense of moral necessity. The enthusiasm of McKibben's young collaborators, who merge a neo-sixties aesthetic with a faith in the power of new communication technologies (and competencies in using those technologies), adds another vital component for an aspiring social movement.

Complicating "the Public" and "the Public Sphere"

Felski (1989), Fraser (1991), and others have examined the limitations of overly idealized, unitary models of the public sphere. Accordingly, some theorists have replaced the notion of a unified "public" with the concept of multiple "counterpublics" that challenge the prevailing social order in diverse and evolving ways (Asen 2000). Asen and Brouwer (2001) noted three aspects of such "counterpublics theories." First, those theories recognize the "multiplicity of the public" and the corresponding complexity of public discourse (Chesters and Welch 2006). Second, they emphasize that boundaries between "public" and "private" concerns are permeable: issues can be reframed rhetorically with consequences for inclusiveness, standards of argument, and sites and forms of engagement. Thus, we suggest that SIU seeks to challenge prevailing rhetorical "boundary work" (Gieryn 1983, 1995) or "demarcation" (Taylor 1996) that isolates global warming in the technical sphere, by remoralizing the issue and thereby broadening public participation. Third, counterpublics theories recognize important "imbrications of state and public" (Asen and Brouwer 2001, 13), that is, direct mutual influences and strong interdependencies between official decision making and public discourse. Habermas' original model (1974) viewed the public sphere as a site for open discussion and creative invention of policy options, which are then passed on to the formal political system for consideration. Counterpublics theories suggest a more direct policy influence on the part of activist groups, and as we demonstrate below, SIU appears to be seeking such a role.

Counterpublics theories also challenge models of "new social movements" (Habermas 1981; Laraña, Johnston, and Gusfield 1994), which emphasize a transition from a material politics of distributive justice (or, for Beck, distributive risk) to a symbolic politics of identity and recognition

(Aronowitz 1992; Fraser 1991; Taylor 1994). Rather than reifying sharp boundaries between the material and symbolic domains, some counterpublics theories explore how symbolism and identity can become tools for addressing the material inequities of capitalism (Cloud 2001). As a focus of political protest, global warming can link material issues of consumption and ecological crisis with symbolic issues of identity, community, and lifestyle. Additionally, global-warming activism is emerging as new communication technologies are—at least potentially—restructuring the possibilities for symbolic action and the relationships among activism, identity, and community (Foot and Schneider 2002; Palczewski 2001; Poster 1997). SIU has explicitly embraced the possibilities afforded by new communication technologies and has made them central to its organizing strategy.

Analysis: SIU as a Social Movement for Action on Global Climate Change

Counterpublics theories offer important insights regarding the changing conditions of contemporary public discourse, and we consider those throughout our analysis of SIU's efforts. Our goal, however, is to assess SIU in terms of its own self-representation as a new social movement. Accordingly, we draw upon a standard social-movements approach to organize our analysis. Stewart, Smith, and Denton (2007) identified six characteristics of social movements: organized collective action, lack of institutionalization, spatial and temporal scope and diversity of actions, commitment to changing societal norms and values, sense of moral struggle, and pervasive use of persuasive strategies and tactics. Expanding on the last of these characteristics, they also identify five "persuasive functions" of social movements: transforming perceptions of social reality, altering self-perceptions of protestors, legitimizing the movement, prescribing courses of action, and sustaining the movement. Stewart, Smith, and Denton's six characteristics provide useful criteria for considering the status of SIU as a "genuine" or "bona fide" social movement.[3] Similarly, their five persuasive functions provide criteria for evaluating the success of the SIU effort, irrespective of its status as a social movement. Here, we present a set of observations organized according to those criteria, based on data collected from the sources we have described above. As Benford and Snow suggested, we view "movement actors...as signifying agents actively engaged in the production and maintenance of meaning for constituents, antagonists, and bystanders or observers" (2000, 613). Our observations also demonstrate how SIU exemplifies some of the challenges posed to the traditional social-movement perspective by counterpublics theories and by changing social, political, and technological conditions.

Characteristics of Social Movements

Organized Collective Action

The organized and collective character of SIU is evident overall, but its mode of organizing is of particular interest. Since the launch of the national Web site in January 2007, SIU's organizers have facilitated what might be called a *mediated social movement*. Although the organizers have expressed surprise that participation grew so rapidly and to such levels, that surprise itself functions rhetorically by suggesting a high degree of success:

> [W]e launched Stepitup2007.org, the first open source, web-based day of action dedicated to stopping climate change. We had no expectations, so we were surprised when people all over the country—many of them new to organizing—signed up to host rallies in their communities. (Step It Up 2007a)[4]

Thus, SIU explicitly presents itself as a Web-based movement, using an "open-source" metaphor to describe its mode of organizing (this volume's chapter 11 discusses the open-source metaphor further). That metaphor has both technological and democratic implications. "Open-source" refers to software made available free of charge to all interested users. Anyone may download the software from the Web and use it without fear of violating copyright laws. Typically, one person or group initially designs the software (or adapts it if there is a copyrighted original version available) and places it on the Web for public use. Often, there is an expectation that it will be developed further communally. In some cases, users donate toward the costs of developing and maintaining the software; in other cases, users double as product testers, reporting errors for the designers to fix. In the case of SIU, the metaphorical "software" consists of instructions for organizing local actions based on the shared theme of global warming, the shared goal of reducing U.S. carbon emissions by 80 percent by the year 2050, and the shared objective of acquiring support from members of Congress and other political leaders. This volume's chapter 8, on SIU's use of new communication technologies, refers to that template as a "cookbook" for local action.

Non-Institutionalization

Stewart, Smith, and Denton (2007) maintained that social movements are distinguished by a lack of institutionalization, operating as alternatives or in opposition to the prevailing system of institutions. Indeed, some critics of the U.S. environmental movement focus precisely on its tendencies toward institutionalization, as grassroots efforts are increasingly supplanted by the work of large, formalized organizations maintaining close relationships with politicians and industry. Shellenberger and Nordhaus (2004) linked the crisis of global warming, specifically, to this alleged institutionalization of environmentalism. Viewed as a new, grassroots presence within the environmental movement, SIU can potentially help transcend such concerns. Alternatively, SIU might be viewed as a phenomenon outside the existing environmental movement, enrolling actors from beyond that movement and acting independently of it to address the problem of climate change. Two cautionary observations are appropriate, however. In the short-term, SIU must balance the benefits of appearing new and different with the benefits of aligning with the existing environmental movement and its political allies. In the long-term, if SIU seeks to sustain itself more permanently—another social-movement characteristic noted by Stewart, Smith, and Denton (2007)—then it must address the continuing risk of institutionalization or co-optation over time.

Spatial and Temporal Scope: Variety of Actions

Obviously, it is too early to evaluate SIU's status as a social movement in terms of temporal duration; thus, we focus here on the geographic scope and variety of actions that have characterized the effort to date. Three months after the launch of the Web site, SIU's "open-source" approach facilitated the organization of a large number of rallies held at sites throughout the United States. By April 14, 2007, the date of those rallies, the Web site reported 1,434 scheduled events; more than 850 local participants have since posted reports. The local user groups had "downloaded" the master template, adapting it to their specific rhetorical situations in a variety of ways detailed throughout this book's chapters. Most of these users were members of preexisting groups with a wide range of identities and missions. In Raleigh, North Carolina, where approximately five hundred people attended, participating groups included the investment company Ameriprise Financial, the Audubon Society, Bleeker Electric Car Company, Code Pink, Community Green Guide, the local open-space activist group Dix 306, Environmental Defense, Environment North Carolina, the Fair Trade Federation, the faith-based group Interfaith Power and Light, Piedmont Biofuels, Southern Energy Management, North Carolina Sustainable Energy, North Carolina Green Power, North Carolina State University's MBA Program, North Carolina Waste Awareness and Reduction Network (NC WARN), and Population Connection. The rhetorical linkages between these groups and SIU's global-warming theme were diverse, ranging from the promotion of educational programs and "green" investment opportunities to the connections among global warming, oil, and militarism to stopping the construction of new coal-fired and nuclear power plants in North Carolina.

At the multiple rallies in western Washington state, most of the more than 2,500 attendees were members of community, environmental, faith-based, or public-interest groups who either traveled to the rally together or met up with fellow members at the sites. According to notes from a planning meeting for the western Washington events, participating groups included Campus Climate Change Coalition, Earth Ministry, Greenpeace, the Interfaith Creation Council, the League of Women

Voters, NetGreen, Northwest Animal Rights Group, the Sierra Club and Student Sierra Club, Sustainable Ballard (a Seattle neighborhood organization), and Washington Public Interest Research Group (WASHPIRG). Other participants, not affiliated with groups, reported hearing about the rallies via National Public Radio or the DailyKos political blog site.

In Kalamazoo, Michigan, students from Kalamazoo College helped stage events, while the First United Methodist Church gathered about seventy-five people, including some from the First Baptist Church, for a ceremony including a reading of a letter from the SIU leaders and a brief religious service. A separate rally in East Lansing attracted approximately 125 people, the majority apparently from Michigan State University; many of those participants reported learning of the event via the social networking site Facebook. The number and diversity of groups and individuals appropriating the vision of SIU can be taken as one indicator of status as a social movement, although it is not by itself a conclusive indicator. We should also note that rallies staged at some other sites were not numerically successful.

Orientation to Changing Societal Norms and Values

Another aspect of social movements involves the participants' orientations to effecting social change. In this regard, statements by local organizers and speechmakers at the Raleigh event sometimes contrasted with comments by individual participants during postevent interviews. Even after the rally, a number of individuals still seemed to lack clarity regarding what SIU was about, or associated it primarily with one of the cosponsoring groups. For some of these participants, the multiplicity of voices contributing to the events may have obscured the intended message of shared identity and purpose. On the other hand, the variety of voices may also convey some of the benefits of "strategic ambiguity," enhancing the possibilities for "unified diversity" (Eisenberg 1984). Each cosponsoring group presented its own ideas for addressing global warming; for example, the antiwar group Code Pink displayed bright pink bicycles with a message linking oil use and the Iraq conflict. Thus, participants could identify with interests spanning business opportunities, environmental protection, faith commitments, antiwar activism, open-space preservation, and resistance to the further growth of nuclear power. If this range of identifications is an indicator, one can contribute by praying, consuming, reducing consumption, writing letters or e-mails, opposing war, bicycling instead of driving, or protecting open spaces. In an increasingly diverse culture with a multiplicity of media messages espousing many viewpoints, this breadth of options might promote a convergence of normally competing or disconnected voices.

Many interview respondents expressed concern for their children or for the future more generally and commented on the importance of democratic participation to effect changes. There seemed to be a general sense of a need to sacrifice, that Americans use and waste too much, and that the nation must move toward a more sustainable balance. Many participants expressed existential frustration regarding suburban and urban sprawl, transportation issues, and a culture of consumption. Despite their attendance, some respondents questioned the power of events such as SIU to challenge the prevailing economic and political system. Nevertheless, most of those participants expressed a determination to engage rather than a sense of hopelessness.

Sense of Moral Struggle

As we have noted above, a number of statements by McKibben and others on SIU's national Web site convey a sense of moral import. This framing is in partial tension, however, with other messages that circulated at local rallies on April 14, mirroring an ongoing debate within the environmental movement regarding radical versus incremental change. For example, a theme reported by many observers at the local events was the promotion of compact fluorescent lightbulbs as a practical tool for energy conservation. At the Raleigh event, volunteers at two tables distributed samples and an authoritative academic speaker observed that North Carolina could eliminate the need for one new coal-fired power plant by replacing four incandescent lightbulbs per household with compact fluorescents. Although this new technology surely has considerable value, some interview respondents commented that "it will take a lot more than changing our lightbulbs" to address the problem of global warming. Their comments, in fact, echo a statement by McKibben that had appeared on the national Web site. Taken together, these comments highlight both the merits of incremental approaches and the need to go beyond them. In a remark that captures additional tensions between incrementalism and moral framing, the same academic speaker applauded the increasing involvement of faith communities in the global-warming issue but added the ironic caveat, "—as long as people don't drive their SUVs to church!"

Pervasive Persuasion

The national Web site explicitly identified a specific, persuasive goal for the April 2007 actions: to persuade the U.S. Congress to enact legislation leading to an 80 percent reduction of carbon emissions by the year 2050. The template provided on the Web site highlighted that message, encouraging local organizers to create a banner and to gather participants to pose with it for digital photographs at each rally:

> On April 14th, the SIU National Day of Climate Action, communities came together in more than 1,400 places, all holding up banners reading: "Step It Up, Congress: Cut Carbon 80% by 2050." On that night, photos streamed in from all across the country, from underwater in Key West to dwindling glaciers in Wyoming to the levees of New Orleans. At schools and on town greens, concerned citizens made their voices heard, and the results speak for themselves. (Step It Up 2007a)

In practice, this persuasive goal became part of a much broader effort to address global warming through a variety of individual, communal, and political approaches. The various rallies exhibited efforts to influence consumption behaviors directly, to inform and educate, to encourage political engagement by attendees, to recruit new members for sponsoring organizations, and to build shared identification and a sense of community. A variety of persuasive approaches was evident; the Raleigh event, for example, featured speeches, musical performances, visual displays, and carnivalesque parodies. The program included a wide range of speakers drawing on diverse sources of authority: an academic researcher, a Native American tribal leader, a state assembly representative, a minister-in-training, and a group of children who made personal statements about the implications of global warming for their future.

SIU's persuasive strategy, overall, exemplifies the new media principles of "co-production" and "mobilization" (Foot and Schneider 2002). As coproduction, SIU's national organizers distributed a general template that local leaders used in innovative and context-specific ways. As mobilization, SIU has employed a two-step form of persuasion in which local participants were persuaded to help persuade political decision makers. This volume's chapter 9 examines this Web-based persuasive strategy in more detail.

Functions of Social Movements

Transforming Perceptions of Social Reality

For SIU, transforming perceptions of social reality begins with transforming perceptions of material reality. Textual descriptions and visual images of the present and predicted effects of global warming have been key elements on the national Web site and at all the local rallies. These materials provide the starting point for what is, essentially, SIU's master argument: global warming poses vast ecological and economic threats and is primarily a product of the social order, including economics, politics, and culture; therefore, social changes are fundamental to addressing the problem.[6] This argument highlights the urgency and consequentiality of the issue, typically placing the implications of global warming in the domain that Cox (1982) called the "locus of the irreparable." It does not, however, directly address the tensions between incremental and radical solutions, and by leaving room for disagreement on that point, it maintains its appeal to a broader audience. Often, the argument identifies issues of environmental justice to link the material and social consequences of global warming, as examined in this volume's chapter 6, titled "Toward Just Climate-Change Coalitions."[7]

At the Raleigh event, examples of this argument in action included the testimony of a Native American tribal leader who serves as director of natural resources and environmental affairs for the Southern Cherokee Nation, who described his recent trip to Alaska to witness the effects of climate change on the Inuit people. An environmental scientist referenced the recently released physical science component of the Fourth Assessment Report of the Intergovernmental Panel on Climate Change (IPCC 2007), a powerful new warrant supporting the case of anthropogenic warming, linking that relatively abstract global perspective to local projections of the impact on coastal North Carolina. A student minister suggested that "some would call global warming a tragedy of biblical proportions." In each case, the speakers connected these material consequences to social consequences: the devastation of Inuit communities, the "loss of biodiversity and [consequently] cultural heritage in Appalachia," and opportunities for developing "human potential" in the fight against global warming.

Altering Self-Perceptions of Protestors

On its Web site and at local events, SIU sought to alter the self-perceptions of participants by highlighting their sheer numbers and emphasizing the unity of the national effort. The local coordinator for the Raleigh event opened the rally by stating that there were 1,377 events planned nationwide (the number listed on the national Web site later increased), with twenty-three actions in North Carolina. Describing a hypothetical job description for "climate superheroes," she proclaimed, "I challenge you to become a climate superhero…We are the solution and we have superhero strength…Now, let's get started with our speakers."

The final speaker, the student minister, invoked both religious and secular heroes: the ecological author and Catholic priest Thomas Berry, a North Carolina native, who wrote of a coming "ecozoic era" of human concern for the environment, and the Apollo 13 mission director, Gene Krantz, who responded to a reporter's comment regarding the space program's "lowest moment" by insisting that, on the contrary, salvaging the mission would be NASA's "finest hour." Merging theology and technology, these remarks linked themes of holism, hope, human responsibility, and a new, enlightened form of progress. Participants at the rally experienced an opportunity to view themselves as agents of positive and historical change, politically empowered and connected to a like-minded community.

Legitimizing the Movement

At the Raleigh event, both the setting and the speakers contributed to a sense of political and moral legitimacy. The rally was held on the grounds of the old state capitol, amidst a rich collection of historical and political symbols: statues of political and military heroes from a range of time periods, cannons, flags, and the capitol building itself. One state assembly member spoke to the participants, conveying the support of another member who had been present earlier in the afternoon. The dean of Duke University's Nicholas School of the Environment lent his own technical authority to the event and invoked the broader authority provided by the recent findings of the Intergovernmental Panel on Climate Change. At the same time, a theme of grassroots democracy was prevalent: the speakers consistently stressed that this action was planned and executed by ordinary people, in the interest of all people, all species, and the planet.

Consistent with a sense of growing legitimacy, the SIU Web site now lists more than eighty political figures who supported the follow-up events in November 2007. Much of this support was mediated in form; many of those cited did not appear in person at the rallies but instead sent representatives or messages. At least four potential presidential candidates expressed their support. Former senator John Edwards and Congress member Dennis Kucinich each attended events, Senator Hillary Clinton sent representatives to at least two events, and Senator Barack Obama sent a videotape and a written message to at least one event. Posted on the national Web site, these figures' images and messages add political legitimacy to SIU's activities.

Prescribing Courses of Action

The central message of SIU prescribes a clear and specific course of action, entreating the U.S. Congress to enact legislation reducing carbon emissions by 80 percent by the year 2050. This message alone is too specific to ground a movement, as it calls for a single policy change—albeit, one that would have far-reaching consequences—rather than a continuing effort to redress a moral wrong and to change society at a fundamental level. Another risk of attempting to found a movement on this message is that the specific plea to Congress may well be unsuccessful; in that case, SIU would need to change or broaden its focal message. As we have seen, the message has already been expanded in many ways to encompass a broad range of approaches for addressing anthropogenic climate change. These broader efforts may prove to be the more consequential outcomes of the events held in 2007 by raising the public visibility of climate change, by offering tangible suggestions for action, and by recruiting new individuals, groups, and organizations to the cause.

At the close of the Raleigh action, the organizers reminded participants to sign up at the coordinators' table and to record the distance they had traveled because organizational sponsors had agreed to subsidize carbon-offset credits for all who attended. As another incentive, the green-branded grocery store Whole Foods had donated a gift bag to be raffled using the sign-up cards. The sign-up cards, of course, provided a useful tool for continued contact with rally participants. At other rallies, including the one in Seattle, pledge cards offered options for individual action, sending messages to policymakers, and involvement in continued organizing efforts. As her final environmentally conscious comment, the Raleigh event host noted that "grounds keeping matters, so please leave the site clean."

Sustaining the Movement

As of June 2009, SIU's national Web site displayed 642 reports of follow-up events held on November 3, 2007. The November action organized by the Raleigh group did not take place downtown but instead at Falls Lake reservoir, approximately seventeen miles distant. Although the number of participants was smaller than in April, again the setting provided symbolic significance, as the city's water supply was visibly low due to unprecedented drought conditions that had plagued the region for months.

Compared to the approximately 850 reports of events held in April, the reduced number of reports for November suggests a decline in support, and the available data also indicate that the November actions were smaller in scale. However, these numbers are only one benchmark for assessing the sustainability of SIU's effort. One outcome of the November actions was a linkage of SIU with 1Sky, another emerging climate-change effort that describes itself variously as a "campaign" and, like SIU, a "movement." According to the 1Sky Web site,

> The 1Sky campaign is dedicated to aggregating a massive nationwide movement by communicating a positive vision and a coherent set of national policies that rise to the scale of the climate challenge before us. It has been born from the collective urgency and determination of leaders throughout the country. Our movement has many strengths, audiences, and organizing strategies—1Sky is aggregating and magnifying our power. (1Sky 2008b)

1Sky has also linked its efforts with those of Power Shift 2007, a project that staged a "national youth summit to solve the climate crisis" in Washington, DC, on November second, the day before the second round of SIU events. Promising to "take the climate movement to new levels," Power Shift identifies three goals not unlike the broader goals of SIU: to "make U.S. Presidential candidates and Congress take global warming seriously," to "empower a truly diverse network of young leaders," and to "achieve broad geographic diversity" representing "nearly every Congressional district in the United States to demand scientifically based solutions from all who represent us" (Power Shift 2007).

By June 2009, the SIU Web site provided links to 103 groups and organizations, including 1Sky and Power Shift, listed under the heading "friends and allies." These included entities devoted directly to climate change, such as Focus the Nation, another student-oriented project planning a "national teach-in" for January 31, 2008, and Global Climate Campaign, which organized a "third international day of climate action" on December 3, 2007. Other groups and organizations, such as green-branded food-and-drink manufacturer Clif Bar and Company, have less-direct connections to the issue of climate change. Some are connected via general commitments to ecology and energy concerns, social and environmental-justice missions, alternative-media operations, or antiwar, antiglobalization, and other progressive political activities. Also prominent on the SIU Web site is a box titled "Find Us On," with links to the social networking and file-sharing sites Facebook, Flickr, and YouTube. These links provide opportunities for individuals to connect directly with others without necessarily engaging with SIU or its allied groups and organizations.

Despite its enthusiastic embrace of networking opportunities, SIU has thus far maintained its own identity rather than disappearing into the larger grassroots effort. The SIU Web site now features information on *Fight Global Warming Now: A Handbook for Taking Action in Your Community*, released in October 2007 and authored by Bill McKibben and the Step It Up team, with six of McKibben's young collaborators listed as contributors. Also featured is Project Hot Seat, a political-influence campaign encompassing letter writing, phone calling, voter mobilization, and public education, described as a joint project of Greenpeace, TrueMajorityAction.org, SIU, and 1Sky. The broad effort to sustain and grow a climate-change movement appears to be embracing the coproduction possibilities afforded by new media, while also honoring the specific contributions of key organizations.

Discussion: Nested Movements, Mediated Action, and Public Moral Argument

We now return to our central research questions in light of the observations provided above. To what extent does SIU represent a genuine social movement? How successful has the SIU effort been? What is distinctive and instructive about the case of SIU? Rather than addressing these questions sequentially, we view them as interconnected and offer three closing observations, noting some implications for climate-action organizers.

First, we regard SIU less as an autonomous social movement and more as part of a broader, rapidly evolving network of organizations, groups, and individuals seeking action on global climate change. The amorphousness of this network, linked innovatively through the use of new communication technologies, raises questions regarding how to define the "climate-change movement" referenced so frequently and ubiquitously by participants. If SIU *is* a movement, then perhaps it is a movement within a movement (which may—or may not—be nested within a still-broader environmental movement). As 1Sky's online brochure remarks:

> The emerging climate movement does not lack ideas, leaders, or organizing capacity. It lacks power...1Sky is working to bring the movement together, and to expand it dramatically by inviting a broad range of key stakeholders outside the environmental groups to the table. The unifying force is a common destination—a vision, a moral proposition, and a communication strategy for real policy solutions. (1Sky 2008a)

Accordingly, climate-action organizers should be attentive to the nuances of rhetorical framing and positioning. In some communities or geographic regions, and in connection with particular issues that have demonstrable relationships to climate change, identification with a broader environmental movement may be a rhetorical asset. For example, efforts to connect climate change with issues of environmental justice at the Raleigh event appeared to be successful. At the same time, organizers should note that the issue of climate change appears to resonate with audiences not typically identified with environmental causes; to appeal to those audiences, climate activists may need to distinguish their goals from the broader environmental agenda. Doing so may be especially important in geographic regions not known for a high degree of sympathy to environmental issues.

Second, we note that the new media principles of aggregation, coproduction, and mobilization are key parts of this communication strategy. As our analysis has demonstrated, the resulting network of climate-change activism exhibits a number of the social-movement characteristics described by Stewart, Smith, and Denton (2007): organized collective action, broad geographic reach and diversity of actions, and pervasive use of persuasive tactics and strategies. The diffuse and decentered characteristics of Web-based communication, which SIU and its allies use so extensively, provide some protection against the risk of creeping institutionalization. A commitment to changing social norms and values is also evident in these efforts, although there is ambiguity regarding how to manage the tensions between incremental and radical change. These characteristics may suggest the emergence of a fairly successful social movement, but if so, it is a rapidly evolving and highly variegated movement. For organizers, questions of how to sustain such a movement over time are paramount. Framing and positioning are crucial for movement sustainability, but should be flexible with respect to regional differences and changing public concerns. For example, subsequent to the Step It Up events described here, the salience of rising fuel prices has increased significantly, presenting a clear rhetorical opportunity; new media communication tools can help organizers respond rapidly to such developments.

Our final observation suggests another way of addressing the sustainability problem and highlights a crucial feature of the climate action effort. We note that the remaining social-movement characteristic noted by Stewart, Smith, and Denton—a sense of moral struggle—also appears in 1Sky's statement and in our observations of SIU. As counterpublics theories suggest, organizations such as 1Sky and SIU provide virtual safe spaces where parties can express their concerns, forge common identifications, and strategize before speaking out more publicly with a shared vision and a sense of moral authority. New media tools, in the hands of a new generation of competent and committed activists, can help promote that moral vision.

Cognitive psychologist Steven Pinker provided a cautionary observation regarding moral framings of climate change:

> The threat of human-induced climate change has become the occasion for a moralistic revival meeting. In many discussions, the cause of climate change is overindulgence (too many S.U.V.'s)

> and defilement (sullying the atmosphere), and the solution is temperance (conservation) and expiation (buying carbon offset coupons). Yet...even if every last American became conscientious about his or her carbon emissions, the effects on climate change would be trifling, if for no other reason than that two billion Indians and Chinese are unlikely to copy our born-again abstemiousness. Though voluntary conservation may be one wedge in an effective carbon-reduction pie, the other wedges will have to be morally boring, like a carbon tax and new energy technologies, or even taboo, like nuclear power and deliberate manipulation of the ocean and atmosphere. Our habit of moralizing problems...and resting content when we feel the right feelings, can get in the way of doing the right thing. (2008)

Pinker's observation is provocative, but we believe he underestimates the degree to which climate change is, inevitably, a moral issue. Technical solutions are surely important in the portfolio of responses to global warming, but they, too, are insufficient: the exigence of global warming demands more fundamental changes in our collective values and practices. Indeed, technical solutions come with their own moral entailments. Policy decisions regarding nuclear energy or manipulation of the ocean or atmosphere are linked to particular visions of the common good and the relationship between humans and nature. Carbon-tax proposals engage debates over free-market capitalism, globalization, and the tensions between individualism and collectivism. Competing approaches to the problem of climate change raise issues regarding the distribution of material risks and economic costs and benefits, as well as questions of social and environmental justice. These aspects place the problem of climate change squarely within the realm of moral conflict (Pearce and Littlejohn 1997) or public moral argument (Fisher 1987).

SIU and its allies have made important contributions by expanding the conversation about climate change and by promoting practical actions that can improve the material situation. A true social movement may, in fact, be necessary to address the problem of climate change creatively and adequately across political, geographic, and cultural boundaries and over the requisite time frame. SIU and the broader network for climate activism may constitute the beginnings of such a movement, distinguished by a unique linkage of moral fervor, recognition of the practical necessity of social change, and astute use of new communication tools. Continued attention to their efforts is well warranted and provides a timely opportunity for fruitful collaboration between scholars and activists.

ENDNOTES

1. Here we use the term *new social movement* in a colloquial sense, as it is used by the SIU organizers, rather than in the more specific theoretical sense discussed later.
2. Details on the national SIU research project and the eight research teams are provided in the introduction to this volume.
3. The concept of "bona fide groups" (Putnam and Stohl 1990) provides a useful analogy for evaluating SIU's status as a social movement, but we do not develop that analogy here.
4. The SIU organizers also referred to this theme of surprise in a telephone interview with contributors to this volume on July 25, 2007.
5. Eder offers another view of the "institutionalization of environmentalism," arguing that as environmental concerns become increasingly common in public discourse, "environmental protest actors" must compete with other parties "in the market of producing and communicating 'green' images" and must "defend their part in the public discourse on the environment they themselves once launched" (1996, 204).
6. More nuanced versions of the argument acknowledge the increasingly ineffective objection that global warming might have a nonanthropogenic component but stress that the problem is primarily a product of human activity.
7. For more on the concept of environmental justice, see Sandler and Pezzullo (2007).

References

1Sky. 2008a. The 1Sky campaign. http://www.1sky.org/wp-content/uploads/pdfs/ 1sky_background.pdf (accessed April 13, 2008).

———. 2008b. About 1Sky. http://www.1sky.org/about (accessed April 13, 2008).

Aronowitz, Stanley. 1992. *Politics of identity: Class, culture, social movements*. New York: Routledge.

Asen, Robert. 2000. Seeking the "counter" in counterpublics. *Communication Theory* 10:424–446.

Asen, Robert, and Daniel C. Brouwer. 2001. Introduction: Reconfigurations of the public sphere. In *Counterpublics and the state*, ed. Robert Asen and Daniel C. Brouwer, 1–32. Albany: SUNY Press.

Barber, Benjamin. 1984. *Strong democracy: Participatory politics for a new age*. Berkeley: University of California Press.

Beck, Ulrich. 1992. *Risk society: Towards a new modernity*. Thousand Oaks, CA: Sage.

———. 1995a. *Ecological enlightenment: Essays on the politics of the risk society*. Atlantic Highlands, NJ: Humanities Press.

———. 1995b. *Ecological politics in an age of risk*. London: Polity.

Benford, Robert D., and David A. Snow. 2000. Framing processes and social movements: An overview and assessment. *Annual Review of Sociology* 26:611–639.

Bitzer, Lloyd F. 1968. The rhetorical situation. *Philosophy and Rhetoric* 1 (3): 1–14.

Chesters, Graeme, and Ian Welch. 2006. *Complexity and social movements: Multitudes at the edge of chaos*. New York: Routledge.

Cloud, Dana. 2001. Doing away with Suharto—and the twin myths of globalization and new social movements. In *Counterpublics and the state*, ed. Robert Asen and Daniel C. Brouwer, 235–263. Albany: SUNY Press.

Cox, Robert. 1982. The die is cast: Topical and ontological dimensions of the locus of the irreparable. *Quarterly Journal of Speech* 68:227–239.

Eder, Klaus. 1996. The institutionalization of environmentalism: Ecological discourse and the second transformation of the public sphere. In *Risk, environment, and modernity: Towards a new ecology*, ed. Scott Lash, B. Szerszynski, and B. Wynne, 203–223. Thousand Oaks, CA: Sage.

Eisenberg, Eric M. 1984. Ambiguity as strategy in organizational communication. *Communication Monographs* 51:227–242.

Felski, Rita. 1989. *Beyond feminist aesthetics: Feminist literature and social change*. Cambridge, MA: Harvard University Press.

Fisher, Walter R. 1987. *Human communication as narration: Toward a philosophy of reason, value, and action*. Columbia: University of South Carolina Press.

Focus the Nation. 2008. Focus the nation: Global warming solutions for America. http://www.focusthenation.org (accessed April 13, 2008).

Foot, Kirsten A., and Schneider, Steven M. 2002. Online action in campaign 2000: An exploratory analysis of the U.S. political web sphere. *Journal of Broadcasting and Electronic Media* 46 (2): 222–244.

Fraser, Nancy. 1991. Rethinking the public sphere: A contribution to the critique of actually existing democracy. In *Habermas and the public sphere*, ed. Craig Calhoun, 109–142. Cambridge, MA: MIT Press.

Gieryn, Thomas F. 1983. Boundary-work and the demarcation of science from non-science: Strains and interests in professional ideologies of science. *American Sociological Review* 48:781–795.

———. 1995. Boundaries of science. In *Handbook of science and technology studies*, ed. Sheila Jasanoff, Gerald E. Markle, James C. Petersen, and Trevor Pinch, 393–443. Thousand Oaks, CA: Sage.

Global Climate Campaign. 2008. International demonstrations on climate change. http://www.globalclimatecampaign.org (accessed April 13, 2008).

Goodnight, G. Thomas. 1982. The personal, technical, and public spheres of argument: A speculative inquiry into the art of public deliberation. *Journal of the American Forensic Association* 18:214–227.

Habermas, Jürgen. 1974. The public sphere: An encyclopedia article. *New German Critique* 1:49–55.

———. 1981. New social movements. *Telos* 49:33–37.

———. 1984. *The theory of communicative action, volume 1: Reason and the rationalization of society*. Trans. T. McCarthy. Boston: Beacon.

———. 1989. *The structural transformation of the public sphere: An inquiry into a category of bourgeois society*. Trans. Thomas Burger. Cambridge, MA: MIT Press.

Intergovernmental Panel on Climate Change (IPCC). 2007. *Climate change 2007: The physical science basis: Contribution of Working Group I to the Fourth Assessment Report of the IPCC*. Cambridge: Cambridge University Press.

Kinsella, William J., and Jay Mullen. 2007. Becoming Hanford downwinders: Producing community and challenging discursive containment. In *Nuclear legacies: Communication, controversy, and the U.S. nuclear weapons complex*, ed. Bryan C. Taylor, William J. Kinsella, Stephen P. Depoe, and Maribeth Metzler, 73–107. Lanham, MD: Lexington Books.

Laraña, Enrique, Hank Johnston, and Joseph R. Gusfield, eds. 1994. *New social movements*. Philadelphia: Temple University Press.

Luhmann, Niklas. 1989. *Ecological communication*. Trans. John Bednarz Jr. Chicago: University of Chicago Press.

McKibben, Bill. 2007. Step It Up: Introducing a brand-new, mass-protest climate movement for 2007. *Grist: Environmental News and*

Commentary. http://www.grist.org/comments/ dispatches/2007/01/08/mckibben/index.html (accessed September 21, 2007).

———. 2008. Gandhi: A man for all seasons—What we still need to learn from the example of Gandhi. *Odemagazine.com*. http://www.odemagazine.com/doc/50/what-the-world-needs-now (accessed April 13, 2008).

Palczewski, Catherine Helen. 2001. Cyber-movements, new social movements, and counterpublics. In *Counterpublics and the state*, ed. Robert Asen and Daniel C. Brouwer, 161–186. Albany: SUNY Press.

Pearce, W. Barnett, and Stephen W. Littlejohn. 1997. *Moral conflict: When social worlds collide*. Newbury Park, CA: Sage.

Pinker, Steven. 2008. The moral instinct. *New York Times Magazine*, January 13, 2008. www.nytimes.com/pages/magazine/index.html (accessed January 13, 2008).

Poster, Mark. 1997. Cyberdemocracy: Internet and the public sphere. In *Internet culture*, ed. David Porter, 201–217. New York: Routledge.

Power Shift 2007. 2008. About Power Shift 2007. http://powershift07.org/about (accessed April 13, 2008).

Putnam, Linda L., and Cynthia Stohl. 1990. Bona fide groups: A reconceptualization of groups in context. *Communication Studies* 41: 248–265.

Sandler, Ronald, and Phaedra C. Pezzullo, eds. 2007. *Environmental justice and environmentalism: The social justice challenge to the environmental movement*. Cambridge, MA: MIT Press.

Shellenberger, Michael, and Ted Nordhaus. 2004. The death of environmentalism: Global warming politics in a post-environmental world. The Breakthrough Institute. http://www.thebreakthrough.org/PDF/Death_of_Environmentalism.pdf (accessed April 13, 2008).

Step It Up. 2007a. Our story. http://stepitup2007.org/article.php?list=type&type=48 (accessed April 13, 2008).

———. 2007b. The Step It Up Team. http://stepitup2007.org/article.php?id=687 (accessed April 13, 2008).

Stewart, Charles J., Craig Allen Smith, and Robert E. Denton. 2007. *Persuasion and social movements*. Long Grove, IL: Waveland.

Taylor, Charles. 1994. The politics of recognition. In *Multiculturalism: Examining the politics of recognition*, ed. Amy Gutman, 25–73. Princeton, NJ: Princeton University Press.

———. 1996. *Defining science: A rhetoric of demarcation.* Madison: University of Wisconsin Press.

Chapter Interlude

Interview with Stephanie Kimball

Phaedra C. Pezzullo

Background note: Stephanie Kimball was the primary organizer of Step It Up! in Bloomington, Indiana, on April 14, 2007. She was born in Washington, DC, moved a lot growing up, and has lived in Bloomington since 2000. She is a European American woman, wife, and stay-at-home mother who homeschools her two children. She has a PhD in education from Virginia Tech. This is an edited version of an interview that took place on October 2, 2007, in a locally owned bakery. Pezzullo, a resident of Bloomington since 2002, met Kimball earlier in the year at a meeting about environmental education and outreach, which Pezzullo was attending as an environmental commissioner for the city. The Environmental Commission was one of the cosponsors of the event.

Pezzullo: How did you first hear about Step It Up (SIU)?

Kimball: I think somebody sent me an e-mail linking me to their Web site, but I can't remember who or how. I looked at their Web site and thought, "Wow. This sounds really interesting." There wasn't anything listed yet for Bloomington. That was in January of this year [2007].

Pezzullo: **From what I understand, this was your first political organizing event ever. What made you decide to not only be a part of but actually to organize Bloomington's SIU events?**

Kimball: I had actually seen Al Gore's DVD [*An Inconvenient Truth*] in, I think it was December [of 2006]. It had taken a lot for me to skewer up the courage to watch it. I knew about global warming—and I kind of didn't want to know any more. My life was pretty stressful. I finally watched it, and I remember announcing to my family at the end of December that this was going to be my focus for the following year at least, that I was going to do everything that I could to join the fight against global warming. I didn't know what that was going to look like.

Then this thing came across my virtual desk called "Step It Up!" and it just seemed like a promising thing to do. I was really hesitant because I didn't see myself as an organizer and I didn't see myself as someone who knows anybody—it's very isolating being a homeschool mother of two living in the country! (Although my husband had just run for county council, so I realized I did know quite a few people, actually, at least in a secondary sort of way.) There was a form on the Web site asking what you are planning on doing. At the time, it seemed to me like they were asking, "What are your credentials? What business do you have taking this on?" And I was thinking, "Ugh." It felt like a job interview! So, I was really being careful about how to word my response, and I set it aside for a couple of days trying to figure out how I could convince anyone.

In the meantime, someone across town was having a similar experience. Maire Quilter probably had seen the same thing on the same day, and she told me later that she was thinking the same thing: "Well, I haven't been in town long, and I don't know many people." But she didn't hesitate as long as I did; so, when I went back to the Web site to submit my form, I saw she had already posted hers. So, she actually was the official organizer. We hadn't met each other before that...It turned out that we go to the same [Unitarian Universalist] church. And when we met for the first time, we realized we had almost identical visions for this whole thing, down to who to invite. It was really weird. Unfortunately, she had some health issues, so she wasn't able to be as active as she had wanted to be.

There is just one other thing I want to add regarding my motivation to organize an action as opposed to just attending, or participating in a smaller way. One of the homeschooling study habits that has evolved for my kids and me is to visit the library and bring home a dozen or so books on a particular topic of interest, then get comfy on the family-room floor and read them together. So one January afternoon, shortly after my own announcement to my family, my son, Hans, decided to do his science-fair project on global warming. I found myself reading book after book about global warming to my then- three- and six-year-old children. Not surprisingly, after a while, it became a bit overwhelming. Finally, I closed the book we were looking at and said, "This stuff is a little scary, isn't it?" They agreed that it was. So I said the only thing I possibly could say in the circumstance: "You know what? We're going to stop global warming. It's going to take a lot of work, but we WILL do it." I am not in the habit of making promises, especially about things that are almost entirely out of my control, but I did. And so far, those words have turned out to be incredibly powerful to me. I've found that they have pushed me past my fear and fatigue when I've had to face new challenges—having to speak to large groups, ask people for things, learn how to build Web sites—and they've stopped me from quitting in the face of despair. I know it is corny, but even so, it's true: I have been pushing myself into roles and responsibilities, many of which I'd just as soon pass up, simply because I made an audacious promise to my precious and beautiful, full-of-life children, that we will, somehow, stop global warming.

PEZZULLO: **Did you ever have direct contact with the SIU national organizers prior to or during the organizing, or were you mostly in communication with them online?**

Kimball: It was all online, mostly generic messages to all the organizers. At some point during the process, a question came up—we were wondering if we should have a petition at the event for people to sign and what were the guidelines—so I went to the organizers' forum on the Web site, where you could talk directly with other organizers, and I posted something there, asking if we should do this. About six or seven people replied, as I recall, and said they wanted to do the same thing, but nobody knew exactly how. At that point, I e-mailed the national folks to see if anyone was working on this, and I think that was my first contact with them. (And, as a side note, they ended up working with MoveOn to produce a petition that was used.) There were a couple of other things like that, where I e-mailed them and had a nongeneric interaction, which was nice...They were actually really good about getting back to me promptly, no more than a day or two, and [they were] very encouraging.

Again, one of the things I guess that shifted for me in this process of this was that I was starting off really imbalanced. I saw the national organizers as faraway people who obviously had some kind of authority in something—I don't know what—but, enough that they could be putting out this call to people everywhere to do something. And I saw myself as really unqualified and kinda scared, but feeling like, "Well, somebody's got to do this." And as the process went on, I came to realize that these are—not to put them down or anything in *any* way, because I *really* admire these people—but, that they're real people and they're working their butts off to do this stuff, just like everybody else. And *they* didn't see themselves on any kind of pedestal. They saw themselves as facilitators and catalysts and whatnot, which *they are*. But that was a really important, I think, kind of shift for me—that we're all in this together in a really basic way.

Pezzullo: The question of authority for you, was it more about being an organizer or about global warming as a scientific topic?

Kimball: Definitely both. Since then, I've *read a lot* and I understand a lot more than I did eight months ago, for sure; but, still, I'm definitely not an expert. I'm aware that there are people in the community, just as there are scientists all over the place, who can rattle off all sorts of information. And there are people that can remember facts much better than I do—

Pezzullo: —And me!

Kimball: But, you know, I do find that I can explain things to my kids. There are different ways that we can all be different parts of the puzzle.

And, definitely, also as an organizer, because I had only ever been a participant before…I was kind of relieved when I saw somebody else had signed up first, and I thought, "OK, I can go along with this and do what I can"—because I've got two kids and an extremely busy husband…oh, [being involved with SIU] was kind of a bold thing to end up doing. I did sort of end up central to the organizing event, and I'm *amazed* that it even happened because of that! [*laughs*]

Pezzullo: Much of SIU involved the Internet. Could you imagine having done this event without it?

Kimball: Um, no. It wouldn't have occurred to me. And if it had been just one call, like, "Hey, everybody, let's have events on April 14," and then we're all on our own—probably not. I mean, there was a point where I was checking that Web site several times *a day*. That's how I would start my day—I probably shouldn't admit this, but I'd have my computer by my bed and when I'd wake up, I'd look to see how many people had signed up overnight. It was scary! [*laughs*]

Pezzullo: What was the biggest challenge to you personally in organizing?

Kimball: The biggest challenge to me—I'd say there were two things that are kinda related. One was just getting things done. It's one thing to sit with a bunch of people and come up with great ideas, and it's another thing to actually get information onto the Web site, make the phone calls, all that kind of stuff. And the related challenge was getting *other* people to do things, getting people to show up to meetings, because *everybody* is doing so many things and is so overcommitted, and I totally sympathize.

PEZZULLO: **Whenever I saw you last spring, you were wearing a pin that said, "Ask me about SIU." Did you find wearing that pin has generated interest in the issues or event?**

KIMBALL: I had several people ask me about it. There's something about wearing something that says, "Ask me about…" that means you'll get a lot of kind of obnoxious people who will say [*in a taunting voice*], "Well, then, tell me about it." But, I'd tell them about it anyway! [*laughs*] Actually, it did help with friends and acquaintances who maybe did not know I was involved in this; it created opportunities for me to talk about the event and the issue.

PEZZULLO: **The rain moved the event in Bloomington from outside at the farmers' market to inside the city council building next to the farmers' market—except for the initial march through the downtown. You had planned for backup in case of rain. Do you feel the weather impacted the number of people who attended?**

KIMBALL: Definitely. It wasn't just rain, it was *cold*. It was almost snow. It was kind of nasty. I think if it had been a nice spring day, there would have been tons of people at the farmers' market and they would have drifted over to find out what was going on. As it was, you had to be pretty die-hard to be at the farmers' market that day. And even if you *were* at the farmers' market, you might not have known that anything was happening inside the building—even though we had signs. So, you had to pretty much know that the event was going on. Given all of that, the fact that we had, I don't know, maybe three hundred people—that was pretty amazing!

PEZZULLO: **You had an amazing schedule and ended up organizing a full day of speakers and performers (see appendix B). Can you explain why you chose to invite both politicians, including the city mayor and Congressman Baron Hill, as well as artists, including local environmental writer Scott Russell Sanders and musicians?**

KIMBALL: As I recall, that intention was there pretty much from the beginning. Maire and I wanted the event to be *huge*. So we asked ourselves, how do you reach people? Music reaches people in a way that nothing else can. But, also, we wanted to include the politicians because they're in a position of power and they can actually get stuff done. Inviting Baron Hill was a no-brainer, though we had no idea if he would participate; so, it was great that he did, especially since there weren't too many congressional representatives around the country that actually did. And we thought he would be a draw. We wanted to get to him, to influence his actions if we could, and also we thought he would draw a crowd if he came to speak.

PEZZULLO: The event also included tables of municipal commissions and local nonprofit environmental organizations. There were free compact fluorescent lightbulbs handed out in Bloomington by our city's Sustainability Commission, information about natural landscaping provided by our city's Environmental Commission, and a tree planting sponsored by the City of Bloomington Department of Parks and Recreation. Was there anything else that you feel was important happening at the tables to help reduce global greenhouse-gas emissions?

KIMBALL: There were a lot of other groups [see appendix C]...We really wanted to educate people, because the dimensions of this problem are just enormous. I think it's easy to focus on certain obvious things, like stop using so much electricity and stop using so much gas. But, beyond that, to start thinking about, where's your food come from and how's it grown? Or the implications of building I-69 [a newly proposed highway in Indiana] and the implications for more traffic that that creates. People don't think about that as a global-warming issue. We wanted to be as broad as we could...It was amazing, actually, a learning process for me on how many groups are working on [global warming].

PEZZULLO: I don't recall, were there any private companies there?

KIMBALL: Well, actually, we had Honda and Toyota bring their hybrid vehicles; so, it was a chance for them to give pitches. Some people were upset about that, that we should not be promoting cars at all. But, we thought people are going to be driving anyway, so, if you're going to drive, drive a hybrid...Bloomington Transit also was there with their hybrid bus.

PEZZULLO: **According to the 2000 census, Bloomington is a small city of about 70,000 and is 83 percent European American. Although the majority of the event lineup appeared to involve and to attract European Americans, there was some racial diversity amongst the crowd that day. Did you make any efforts to reach a racially and/or economically diverse audience?**

KIMBALL: It definitely came up in the discussions. That's something that's really important to me. I think it's easy to talk with the crowd you're familiar with and, in this town, I think that key topics are mostly addressed to the university or university-related crowd...We did have a little bit of racial diversity in our turnout...There's another SIU event in November, and I want to do more to get our message out to a wider range of people...In terms of socioeconomic status, it doesn't make much sense to hand out [compact fluorescent] lightbulbs to people who would buy them anyway.

PEZZULLO: **Did SIU Bloomington have an impact?**

KIMBALL: The mayor of Bloomington did do a proclamation: he did declare it SIU Day. [*laughs*] I think that was more of a pat on the back of the organizers than something that really made a difference. But, I do feel like SIU made a difference at various levels, locally, statewide, and nationally. The process of making it happen made as much of an impact as the event...Particularly the 80 by 50...When I first read that Bill McKibben and his team were calling for 80 percent cuts in carbon emissions by 2050, I thought that was pretty extreme. Actually, I assumed they were purposely setting the goal really high, in hopes of getting *something*. But then the IPCC [Intergovernmental Panel on Climate Change] reports started coming out and it became clear that 80 by 50 is what's needed, not an exaggeration. By April 14, that language was pretty common. I think SIU is part of why this is being talked about more, nationally and locally...I think that another thing that happened is that a lot of groups were made aware of each other that maybe weren't before, and maybe in different ways...

For SIU2—I'm don't mean to capitalize on the language too much, but we're trying to "step up" our efforts. I think we're kind of done with the lightbulbs. We need to ask ourselves, really and truly, how are we going to reduce carbon 80 percent by 2050? We have to stop Duke Energy from building another coal-fired power plant [in a nearby county], and if we don't, then we can change all the lightbulbs we want but the savings will be wiped out in one month of that power plant's operations. Stuff like that. We've got to figure out how to stop driving...There's just such a massive mind shift that needs to happen...

At SIU1, there was sort of this tone of, "Global warming is real and it's scary, but don't worry, it's OK because you don't have to change too much." Our MC was so wonderful, but he made a point of saying, "You don't really have to change your life that much." And I think we *do* have to change our lives that much. I know there are lots of people who think you just can't say that to people because it will turn them off, but my feeling is, well, if they're not going to do anything because they're mad or they're not going to do anything because they're complacent, well, either way they're not going to do anything. I think there are plenty of people who can handle a stronger message and will do more if they understand how much more needs to be done.

Pezzullo: How has organizing this event changed you or your life?

Kimball: It has made me much more willing to take on projects, and I'm noticing people seem to expect it. Now there are things I'm planning that I hadn't even imagined before. For example, I'm working on a running relay from California to Delaware...I guess I'm thinking on a bigger scale than I ever would have before. And in my own life right now, I'm thinking of much bigger changes...

Pezzullo: Is there anything else you would like to add?

Kimball: I hope the Vermont team will keep doing this. I know when I've spoken with them, they've been really concerned with not becoming a top-down sort of movement...which I think is absolutely valid, but I also really think their leadership is critical in going forward. Last spring, as April 14 approached and the weather forecast was looking worse and worse, it made a huge difference to me to know that no matter what happened with our local event, nationwide, SIU would be a success. That evening, it was amazing to watch reports coming in from all over the country. It's so easy to feel isolated and insignificant in a fight as big as this; feeling solidarity with 1,400 communities across the country…that was something.

I don't know how SIU will keep the energy going. I'm guessing our November 3 event is going to be smaller than last April's. I wonder if the Vermont team had waited until next spring to do another event, whether the energy would be back up. Then again, this is so urgent. We can't take a year off. This has to be in our faces all the time.

PEZZULLO: So, from this interview, it sounds to me like, before SIU, you would have identified as an environmentalist, but not an activist—is that fair?

KIMBALL: Yeah, I think so. Actually, I've thought a lot about this, because I wonder where on earth I've been. I'm *really*, *really* mad at myself. I remember hearing about global warming like everyone did, some twenty years ago, and I remember thinking, "Good grief, what are we going to do now?" But, you know, I composted, I used my clothesline, I drove very little. I figured I was doing my part and so they weren't really talking to *me*. [*laughs*] And so I'm really grateful to Al Gore, because I know that his work has the same effect on an awful lot of people—to just kind of wake us up that it doesn't matter how much you are recycling, because there is more that you can do and need to do. So I don't know if I would have claimed environmentalist… because I've been sitting on my butt!

PEZZULLO: Any advice for future organizers if they have never organized one of these events before?

KIMBALL: I guess the only thing I would say is just do it. If it's going to be small, it doesn't matter. Let it be small, let it be big, let it reflect your community. Think as big as you can.

Part III

Practices of Citizenship

Chapter 8

New Media, New Movement?

Jodi M. Minion, William J. Kinsella,
Chad O'Neil, and Tarla Rai Peterson

Environmental activism in the United States isn't what it used to be. A decade or more ago, participating in a protest meant we marched on Washington, picketed, initiated consumer boycotts, joined in a tree-sit or other forms of civil disobedience, or even, perhaps, engaged in the occasional act of ecotage. We formed tight-knit groups and planned actions using local resources; we gained new members through word-of-mouth and mass mailings. Movements took hold when charismatic leaders brought grassroots concerns to the national spotlight (Shils 1982). These types of activism still happen, but the opportunities to participate in environmental decision-making processes have changed. We now have new forms of media that have developed into public forums: e-mail supplements snail-mail and chat rooms, and blogs replace face-to-face meetings; we download theme music instead of spin it. Activists use new media to communicate about important issues, to build group and movement membership, and to organize virtual and on-the-ground actions. In short, new media change how we communicate about and do activism; they change how we participate as citizens in a democracy.

So, what exactly are new media? We define them as interactive media that are produced and distributed using computer technologies and whose data are accessed at random (Manovich 2002). Examples of new media include the Internet, Web sites, and DVDs. Unlike print text or imagery, we can experience any part of new media at any time in almost limitless quantities as long as we have the technology. For instance, a lone picture of two fingers held high above a crowd in the shape of a "V" shown in a print magazine is now supplemented with thousands of digital photos available for searching and downloading, in any order and at any time. New media influence activism by changing how movement symbols and ideas are communicated and to whom.

Throughout this chapter, we maintain a broad definition of public participation: citizen input into decision-making processes. In a post-9-11 context, public participation in the United States and other nominal democracies has been compromised by an increased emphasis on national security. This emphasis threatens democracy's vitality by removing legitimate venues for dissent. As Robert L. Ivie (2005) and Chantal Mouffe (2005a) have argued, democracy requires vigorous participation by an active and informed citizenry to thrive. When dissent is no longer allowed, both participation opportunities and message content become suspect.

New media suggest ways activists may adapt to contemporary political constraints by inventing new forms of participation, such as virtual marches and petitions. These forms of participation may reach audiences that are insulated from traditional protests, whether by conscious choice or by the more insidious power of any status quo. New media saturate the everyday lives of U.S. citizens, thus encouraging development of new social movements that focus on participant values, ideologies, and identities (Schutten 2008). Further, Internet activism gains saliency as Web-based groups such as MoveOn.org and StopGlobalWarming.org penetrate the mainstream. These groups combine online tools with more traditional on-the-ground organizing and activism. For example, MoveOn.org made headline news after placing a controversial antiwar advertisement in a major newspaper. StopGlobalWarming.org organized a college-campus tour featuring Sheryl Crow to stir up climate action. Both organizations have large Web-based memberships and regularly submit e-letters and e-petitions to Congress. It is clear new media have changed how we communicate about and participate in political processes. What is less clear is how they affect (1) who participates, (2) how they participate, and (3) the practice of democracy in general.

In this chapter, we use Step It Up (SIU) to interrogate what it means to participate in Web-based activism, and how global-warming activists have used new media to build a movement. We look at how the Internet facilitated and constrained different aspects of public participation for the SIU movement-building effort and suggest what future environmental activists might learn from SIU organizers' efforts. In our first section, *Finding a Democratic Space for Dissent*, we use the work of Chantal Mouffe to demonstrate the importance of legitimate venues for dissent when freedom of speech is compromised. In *Public Participation in U.S. Environmental Policy*, we briefly review traditional and nontraditional forms of participation in U.S. environmental decision-making processes. We then highlight how basic tenets of democracy are affirmed in a new media setting. In *How SIU Organizers Capitalized (Or Not) on New Media*, we explore how SIU used new media in different ways to help start a movement. We conclude our argument with *What New Media May Offer Environmental Organizers* by suggesting SIU used new media to (1) cast a wide net and increase opportunities for ordinary citizens to participate in the political process, (2) offer new venues for dissent, allowing those with low risk-tolerances access to participation in building a social movement, and (3) shift power away from the federal government to the local grassroots level. Finally, in *Lessons Learned*, we offer suggestions for how both scholars and activists can make the most of new media technologies while researching or doing environmental activism in the post-9-11 context.

FINDING A DEMOCRATIC SPACE FOR DISSENT

The most commonly used model for analyzing public participation in democratic regimes—the public sphere—is derived from the work of Jürgen Habermas (1984, 1991) and emphasizes rationality, sincerity, shared interests, and a mutual effort to reach consensus through public deliberation. The practical problem with this perspective is that it does not grapple with power and influence and minimizes the importance of difference. Idealizing democratic politics as rational, clean, efficient, and reliable can have unforeseen negative consequences (Ivie 2005; Peterson et al. 2006). Apathy, cynicism, and despair grow when citizens discover that the political process is messy, highly interest-based, and able to be manipulated through language. Acting under the guise of rationality, elites rhetorically construct their own actions as right and truthful while strategically curtailing citizen rule. This is seen in the case of global warming when the Bush administration effectively curbed development of carbon-emission policy by defining climate scientists and global-warming activists as alarmist, extremist, and agenda-driven, as opposed to the supposedly moderate position of the administration.

Chantal Mouffe (2000, 2005a) offered an alternative model of democracy, arguing that dissent is far more important than rationality, sincerity, and consensus. To gain legitimacy, the public forum must encourage divergent voices in the decision-making process. Mouffe calls her space for legitimized dissent *agonistic pluralism*: *agonism* implies legitimacy of oppositional messages exchanged between dissenting parties, and *pluralism* implies multiple voices allowed into the decision-making process. M. Nils Peterson et al. explained,

> Meaningful engagement in argumentation requires participants to fully explain their own perspectives to those with opposing views, as well as to actively listen to opposing viewpoints with the goal of understanding those perspectives. Participants need not like each other, but they must respect each other as adversaries worth arguing against. (2004, 758)

The point of agonistic pluralism is to negotiate among differences rather than erase them. Not only is dissent important to prevent democracy from collapsing in upon itself, but it also may facilitate development of new ideas and unveil alternatives to current policies and perspectives.

Strong democracies involve active contestation in the context of difference and division (Mouffe 2000; Ivie 2004). Robert L. Ivie (2004) explained that current regimes weaken democracy by relegating difference to a space of "undecidability" and presenting conflict negatively. He argueed for replacement of the current culture of conformity with a culture of dissent. Finding appropriate forums for dissent is increasingly difficult in the "us" versus "them," "friend" or "enemy" discourses promulgated in the United States (Bush 2001). This sharp distinction between friends and enemies attacks the moral standing rather than merely the politics of dissenters, thus delegitimizing both specific opponents and the concept of opposition in general (Mouffe 2005b; Peterson et al. 2005, 2006). Dissenters are framed as unpatriotic and dangerous (Ivie 2005). This is seen in the Eugene, Oregon, trials of six environmental activists accused of arson between 1999 and 2001; participants in direct action now stand trial as terrorists, under an expanded definition that includes vandalism (FBI 2001; Rasmussen 2007).

Tarla Rai Peterson et al. (2006) added that it is not enough to provide an open space for dissent, but rather, democracy requires an inclusive participatory process where information is shared and the citizenry is informed through collaboration and learning. Access via an inclusive process and an informed citizenry are key tenets for effective public participation that provides legitimacy between differing views. At the same time, Peterson et al. argued that participation in a democracy is not, and should not be, comfortable. As environmental activists use new media, it is important to ask whether reliance on the relative safety provided by the anonymity of the Internet and forgoing on-the-ground action may compromise the democratic struggle. On the other hand, careful application of new media may complement embodied democratic participation, thus revitalizing democracy.

PUBLIC PARTICIPATION IN U.S. ENVIRONMENTAL POLICY

The National Environmental Policy Act of 1969 (NEPA) institutionalized public participation by mandating public information and involvement activities such as environmental impact statements, public meetings, and citizen-comment procedures. A potential problem with NEPA is that its approach is consultative (via governmental employees and interest groups) rather than collaborative, providing an advisory rather than policy-making role for the public (Cox 2006). In cases where these traditional forms of public participation do not exist (e.g., the lack of U.S. climate-change policy), activists protest within the legal system (e.g., suing the federal government for violating tenets of NEPA), or they stage protests outside of the establishment's institutions, calling for social or political change. Traditional forms of protest include letter writing, boycotts, and face-to-face encounters such as demonstrations (e.g., the "cracking" of Glen Canyon Dam), rallies (e.g., Earth Day), and marches (e.g., the Great Peace March). New media have the potential to facilitate broader, more inclusive forms of participation by increasing the number and types of participation opportunities (e.g., blogs, forums, electronic petitions, public comments), distributing those opportunities across space and time, increasing overall access to information, providing new systems for managing information, and linking individuals and groups (e.g., via e-mail lists, social networking, and Web site links).

Several scholars have claimed the Internet allows anonymous and thus unconstrained political dialogues, enabling inclusion of those unwilling to dissent in the physical realm (e.g., Papacharissi 2004; Poster 2001). However, we think it makes more sense to argue that the Internet provides different venues for dissent, including some that are less risky than embodied protest. For example, blogs allow citizens to put forth ideas in an editorial-like form without real-time response and without having to divulge personally identifying information. Nonprofits like SIU and MoveOn.org mix virtual activism with on-the-ground action by asking ordinary citizens to organize local rallies. These organizers discuss and brainstorm ideas for events within designated organizer forums on the

Internet. These forums provide conversation opportunities where participants reveal information to one another about themselves and their locations. SIU and MoveOn.org also act as facilitators for organizers to meet event participants online and eventually, face-to-face, on the day of the action.

When providing low-risk opportunities for dissent, activists must consider the quality of the message. John Peters (2005) argued that one reason political expression matters is that participation with the body is required for an effective citizenry. In cases where democracy and civil liberties are threatened, ordinary citizens who protest in the physical sphere place their bodies at risk. Peters (2005) and Phaedra Pezzullo (2007) suggested the risk of witnessing with the body provides resonance needed to gain saliency for political issues. Dissenting with the body is also thought to increase legitimacy for ideas and actions, a central tenet of participation under Mouffe's agonistic pluralism. As cyberactivism expands, the question of whether a social movement requires physical or material movement as well as virtual movement is very real.

How SIU Organizers Capitalized (Or Not) on New Media

In January 2007, Bill McKibben and six Middlebury College students responded to the minimal media coverage of their five-day walk across Vermont for climate action by launching a Web-based campaign at http://www.stepitup2007.org. SIU asked ordinary people across the nation to organize climate action rallies in their hometowns and at America's iconic places. On April 14, 2007, all rallies carried the same message: "Step It Up, Congress: cut carbon 80 percent by 2050." Organizers from SIU headquarters explained in an interview that they were available to local organizers via phone and e-mail and that they developed their Web site based on organizer needs.

Stepitup2007.org provided local organizers with ways to post information about their event and for participants to contact local organizers. When SIU headquarters became aware that many local organizers were concerned citizens who had never even attended a rally, let alone organized one, they provided all the components needed to organize a rally (McKibben et al. 2007).

We call what SIU provided to local organizers a "cookbook" for how to do a rally; local organizers only needed to add flavor and, if needed, a permit to gather. The SIU Web site provided information on how to contact elected officials and how to best communicate with the media; it also provided downloadable fact sheets, press releases, and posters. The Web site provided bannermaking 101, action ideas, and means for local individuals or groups to collaborate with local and national organizations. We consider all of these resources as part of the rally cookbook. The Web site also provided basic information on SIU and the headquarters team, a blog, a place to download and listen to movement music, and links to climate-change and SIU media coverage. For this chapter, we analyzed how the cookbook of Internet-mediated information provided to local organizers and participants directly influenced participation in SIU.

We archived all texts, including working pages, links, and pictures provided on stepitup2007.org from January 4, 2007 (the start of SIU), to May 14, 2007 (one month after the April 14 events). We examined the texts using critical discourse analysis (CDA), which facilitated exploration of relationships between the Internet and public participation and suggested the implications of new media on democracy (Fairclough 1989; van Dijk 2001). We then explored the relationship between the cookbook and images available on the site. To get a sense of how the National Day of Climate Action moved from Internet-mediated communication to on-the-ground action, we participated in SIU events in San Antonio, Texas, and Raleigh, North Carolina.

In addition to CDA, observing the events as participants allowed us to interact with and experience the climate-action culture that was represented at the events. Further, experiencing SIU in person allowed us to see how the events moved from the virtual to the material realm. We worked from field notes and thick descriptions of the events (Geertz 1973). This enabled us to root our story in social and political contexts, time and space, and location and view. Consistent with the goals suggested by CDA, we organized our observations according to themes related to participation, power roles, and movement struggles. Since observations made at the different events are used to inform findings drawn from our examination of SIU texts, we do not attempt to generalize SIU by location, event type, or participant experience.

Analysis of the SIU Cookbook

The SIU cookbook revealed basic tenets for what it meant to effectively participate as an SIU local organizer. McKibben and the SIU team asked local organizers to:

- hold an event on April 14, 2007, in their hometown or at an American iconic place;
- get as many people to attend as was appropriate for the action type (e.g., scores of folks for a rally, a handful for a barbeque);
- be creative with who was invited to attend and the action type;
- contact the press;
- use a banner that carried the SIU message; and
- report back (with a group photo, if possible).

SIU encouraged people to use and pull from established interest groups; if no group existed, SIU provided individuals with networking tools to either form a new group or connect with others at local or national levels. The cookbook offered concerned citizens, all acting in different ways to affect climate change, a unified message and the push to move climate policy from action at the individual level (i.e., changing lightbulbs) to the halls of Congress.

Throughout the SIU cookbook texts, SIU maintained an inclusive stance on who participated and how; it considered potential socioeconomic barriers and encouraged participation that allowed those with little knowledge of global warming or those with few resources access to membership in the movement. SIU also provided basic information on the relevant political situation along with links to information resources, thus encouraging development of a more informed citizenry. Further, SIU empowered individuals and local groups to participate in the movement in their own way, possibly encouraging a power shift away from a centralized elite to a local, grassroots level. In what follows, we explore each of these ideas in detail to illustrate how SIU used the Internet to help start a movement.

Access

Stepitup2007.org encouraged everyone to participate. To facilitate this process, the Web site encouraged local organizers to involve those they knew were interested in climate action, and to cast a wider net by inviting everyone on their e-mail lists and even people they did not know. The Web page titled "Involving More People" encouraged readers to

> invite all your friends (and) involve people whom you don't know: communities who live close to dirty energy sources, who are dealing with pollution in their daily lives. Global warming affects us all, even those people whom you don't think will be interested;...find out what matters. (Step It Up 2007d)

Barriers to participation due to limited resources were considered as well. The Web page titled "Bannermaking 101" explained how to make a banner for those who did not know. The instructions were clear and did not require that organizers purchase materials (Step It Up 2007c). Rather, SIU encouraged people to use recycled materials and to be creative.

The SIU cookbook provided the basics for those who had never organized or even attended a protest event, such as the local organizer for SIU San Antonio. Following the suggestion to select an iconic location, local organizers held SIU San Antonio in Alamo Square amidst numerous shoppers and tourists on a blustery Saturday afternoon. All aspects of the SIU cookbook were seen at the rally: the banner, flyers, stickers, prepared speeches, media, and over one hundred concerned citizens. Participants at that event comprised an array of local grassroots organizations, including the Alamo chapter of the Sierra Club, Texas Democrats, and local activists attempting to thwart plans to fast-track NEPA. Participants ranged from toddlers to ninety-somethings. Most were non-Hispanic whites, despite a thriving Hispanic culture that characterizes south-central Texas. Organizers pulled from local resources as rally-goers watched belly dancers perform in the unseasonable cold wind and later played environmental Trivial Pursuit. Connection with the movement was clear as participants chanted the unified message: "What do we want?" "80 percent!" "When do we want it by?" "2050!" A group photo was taken in front of the Alamo and became one of the scrolling images seen on the SIU Web site homepage.

Participating as a local organizer is framed on the Web site as easy and not time-consuming; in fact, a national SIU organizer explained in an interview that the SIU team absorbed a lot of the time required to organize an event by working around the clock to support local efforts. For those needing to ask questions via telephone, SIU team members were available. For those who preferred to use tools offered online, the information provided was "as simple as 1-2-3." This idea is illustrated in the interview with the local organizer for SIU Bloomington (see chapter interlude between 7 and 8), who recalls her interactions with the SIU team as prompt and "non-generic." She later added that the SIU team was "real people and they're working their butts off to do this stuff, just like everybody else." SIU also provided detailed instructions for how to effectively work with the media:

> When you call reporters, always ask if they have a minute to talk. Often they will be on deadline and will not be able to talk to you then. If they are busy, ask when a better time would be to call them. The reporter will appreciate this greatly and be more receptive. (Step It Up 2007e)

This information reduces potential challenges involved with organizing. Further, it provides new organizers with a short lesson in media etiquette, which helps the movement look professional and may indirectly increase media coverage.

The point of access to SIU was the use of the banner carrying the message of 80 percent cuts in carbon emissions by 2050 to Congress.

The banner was the unifying symbol needed to start a movement. Not only did its message convey expectations to elected officials, it united thousands of fragmented citizens for a couple of hours on one day. Classical movement scholars stress the importance of unifying symbols that have meanings of which all movement members are aware (e.g., Shils 1982; Johnson 1982). Without a symbol, there is no movement. In the case of SIU, the Internet facilitated the process of connecting a highly disconnected citizenry by providing the symbol that was then used to demonstrate the material presence of the movement in (hopefully) iconic locations throughout the nation.

SIU increased access for a wide range of rally participants with different perspectives on activism. When considering participating in direct action, some may fear being arrested or being seen with extremists. Some citizens hold professional and social positions that do not allow them to safely participate in actions that protest the establishment. The speakers at SIU San Antonio placed participants at ease with lighthearted messages that did not put participants at risk of arrest or jeopardizing political positions. At the Raleigh event, five of the event speakers were between the ages of eight and twelve years old. On the reverse side, some prefer direct action. For motivated individuals with high tolerance for personal risk, SIU suggested participating via extreme sports (that retain a pro-American cast) such as deep-sea diving or mountain climbing. SIU set the stage for Mouffe's agonistic pluralism by maintaining a respectful, nonthreatening dialogue when dissenting from the establishment while calling for bold and immediate action to effect change. They offered opportunities for face-to-face argumentation while providing cybervenues for sharing ideas and tactics for dissenting in the physical realm.

Informed Citizenry

SIU provided information to educate Web site visitors on the more general climate issue and specifically on movement building. The home page included a brief introduction from Bill McKibben—"Friends, this is an invitation to help start a movement" (Step It Up 2007b). The page also included pictures of events that were taken in advance, links to climate-movement music and various collaborators and sponsors, and a blog. The Web site provided information on how people can act at the individual level. This idea carried through to the actual events, as local organizers offered ideas on how rally participants could reduce their carbon footprints. For example, at SIU San Antonio, rally organizers planted the answers to the trivia questions in the audience so the crowd seemed informed and knowledgeable about the topic:

> SPEAKER: Now, what [do we do] about water protection projects?
> CROWD: Rainwater capture! Low-flow!
> SPEAKER: Low flow toilets…did anybody say that? Somebody was supposed to say low-flow toilets. Thank you. Oh, okay. Now, what do we do to decrease energy consumption in our home?

At the SIU global climate-change rally in Raleigh, North Carolina, a speaker linked the local event with information found on faith-based climate-change Web sites:

> I think it's fine to believe that climate stewardship reflects God's orders for us to be stewards of the planet, but probably not if you drove an SUV to church. You know, there needs to be some kind of consistency in what we believe and our behavior as a result of that.

SIU collaborated with mainstream and grassroots climate-action groups to position its Web site as both a networking hub and a source for climate-change and action information. The point was to not only educate Web site visitors or rally participants but to offer information in a context that allowed and encouraged them to use it to take action.

The SIU blog provided great potential for climate-change activists seeking to develop an informed citizenry. Blog postings were written by the SIU team and nationally known climate-action figures such as Winona LaDuke and Laurie David. The blog followed local policy issues and kept movement members informed of the progress of SIU events. Certain actions were highlighted; those with extreme events (e.g., aerial photos taken in Park City, Utah; deep-sea diving off Key West) gained the spotlight, but so did a community gathering in Wyoming. The blog also offered links to scientific reports, media coverage, YouTube videos, and other groups working on climate action. For those interested in climate action, the blog was rich with potential for gaining information on the issue and then provided links and tools for those interested in creating change.

Information provided by SIU was not only made available on its Web site but it was also delivered in a safe, familiar manner, making it accessible and acceptable to Web site visitors who were unsure about participating in Web-based activism. The Web page titled "About" provided information about the climate-action movement and the SIU team, and a letter from Bill McKibben (Step It Up 2007a). The SIU team members were framed as friendly, everyday folks seeking to join with others to effect change at the national level. One SIU team member discussed his dream to bake bread and work in grassroots activism but stated that he was taking time off from his dream to affect national policy. Web site visitors could also choose to view answers to frequently asked questions such as "Why not just have a march on Washington?" or "Will this thing work/will Congress be listening?" (Step It Up 2007f). The SIU answers reduced barriers to participation by those unsure of Internet-mediated activism. The Web site also provided information on how to agitate for change. Bowers, Ochs, and Jensen (1993) suggested movements form when a group of suppressed citizens becomes fed up with social or political conditions. Agitation then increases to the point of action at scales that go beyond individual households. An SIU rally speaker in San Antonio informed the public that participating in the national action was a critical step in being an informed citizen:

> Individually, you may think, "Oh, I can't do anything much"... you can do something. "I can't do a whole lot." You can do something. We can ask our Congress to step up to the plate. It's not too much to ask that we breathe clean air. Is it?!

Bill McKibben provided another example in his letter to Web site visitors:

> Those of us who know that climate change is the greatest threat civilization now faces have science on our side; we have economists and policy specialists, courageous mayors and governors, engineers with cool new technology. But we don't have a movement—the largest rally yet held in the U.S. about global warming drew a thousand people. If we're going to make the kind of change we need in the short time left us, we need something that looks like the civil rights movement, and we need it now. Changing light bulbs just isn't enough...[T]his action needs all kinds of people to help out. We can't make it happen—it has to assemble itself...The recent elections have given us an opening, and polling shows most Americans know there's a problem. But the forces of inertia and business-as-usual are still in control, and only our voices, united and loud, joyful and determined, can change that reality. (McKibben 2007)

In the case of SIU, it is not enough to simply know how to mitigate an individual carbon footprint, but rather, an informed citizen holds greater responsibility, which involves acting within the public realm in some directed, deliberate way. New media, therefore, offer endless opportunities for activists to build and maintain an active, informed citizenry.

Shift to Local Power

Movement building involves changing power relations between an oppressed citizenry and the powerful elite. As noted earlier, public participation in direct action faces renewed barriers in the post-9-11 political milieu. Further, political leaders and the fossil-fuel industry have maintained power by discrediting climate science, labeling those who support it as radical or misinformed. The SIU cookbook attempted to rework the current power structure by reminding Web site visitors of their responsibility to act as citizens and by creating a new group of local climate-change leaders. It also offered internationally renowned scientists and scientific panels—and the actions of other developed nations—to justify the need for action.

SIU framed climate-action-movement members against their opponents in a David-and-Goliath-like manner:

> The other side in this battle has essentially unlimited resources: ExxonMobil, for instance, made more money than any company in history last year, and according to a recent report they spent a nice chunk of it spreading disinformation about climate change. But we can beat them if enough of us mobilize—our sheer numbers can outweigh their special influence. (Step It Up 2007h)

This example carries a sense of Gandhi-like mobilization by an oppressed (but free to assemble) citizenry. Oppression, in this case, is manifested in the notion that the Bush administration has compromised scientific integrity in its fervor to avoid appropriate action. The SIU Web site suggested a potential shift in power when citizens demand that their government listen to them instead of to the fossil-fuel industry. This power shift is illustrated at the Raleigh event, as a child speaker explained:

> It's like when my mom tells me to go up and clean my room and I just shove everything under the bed. Usually, I feel guilty enough that I at least have intent to do a better job. But I think I have an idea why this is different. Ignorance—people are just shoving this issue under the bed. After all, if I can't see it, it can't see me, right? That seems to be Bush's theory. But, are we the only ones who see a flaw in their lobby?

Step It Up San Antonio focused on local issues and interests as they played out in the national and global arena. Organizers and speakers positioned local activists as holding the power to do something about global warming. They encouraged individual action along with activist networking to effect national change, as is illustrated in a speech from San Antonio:

> If we're not careful, we're going to build multilevel parking garages into eternity when you and I know we could do something better if we just would power up to the plate and make sure that we produce the kind of a thinking eclectic...If an army marches on its stomach, a democracy marches on its education and information, and I submit to you that's the glory of this moment here, in

> front of the Alamo, where we talk about what we can do, what we should do, what we must do. Let's step it up and make it happen!

By redefining climate action as a mandate for American citizens rather than corporate or political establishments, SIU strategically created a new group of local leaders, shifting power away from the federal government to grassroots and individual levels.

SIU also provided opportunities to explore hidden power in new media. The organizers used tools and networking opportunities provided online to work against the establishment, while identifying themselves with a single nationwide day of action. For example, an SIU team member noted, "If I could manage to separate the mouse from my hand, I'd be raising my fist in solidarity with all of you" (Step It Up 2007g). The extent to which new media allow activists new ways to communicate the diffusion and extent of movement membership is seen on the SIU Web site:

> As people gather, we'll link pictures of the protests together electronically via the web—before the weekend is out, we'll have the largest protest the country has ever seen, not in numbers but in extent. From every corner of the nation we'll start to shake things up. (McKibben 2007)

In this example, SIU linked distinctive characteristics of new media, such as their fast message-delivery and broad distribution capabilities, with the filmic technique of combining multiple images to create a new visual experience. In film, collage can recreate important narrative moments, achieve visual contrasts, or even attempt to portray the experience of memory (Manovich 2002). For SIU, the rhetorical possibilities of new media brought these filmic techniques together in new ways. For example, the scrolling action of the images (see figure 9) enabled by new media technology built a sense of anticipation in the viewer and suggested a national presence (we see many cities and states from across the nation represented), thus appealing to decision makers.

SIU also used new media to encourage networking between activist groups, as exemplified by a brief appearance of the "Raging Grannies" at the Raleigh rally. The Raging Grannies comprise a loosely organized

FIGURE 9. Scrolling photos from SIU action reports.

network that originated in Victoria, British Columbia, in 1987 and now includes members throughout Canada, the United States, Australia, Greece, and the United Kingdom.

At the Raleigh event, the Raging Grannies of Orange County (a local "granny gaggle") appeared, dressed in pale green Lady Liberty costumes, to sing a series of patriotic songs reworked with global-warming lyrics (see figure 10). Exactly as suggested in the "granny starter kit" on the international Web site, the members began by using a pitch pipe to get the members in tune. At the climactic conclusion of their final song, the grannies all raised the lamps they had been holding in their right hands; as the lamps lit simultaneously, they were revealed to the audience as compact fluorescents. Linking a standard organizing template from the Web with their ironic renditions of traditional, patriotic songs, the grannies paired a sense of shared community with a timely and practical suggestion for combating global warming (compact fluorescent lamps were a recurring theme throughout local events across the country). They did so within the framework of the larger organizing template provided to the local rally organizers via stepitup2007.org. SIU provided the information, resources, and networking capabilities to reach well beyond a single day of action and form an actual movement. And indeed, the movement continues to progress, as a second day of action occurred on November 3, 2007; following the November event, the SIU team helped to kick-start the Web-based climate action hub 1Sky.org. The SIU team now works together on a global climate-action project called 350.org.

Our analysis indicates that SIU used new media to provide ordinary citizens information and components needed to organize an event. The cookbook provided those unfamiliar with or unsure about participating

FIGURE 10. Raging Grannies at the Raleigh SIU action.

in building a climate-action movement new venues and networking opportunities to reduce barriers to participation. Other aspects of the Web site complemented the cookbook by providing additional information and networking resources needed to develop an informed citizenry and to further extend a unique assemblage of local actions into a national movement. The events in Raleigh, North Carolina, and San Antonio, Texas, illustrate the interaction between cookbook suggestions, such as providing networking opportunities, carrying the unifying message, and development of an informed citizenry and local sensibilities, with appeals to regional issues.

What New Media May Offer Environmental Organizers

SIU used new media technologies to provide opportunities for ordinary citizens to participate in climate action. These opportunities enabled those concerned with the risk of participating in the physical realm, marginalized groups, and those who did not consider themselves as activists to join with other concerned citizens to affect climate change. Stephanie Kimball, the local organizer for SIU Bloomington (see interlude between chapter 7 and chapter 8), felt that the April 14 event (or at least her participation in it) would not have happened without the use of the Internet. In the case of SIU, new media provided venues for legitimized dissent, venues that catered to a relatively mainstream, proestablishment means of opposition, venues for activists seeking to work with the establishment rather than against it. This could hinder antiestablishment discourses that are needed to build agitation to levels that allow messages to resonate in ways that penetrate decision-making processes. On the other hand, the participation of the Raging Grannies at the Raleigh event indicates that antiestablishment groups can work together effectively with more mainstream citizens to promote change.

One product of the events is a steady stream of photos to Congress (material) and to Web site visitors (virtual); this stream symbolizes the national extent of the SIU campaign. SIU used the Internet to mediate a shift of power away from the federal government to the local grassroots level. Participation in SIU meant networking and collaborating at scales never seen before; stay-at-home moms connected with Greenpeace, and local chapters of the Sierra Club united to create change at the community level. The attempt to shift power from the federal government to local residents is at the very heart of what lies behind all social movements: an oppressed citizenry rising up to challenge the establishment. Web-based social movements may offer the citizenry new tools to complement and/or replace traditional agitation techniques. New media tools hold the potential to help organizers negotiate the post-9-11 political scene by offering venues for dissent and information dissemination that allow messages to reach the halls of Congress while reducing the risk of being arrested and charged with terrorism.

Lessons Learned

Our analysis indicates SIU organizers successfully used new media to increase agitation and to shift power away from the federal government to the local grassroots level. This shift in power holds promise for environmental activists. Increased local power could lead to the invigoration of local economies and more accountability for environmental actions and ecosystem sustainability. Marginalized groups may gain power and momentum through national, Web-based campaigns; further, collaborative environmental conflict-management practices may include stakeholders who are not "local" but who are easily accessible online.

Activists seeking to develop social movements that challenge mainstream suppression of scientific information about the environment can use new media as a unifying tool to provide a simultaneously frustrated and apathetic citizenry with a message that can be used to effect change. The relative success of SIU suggests that organizers can use new media to help negotiate the fragmented political landscape in the following ways: (1) *Cast a wide net*—the Internet provides opportunities to use search engines to meet folks who think like you, to network and collaborate via mass e-mails, to reach out to affected communities, and to get early media coverage; (2) *Provide a safety net* for those who are unsure by keeping the tone of the action light and by mediating communication between organizers and event participants, for both pre- and postevent communications; (3) *Provide simple and clear instructions* on how to do the type of event you propose (if possible, provide instructions in multiple languages); (4) *Establish media relationships* and encourage local activists to do the same by providing, either online or via DVDs, premade press releases and informational fliers that may be (but do not have to be) creatively adapted to local conditions, to increase dissemination of the materials and reduce barriers to participation due to time shortages or geographical access; (5) *Empower leaders at the local level*—encourage grassroots and city-to-city networking and the development of local leaders by building, enhancing, and maintaining Web-based organizer hubs.

References

Bowers, John W., Donovan J. Ochs, and Richard J. Jensen. 1993. *The rhetoric of agitation and control*. Prospect Heights, IL: Waveland Press.

Bush, George W. 2001. Address to a Joint Session of Congress and the American people. http://www.whitehouse.gov/news/releases/2001/09/20010920-8.html (accessed October 7, 2007).

Cox, Robert. 2006. *Environmental communication and the public sphere*. Thousand Oaks, CA: Sage.

Fairclough, Norman K. 1989. *Language and power*. London: Longman.

Geertz, Clifford. 1973. *The interpretation of cultures*. New York: Basic.

Habermas, Jürgen. 1984. *The theory of communicative action, volume one: Reason and the rationalization of society*. Trans. T. McCarthy. Boston: Beacon Press.

———. 1991. *The structural transformation of the public sphere*. Trans. Thomas Burger. Cambridge, MA: MIT Press.

Ivie, Robert L. 2004. Prologue to democratic dissent in America. *Jasnost—The Public* 11:19–36.

———. 2005. *Democratic dissent and the trick of rhetorical critique*. Bloomington, IN: The Polynter Center for the Study of Ethics and American Institutions.

Johnson, Chalmers A. 1982. *Revolutionary change*. Stanford: Stanford University Press.

Manovich, Lee. 2002. *The language of new media*. Cambridge, MA: MIT Press.

McKibben, Bill. 2007. A letter from Bill McKibben. StepItUp.org. http://april.stepitup2007.org/article.php?id=78 (accessed February 25, 2008).

McKibben, Bill, Phil Aroneanu, Will Bates, May Boeve, Jamie Henn, Jeremy Osborn, and Jon Warnow. 2007. *Fight global warming now: The handbook for taking action in your community*. New York: Holt.

Mouffe, Chantal. 2000. *The democratic paradox*. New York: Verso.

———. 2005a. *On the political: Thinking in action*. New York: Routledge.

———. 2005b. Schmitt's vision of a multipolar world order. *South Atlantic Quarterly* 104 (2): 245–251.

Papacharissi, Zizi. 2004. Democracy online: Civility, politeness and the democratic potential of online political discussion groups. *New Media & Society* 6 (2): 259–283.

Peters, John D. 2005. *Courting the abyss: Free speech and the liberal tradition*. Chicago: University of Chicago Press.

Peterson, M. Nils, Markus J. Peterson, and Tarla R. Peterson. 2005. Conservation and the myth of consensus. *Conservation Biology* 19:762–767.

Peterson, M. Nils, Stacey A. Allison, Markus J. Peterson, Tarla R. Peterson, and Roel R. Lopez. 2004. A tale of two species: Habitat conservation plans as bounded conflict. *Journal of Wildlife Management* 68:743–761.

Peterson, Tarla R., M. Nils Peterson, Markus J. Peterson, Stacey A. Allison, and David C. Gore. 2006. To play the fool: Can environmental conservation and democracy survive social capital? *Communication and Critical/Cultural Studies* 3:116–140.

Pezzullo, Phaedra C. 2007. *Toxic tourism: Rhetorics of pollution, travel, and environmental justice*. Tuscaloosa: University of Alabama Press.

Poster, Mark. 2001. *What's the matter with the Internet?* Minneapolis: University of Minnesota Press.

Raging Grannies. 2007. http://www.geocities.com/raginggrannies/ (accessed February 25, 2008).

Rasmussen, Matt. 2007. Green rage. *Orion*. http://www.orionmagazine.org/index.php/ articles/article/6/ (accessed February 1, 2007).

Schutten, Julie K. 2008. *Coming out of the coven: Contemporary witches, hidden populations, and new social movements*. Saarbrücken, Germany: VDM Verlag Dr. Muller.

Shils, Edward. 1982. *The constitution of society*. Chicago: University of Chicago Press.

Step It Up. 2007a. About. http://april.stepitup2007.org/section.php?id=8 (accessed February 25, 2008).

———. 2007b. April 14, 2007—National Day of Climate Action. http://april.stepitup2007.org/ (accessed February 25, 2008).

———. 2007c. Bannermaking 101. http://stepitup2007.org/article.php?id=686 (accessed February 25, 2008).

———. 2007d. Involving more people. http://april.stepitup2007.org/article.php?id=71 (accessed February 25, 2008).

———. 2007e. Media attention. http://april.stepitup2007.org/article.php?id=72 (accessed February 25, 2008).

———. 2007f. Organizer FAQs. http://april.stepitup2007.org/article.php?id=76 (accessed February 25, 2008).

———. 2007g. The Step It Up team. http://stepitup2007.org/article.php?id=418 (accessed February 25, 2008).

———. 2007h. Tell a friend. DemocracyInAction.org. http://www.democracyinaction.org/dia/organizationsORG/SIU/tellafriend.jsp?tell_a_friend_KEY=1251 (accessed February 25, 2008).

U.S. Federal Bureau of Investigation. 2001. Interagency domestic terrorism concepts of operation plan (CONPLAN). http://www.fbi.gov/publications/ conplan/conplan.pdf (accessed December 1, 2007).

Van Dijk, Teun A. 2001. Critical discourse analysis. In *The handbook of discourse analysis*, ed. D. Schiffrin, D. Tannen, and H. E. Hamilton, 352–357. Oxford: Blackwell.

CHAPTER 9

STEP IT UP AND IMAGE POLITICS IN THE PACIFIC NORTHWEST

Leah Sprain, Todd Norton, and Tema Milstein

> We created a visual message in the form of a family friendly carbon-offsetting "Marching Forest." It is said that a picture is worth a thousand words. We hope our forest serves to stir people, no matter how old or what language they speak...We need all voices...When people see that others are putting themselves on the line about this issue, it helps give them courage to join in. The more people we can reach, the easier the rest of the work will be...We want the whole world to see that we citizens in the U.S. mean business about global warming...Let's take to the streets! If we all stayed in our homes who would know?
>
> —Linda Stein
> "The Marching Forest of Shoreline Takes it to the Streets"

A picture is worth a thousand words. This popular cliché points to the widely held notion that images are more valuable and more powerful communicators than words alone. The growth of predominately visual forms of mediated communication—television, YouTube, interactive Web sites—only serves to reinforce our collective preference for pictures over words.

The power of visuals is similarly demonstrated in the above quote from the Marching Forest of Shoreline's Step It Up 2007 (SIU) action report (Stein 2007), which described how the Marching Forest sought to create a visual message. The group's report suggested that public, visual actions can help build movements by telling others about concern for global warming and giving new people "the courage to join in." The action report indicated a clear attempt to use visual images as a movement-building strategy; the effectiveness of the Marching Forest, however, is unclear and its effect uncertain.

In this chapter, we examine how the power of images aids movement building by examining multiple SIU actions held in the Pacific Northwest as *image events*, spectacles designed to gain media attention and widely disseminate visual messages as a means of building a social movement (Johnson 2007; DeLuca 1999). We offer a detailed account of seven distinct events while also making an argument about the possibilities and challenges of staging image events as a movement-building strategy in the twenty-first century. This chapter provides readers insight into the dynamics of Pacific Northwest SIU actions, the concepts of image events and the public screen, and a method for evaluating the usefulness of image events for building future movements.

We begin by introducing the concept of the image event. Research helps us understand how and why staged image events function as effective strategies for movement building. With this background, we describe seven SIU actions in the Pacific Northwest, where we had researcher participants. Then we discuss how these actions do and do not function as image events. This spectrum of examples provides lessons on the challenges and possibilities of using image events in movement building. In the conclusion, we offer practical suggestions for movement organizers who may want to use image events to build future movements.

The Theory Behind Image Events

At a national level, one of SIU's core objectives was to invite individuals and local groups to gather at iconic places and take pictures of everyone gathered with a banner reading, "Step It Up, Congress: Cut Carbon 80% by 2050." This objective suggests the possibility that SIU actions could be usefully viewed as *image events*.

Creating image events has been an effective tactic for environmental and social-movement groups aiming to challenge conventional thought about social issues and practices. For example, environmental-justice group Allegany County Non-violent Action Group (ACNag) had six elderly people bundled up in lawn chairs and handcuffed to a chain traversing a bridge to protest building a nuclear reactor in their town (DeLuca 1999). Through photojournalistic press coverage, this tactic drew attention to its movement by highlighting the identities of the participants involved (elderly citizens) and the local scene of which they were a part (the bridge and river). The tactic was successful because the act of being handcuffed to the bridge itself had symbolic significance—the image of elderly citizens on a bridge was a persuasive, jarring reminder of what was at stake when building a nuclear facility. The same impulse helps explain why radical environmental groups elect to chain people to trees and bury themselves in roads to prohibit logging and drive boats between whales and whalers' harpoons to protest whale-hunting. These events create powerful images that help communicate a movement's message.

Part of the reason image events are an effective tactic in movement building is that we live in an era when much of public discussion is televised, appearing in the form of sound-bytes and pictures. Kevin DeLuca and Jennifer Peeples (2002) have referred to this recent change as the *public screen*. Traditionally, the bulk of political discussion occurred between individuals in public spaces like coffee shops or those provided by print journalism. Jürgen Habermas (1984, 1991) famously called these spaces the *public sphere*, referring to participation in the politics surrounding important social concerns rather than the business sector. The public sphere thus highlights conversation as a site where rational participants meet to develop consensus on a particular issue. To achieve political change in an era of public-sphere discussion, a social movement would need to engage publics in conversation about a particular issue, particularly appealing to norms of rationality and consensus that characterize the public sphere.

In contrast, the contemporary public screen foregrounds the function of swift dissemination, dissent, multimedia, and publicity. Within such a system, social movements benefit from creating images that shock people and therefore attract media attention. During the 1999 WTO protests in Seattle, for example, print and television news coverage significantly increased when anarchists used violent tactics (DeLuca and Peeples 2002). In this way, violence effectively propelled the protests (violent and nonviolent alike) onto the public screen.

Together, the concepts of the image event and the public screen suggest why creating a spectacle might be a successful tactic for environmental movement building. As the examples above demonstrate, environmental groups have successfully used image events to attract media attention and present their message to a wide audience. Nonetheless, critics have questioned whether image events can be used to maintain a sustained, coherent movement. In the next section, we introduce seven SIU actions held in the Pacific Northwest, before considering the extent to which these actions function as image events.

Step It Up, Pacific Northwest

On the days following April 14, 2007, event organizers and participants posted photographs and reports from twenty-four events held in Washington, fifteen held in Oregon, and two held in Idaho on the SIU Web site. These images were posted as action reports and on the SIU home page. National organizers made prints of the photographs for the local organizers to give to their U.S. senators and representatives to demonstrate the local demand for political action on climate crisis.

In addition to reports by local organizers and participants, the research team had participant-observers at six actions in Washington and one in

Idaho. These events ranged in location from northwestern Washington to Northwestern Idaho.

Deception Pass State Park: A nature walk with seven participants highlighted the potential impact of climate change on the waterfront state park on Whidbey Island. Two park rangers sought to create a "dialogue" between participants rather than focus on the national SIU campaign.

Seattle's Golden Gardens: A neighborhood group, Sustainable Ballard, staged a morning event at the beach at Golden Gardens Park. Participants stood in a line paralleling the shore about twenty-five feet from the water, holding nine-foot-tall bamboo poles with yellow caution tape tied to the tops, blue ribbon connecting the poles about three feet up, and pink plastic tape at the bottoms. Each ribbon represented ocean levels that have changed or are forecasted to change due to global warming.

Downtown Seattle: People gathered in local neighborhoods to travel together to join in a rally, march, and solutions fair. For example, the Marching Forest of Shoreline met at a northern Seattle grocery store dressed as forest creatures and rode a bus downtown. Simultaneously, students rallied at Antioch University and the University of Washington before traveling to the central rally to kick off the march. The march filled several blocks with an estimated two thousand participants shouting chants, waving posters, and walking in the rain. At the end of the march, people gathered at a waterfront park to listen to speeches from political, religious, youth, and labor leaders, and to learn about climate-change solutions from environmental organizations and businesses.

Cheney: Despite pouring rain, residents from Cheney and Spokane joined members of activist organizations from Eastern Washington University for a day of climate action. Though small by comparison to other sites in Washington, with approximately fifteen participants, the event received coverage from several local press outlets.

Spokane, Canada Island, in Riverfront State Park: Members of local environmental groups, specifically the Lands Council, which organized the event, gathered on Canada Island in Riverfront State Park. The event was dramatic, given the natural space within which it occurred, a small island near downtown surrounded by the Spokane River with its dramatic waterfalls. Participants formed a human banner spanning one of the island's bridges, wearing shirts that collectively read, "Step It Up, Congress: Cut Carbon 80% by 2050." After taking several pictures for press releases, the event quickly changed as participants removed these shirts and gathered for a series of speeches.

Moscow, Idaho: About seventy people gathered at East City Park for the Step It Up Moscow bike parade. After riding around the edge of town, the parade made its way up Main Street, where an additional seventy or so people waited in the outdoor plaza under the "Step It Up, Congress!" banner. Participants gathered under the banner, holding signs, children, dogs, and each other to take photos for a press release. This event had a dual-city association. Moscow is the home of the University of Idaho and is located seven miles east of Pullman, Washington, the primary campus of Washington State University. Local residents refer to this entire area as the "inland Northwest," and it is common for people to live and work across the state line. Moscow is more activist-oriented than Pullman, which is likely why SIU happened in Moscow rather than Pullman.

We observed these actions, conducted interviews with participants, transcribed and analyzed speeches, and took pictures. We supplemented these field materials with an analysis of press coverage of events and action reports submitted by organizers.

Our central question is, what do these actions tell us about the use of image events in movement building? To answer this question, we begin by analyzing how the events we observed functioned (and did not function) as image events. Our analysis provides insight into what image events are, how they function within this national global-warming action, and how they relate to other types of movement discourse, such as speeches. In the section following our analysis, we draw upon this description to address the possibilities for and the challenges of using image events for movement building.

SIU Actions as Image Events

As participants in the downtown Seattle action marched to a waterfront park for the solutions fair, they passed a polar bear swimming in Puget

Sound, waving a sign that read, "Humans your habitat is in danger too!" (The other side read: "Step It Up, Congress! 80% reduction by 2050."). A local environmental think-tank president exclaimed, "Now, that's a great idea," while parents pointed out the polar bear to their kids.

The swimming polar bear was actually Randolph Sill, a local artist decked out in a dry scuba suit underneath the fuzzy polar bear costume. According to his press release, he donned the costume to "bring attention to global warming and encourage Congress to 'Step it Up'" (Sill 2007). His action functioned as an image event on several levels. First, for the participants walking by, his presence was jolting and unexpected. The image of a bear in the water caught attention, leading people to then read the sign that shifted the typical human-nature relationship. Here an animal was using human terms (e.g., *habitat*) to provide advice. Moreover, participants and news media alike photographed the bear, making the performance a central image that represented the larger SIU events in Seattle in newspapers, blogs, television, and Flickr (a photo-sharing site). The bear was unexpected, cute, and quirky, and the message clearly stated the point in case people had missed it.

The polar bear was not the only image event in the downtown Seattle area. Following the nature theme, the Marching Forest of Shoreline comprised members of the Shoreline neighborhood in northern Seattle who dressed as trees, squirrels, the sun, and other forest-themed entities. The leader instructed the Marching Forest to stay together in order to be a "presence." As indicated in the introduction, the Marching Forest was designed to send a visual message about climate change. In particular, the organizer highlighted that the action "celebrated the benefits of urban trees" (Stein 2007). In interviews with researchers, several Seattle participants reported seeing people dressed as forest creatures, but none of the major news media focused on the Marching Forest as representative of the SIU action. Although the forest did not receive press attention during the rally, the local Shoreline newspaper ran photos of the group making the costumes at the organizer's home in preparation for the event.

The Marching Forest was an attempt to create an image event. It was an attempt at "voice," to quote Linda Stein, through the visual of a marching forest (2007). Unfortunately, the message was less developed than the ornate costumes worn by marchers. Fellow marchers recounted observing people dressed in forest costumes, but did not attribute a broader message or specific meaning to this action other than attracting attention to the issue of global warming.

The Marching Forest participated in the larger Seattle march that filled several blocks: people waving signs, groups occasionally chanting, all the while plodding forward in the rain. A group of children led the march, holding a banner. The march itself created a visual display of people calling for change on climate-change policy, and both news media and participants photographed the march. The *Seattle Times*, which ran an aerial photograph of the march, characterized the march as "placid" in the headline for the accompanying story (Lacitis 2007). As an image, this group seemed approachable and accepting, a family group, not anything dangerous (especially compared to the 1999 WTO protests that shut down Seattle). One researcher's observations of the organizing meetings indicated that Seattle organizers desired positive messages—including climate solutions, involving children, and building for the future—and new people to engage the issues. But *placid* was probably not the intended message, given the overtly political nature of the event and the urgent demands for legislative and personal action to stem climate change.

Once the Seattle marchers passed the polar bear and arrived at the march's endpoint, an urban waterfront park, people were invited to collect promotional material from local organizations and businesses and listen to speeches from local politicians, community leaders, and a guest representing the Marshall Islands. The solutions fair offered more traditional forms of discourse, particularly the information booths and the eleven speeches given by the MC and presenters, which lasted almost an hour.

Notably, the MC started his first speech by referencing an image event:

> We have read a lot of scary news in the newspapers lately about what is going on with global warming. But that's not really news, is it? We've been hearing the science for the better part of

> twenty years. That's not the news. The news is not that the water is rising. The news is that this morning a tsunami, a tidal wave engulfed lower Manhattan. It was a sea of people wearing blue shirts. A sea of people who came out to a Step It Up event just like this one. And they formed a line in lower Manhattan to where the water would come to if the sea level rose ten feet, including very near the site of the World Trade Center. A sea of people is rising faster than the sea level is rising. And you are part of that sea.

This rhetorical use of an image event contrasted with the heavy reliance on spoken and, to a lesser extent, written discourse at the solutions fair. The remaining speeches made several allusions to the rising tide of support and tsunami of people, but much of their content was focused on envisioning Seattle as a leader in efforts to stem climate change, particularly by creating new technology and innovating legislative solutions.

Earlier that day, a rising-tide event took place at Golden Gardens, a waterfront park in Seattle's Ballard neighborhood. The action plan posted on the SIU Seattle Web site stated:

> 30 people 10' apart (spread out in a line 300' long at high tide mark along the beach just north of the bulkhead entrance to Shilshole marina boat moorage) will hold 30 poles 20' in height, with flagging indicating what a water level 5 or more feet higher would look like, just to help people visualize what global warming COULD mean.

A neighborhood group called Sustainable Ballard organized the event. While about forty people, primarily from Sustainable Ballard, turned out to participate in the event, the cool early morning at the beach park may have discouraged spectators. Only a handful of people walked by during the event and, of those, only two or three asked about it. Although event participants identified another photographer as working for *Real Change*, the city's homeless newspaper, images of the event did not appear in *Real Change* or any other media.

The event, however, was organized specifically to be visual. One organizer told the group during an informal speech,

> We're here because I get tired of marching and listening to speeches. I thought it would be interesting to do a visual. It's too bad we weren't able to be later in the day—but we were preempted by the events downtown—because I think more people would have seen it. I really am just delighted that all of you showed up. I'm particularly delighted that the little ones came, because we are doing this for you.

Several months after the event, one of the event organizers told a researcher, "My original goal was to get some good visuals for the media, but it turned out more to be just a moment of reflection for the people who did turn out." In interviews, some participants spoke about their reasons for taking part in the event in terms of the imagery it provided and the action that might unfold from being exposed to that imagery. One participant said she wanted

> to help people *see* what is happening—and this particular project is very visual. I think it gets confusing for people because these things are happening but they're very abstract. This can help them get the idea. More and more awareness will lead to action. The idea is to get people to see it first and then to get policy changes.

In addition to the primary image of the rising ocean levels created by the bamboo poles and ribbons, some participants engaged in other forms of visual communication. For instance, one organizer taped a red-and-white plastic "For Sale" sign to herself, and at the top taped a sign she had handwritten that said, "Earth Not." She spent the event wearing the "Earth Not For Sale" sign on her body. Still another organizer, a woman in her seventies, wore a stuffed starfish that surrounded her face, from which she gave her two informal speeches.

While there was no media coverage of the Ballard event, the experience was visually striking. The sea levels represented by the tape and ribbons on the bamboo poles, some fluttering far above people's heads, were visually enacted and embodied on the sandy beach in front of the backdrop of the Pacific Ocean's Puget Sound. While seemingly not focused on mediated aspects of their image event, participants did seem aware of its potential. For instance, during one informal speech, an organizer told the group, "There's a couple of photographers who want us to clump at each of the poles. Let's sort of step up into a clump to do some photography." After some talk and laughter, a few photographs were taken and the organizer asked, "So where are our photographers? You're good? You're good. Meeting your photo needs? All set? OK."

At the end of the event, organizers asked participants to go to the Seattle downtown march to "recreate this very scene." One organizer stated,

> We're planning to bring the poles down and then, when we hold them in our hands, the wave will be at least above our heads. We'll have at least two waves, hopefully, going on. Then the fish will be swimming; the starfish will be there. We'll make lots more of our signs as we go downtown. Hopefully we'll have a big Sustainable Ballard presence at the march.

Researchers spoke to members of the group downtown but did not observe a reconstructed scene. Organizers later told a researcher that the group marched with their poles, "but their intent could not be seen in the mob."

Spokane provides an interesting case because the event both was and was not an image event. The rally in Spokane was divided into two separate sections. It began with an image event as organizers arranged participants—wearing white shirts over their daily clothes—to spell out the event's message: "Step It Up, Congress: Cut Carbon 80% by 2050." The participants spanned a bridge, and a man with a camera hurriedly snapped several pictures of the human banner.

This portion of SIU Spokane clearly constitutes an image event: there was no dialogue about the problem of global warming, SIU's stated demand to Congress, or what actions people might take to better manage the human and natural complexity of global warming. The objective of the picture was primarily directed to external entities—local and regional press releases, the SIU Web site, local and regional environmental Web sites.

After taking the photographs on the bridge, the entire crowd made its way to a small, open amphitheater. Marking a significant shift in tone, the participants of the human banner removed their banner shirts as they entered the pavilion area. In contrast to the image event, the second portion of the schedule included a series of speeches and dialogue among members. No one appeared to take pictures or record the speeches. Beginning with speeches from the most prominent attendees (such as the then- Spokane mayor and the Lands Council president) and working toward a round of impromptu messages, the morning made a clear transition from image event to dialogue among like-minded individuals. The primary organizer for the day spoke briefly and then introduced the mayor. Members of local environmental groups followed with consistently briefer messages—one person, for example, said simply, "Think global, act local." After a few seconds of silence, people politely clapped, the speaker sat, and another speaker rose. As these speeches ended, people transitioned into informal conversations in small groups to discuss global-warming issues: a researcher talked to a woman in her twenties who had just returned from a camp to be certified to deliver Al Gore's *An Inconvenient Truth* speech to local audiences; two men in their forties complained about President George W. Bush's energy policies; moments later, a researcher spoke at length with participants about green-building supplies.

In contrast to the image event just thirty minutes earlier, these speeches and conversations did not directly orient around the national SIU campaign—the specific demand to reduce carbon emissions by 80 percent—and were focused entirely on local networks and regionally specific issues of climate change. This provided a notable contrast to the image event just moments before, when, as one participant put it, "we need to take a picture for the [SIU] Web site." In contrast to that need, participants now mingled casually and talked about climate-change issues. The only explicit reference to SIU came within the context of people retrospectively talking about how slow people were to commit to the day's events and how quickly the Lands Council brought things together working with preexisting e-mail lists. Beyond these brief notes, researchers did not hear a single reference to SIU or its goal ("Step It Up, Congress: cut carbon 80 percent by 2050") in over an hour of conversation.

In this way, Spokane demonstrated a clear split between the first and second portion of the day. The first section clearly focused on embodying the central SIU message by creating a human banner of that message. The second portion of the day served an internally affirming function for the organizers and participants.

The Moscow/Pullman event shared many characteristics with the Spokane event. In Moscow, participants gathered early in the morning to take part in a bike ride down the three-mile Main Street. The event was intended to embody or represent a solution, in that bike riding provided a specific example of the sorts of transformations necessary to avoid global-warming impacts. As riders made their way through the streets of Moscow, participating adults (some presumably parents) placed children at the front of the collective and also explicitly spoke about them as embodying a solution to global warming. One participant told a researcher, "Look, aren't they cute? They are riding bikes and making a difference." Another person foregrounded the role of children, saying, "Kids are the ones who are going to make a difference."

The function of this image was to implicate both the impact on and solution-potential of children. The bike ride functioned largely as a relatively mild image event that was pleasant, contained no violence, and focused primarily on a solution orientation. The bike rally ended at the historic 1912 center, where they posed "for a group photograph which will be posted on the national Step It Up website" (Hilding 2007, 2). The group waved, shouted, and wielded banners with various SIU messages, including the movement's primary mantra: "Step It Up, Congress! Reduce carbon emissions by 80 percent by 2050." After the photos were taken, the event changed quickly as select public officials, such as University of Idaho's president and an Idaho State representative, addressed the crowd concerning sustainability efforts, and youth members of local environmental clubs sold food. The image-driven aspects of the earlier parade and group picture shifted to a public address event and, later, to a small group space as people listened to the local band Plasma Soul while conversing with one another. In this way, the SIU event in Moscow illustrated the dual function similar to both Spokane and Seattle events, where the early portion of the day was an externally focused image and the second half involved speeches and conversation among participants.

In contrast, the event at Deception Pass State Park was not an image event. People who followed details posted at the SIU Web site for a forest cleanup and protest on the elevated highway bridge would have found nothing; three researchers tried. Only after repeatedly asking an attendant at the ranger station about SIU were the researchers directed to a small sign advertising a "Disappearing Park" nature walk. The two park rangers who led the walk explicitly separated the activity from the national SIU campaign during their discussion with participants.[1] The rangers said they wanted to create a dialogue with participants rather than a protest event. They told participants that in the process of preparing for the walk they realized *for the first time* the extent of the impact global warming could have on this place they knew and loved. The list of impacts was long: the freshwater lake would become part of the ocean, Whidbey Island would be multiple islands, La Connor would be gone, freshwater-fish species in Cranberry Lake would be gone (or change dramatically), and the park would also lose beaches, shoreline, roads, and historic structures that are irreplaceable. During the nature walk, seven participants (two rangers, two local residents, and three researchers) discussed these impacts and learned about the park ecosystems. No one took photographs. No one contacted the media. No spectacle occurred.

The SIU actions we observed represent a range of approaches to action, from successful image events (the polar bear) to attempted image events with no dissemination (Golden Gardens) to mixed events (Spokane, Moscow, Seattle) to events designed to cultivate dialogue between participants that forwent images entirely (Deception Pass State Park). These examples suggest both potential challenges and possibilities of staging image events as a movement-building strategy.

The Challenges and Possibilities of Image Events

The logic of the public screen suggests that image events are effective when they attract media attention and receive wide dissemination. That social movements want to get mainstream media coverage is not a foregone conclusion. The Internet and alternative-media networks (such as Indymedia) provide alternative means of disseminating information (Best 2005). Nonetheless, many social movements do seek mainstream-media coverage. Press releases and media-relations guides written by and for advocates demonstrate that activists are often interested in receiving media amplification of their movement's message. As chapter 8 discusses, gaining media coverage was an explicit goal of the national organizers, who provided tips for attracting media attention in their "cookbook." Organizers recognize the inherent power of media dissemination. If a movement seeks to educate the public or raise awareness, national media coverage can almost instantly introduce the mass public to the movement's message and help construct a different way of thinking and talking about an issue. Mediated images, in particular, are powerful because they are representations of action and they invite readers and audiences to construct their own meaning of the message (Gamson et al. 1992).

With the allure of media dissemination in mind, this section discusses the challenges and possibilities of using image events as a social-movement strategy to gain public attention and build a movement. Sometimes these goals—attracting media attention and movement building—are synonymous. Other times, they can be at odds, such as when attracting media attention comes at the cost of appealing to movement participants. Understanding these lessons from SIU may enable movement organizers to make informed decisions as to what event strategies may be most appropriate for their goals.

Challenges

1. *Sometimes no media appear.* Organizers stage image events as a means of attracting media attention, hoping to get some airtime on the public screen. Unfortunately, not all image events successfully gain media attention. For example, the Golden Gardens event did not receive media coverage despite its visually striking quality, the visual orientation of its organizers and participants, and the promise it held to make the issue of

rising sea levels controversial. Sometimes the lack of media coverage can be attributed to characteristics of the event or a lack of publicity. In the case of SIU, having so many events in one day, which forced the Golden Gardens event to be early in the morning to avoid conflicting with the main Seattle downtown event, may have diluted the effect of possible SIU image events, as the media allotted to cover the event could only focus on so many visual events in one day. The Marching Forest of Shoreline, for instance, received local newspaper coverage prior to the rallies but no press attention during the downtown event itself. When image events do not attract media attention, their impact is reduced to those who directly observe or participate in them. From a movement-building perspective, image events with no press coverage rarely gain notoriety, realign conventional values, or bring widespread attention to the movement's goals.

2. *Attracting media attention may require extreme action.* The difficulty of attracting media attention poses the second challenge of staging image events as a movement-building strategy. Communication research demonstrates that extreme actions such as violence amplify the amount of media coverage an image event receives (DeLuca and Peeples 2004). Yet, extreme actions—in particular, the strategic use of violence—are not without costs. Event organizers must decide how these actions fit into the larger movement they are trying to build. An internal document on the messaging of the downtown Seattle organizers stated,

> With the goal of cutting emissions by 80% by the year 2050, we will use this event to set us on a path to healthier communities, a healthier economy, a healthier environment, and a higher quality of life. Our goal is to both highlight success stories and encourage responsibility.

The risk of using violent images is that it may undermine the possibilities of creating "healthier communities." Extreme, violent actions draw attention to problems, but they also may depict the movement as radical, thereby discouraging new people from joining or supporting the movement. For example, the Golden Gardens event participant who identified herself as "not really your out-there environmentalist" is not likely to chain herself to a tree or set fire to a ski lodge; she wants to be mainstream *and* participate in SIU. Organizers often must choose different tactics to cater to mainstream and radical participants.

3. *Not all movement actions have a clear enemy to confront.* Image events often confront powerful people or industries and conventional ways of thinking about human-nature relations. Staging extreme actions is easier when there is a direct target (e.g., bulldozers to stop, delegates to blockade, corporate buildings to trash). But sometimes it is difficult to identify an enemy. The Golden Gardens event, for example, confronted no concrete person or industry, though President Bush was mentioned informally by some participants as a specific villain when it came to global warming. Organizers of the downtown Seattle event wanted to be positive, choosing to focus on climate-change solutions rather than blaming individuals or even government for inaction. The lack of an opposition target can make it more difficult to create an image event that highlights injustice, leaving organizers to visualize abstract systemic concepts like climate change that do not have a single cause.

4. *An image must tell the right story to be successful.* Some of the participants' tactics were more akin to protest-style visual communication, such as taping signs to one's body or wearing animal representations on one's head, and, therefore, were conventionally recognizable as resistant visual discourse. Other visual representations were less self-explanatory and perhaps a bit too esoteric to function effectively on the public screen. For instance, people holding poles with tape fluttering above their heads may be evocative but need some explanation. To be successful, image events need to tell the movement's story. To do so, they need to resonate with available cultural meanings or work hard over time to establish new frames of reference for their images (as accomplished by tree-sitter Julia "Butterfly" Hill, who lived in a redwood to protect it from logging). Organizers of Shoreline's Marching Forest intended to proclaim the importance of urban forests as carbon sinks and neighborhood habitat. As the forest marched down the street, the costumes were striking, but the intended political and environmental message was not immediately clear.

5. *Sometimes images only tell a portion of the story.* As requested by national organizers, the Spokane organizers took a photo of event participants and the central message of the event. A human banner reading "Step It Up, Congress! Reduce carbon emissions by 80% by 2050" drew attention to the dynamics between Congress and citizen groups. It captured people physically standing to demand political action on climate change. This image does not, however, say much about global warming. For people who do not already support political action on climate change, this image is not likely to sway opinion or compel people to rethink their carbon footprint. Instead, the image captures existing support for action; it does not make a case for why this action is necessary.

6. *Image events may alienate participants when they cater to an outside audience.* At the Spokane event, organizers managed logistics for the photo, yet none communicated to other participants standing in the rain what was going to happen, when, and why. Those not involved in the picture stood around talking about whether they should get behind the banner or stay entirely clear of the picture. Only when a researcher read the human banner as it was being photographed did he realize what it stated. A few directions from the event organizers would have enabled all participants to feel like part of the event as opposed to separate from it, watching it from the sidelines. In this case, the importance of creating an image overshadowed the experience of participants and may have alienated existing members who did not feel included or appreciated. By focusing on external audiences, image events can miss the opportunity to build community among movement participants. Richard Gregg (1971) identified what he called the ego-function of protests as uniting participants by reinforcing their participation as important and worthwhile. This function can be important for building commitment from movement members. Some image events may forsake this crucial ego-function, leaving participants to feel their participation is unvalued and unimportant.

7. *Image events are difficult to stage.* Beyond the creative challenges of imagining the event, natural settings can pose logistical challenges. In the Spokane event, the photographer was perched atop rock outcroppings

upriver from the human banner. The Spokane River undoubtedly provided an iconic scene for this image event. But it was early spring in the Northwest United States, and the melting snowpack from Idaho and Montana created a raging river. As a researcher stood on the bridge watching the human banner collectively pose for shots, he watched the photographer yell directions at the hapless human banner. None of the participants could hear the directions over the roar of the river, however. Without the photographer, the image potential of this event—the possibility to embody the political message of the day—would have been lost. The frustration and time wasted trying to manage the photo shoot could have been minimized had someone planned for the environmental challenges the natural space provided. In the case of Spokane, a two-way radio or even cell phone would have simplified and streamlined the image event. Even the relatively effective swimming polar bear had to overcome logistical challenges, namely, how to stay warm in the chilly Puget Sound (a dry suit did the trick).

Possibilities

1. *Images are memorable.* Despite the challenges and potential problems of staging image events as a movement strategy, such events have the potential to be a vital tool for movement builders. Image events are effective, in part, because people remember the images even after an action or a newscast. As Seattle participants walked by the swimming polar bear, they pointed and exclaimed. But they continued to talk about it later, too. At a meeting with a U.S. Congressional representative to present the group photo from the action, one of the participants referenced the polar bear in describing the meaning of SIU. Though the politician did not know what the participant was talking about, the other SIU people all nodded their heads. This image stuck with them, representing SIU as a whole.

2. *Image events can translate reality through powerful metaphors and anecdotes for movement building.* When image events create a lasting picture that stays with audience members, that picture can continue to remind people about the message of the movement. Image events can

also function as rallying points for movement participants. During his opening speech, the Seattle MC started his speech by referencing a tsunami of SIU participants in Manhattan. By referencing another image event, the MC was able to conjure the visual impact of the other event in his own speech. He then told Seattle participants they were part of that sea, using the image event to unite disparate actions and highlight the force of SIU. Similarly, participating in extreme image events like chaining oneself to a tree or dressing as a polar bear and swimming in the sea can become mythic experiences that perform the ego-function of reminding participants how much they have sacrificed or risked to commit themselves to a cause.

3. *Images communicate efficiently.* SIU demonstrates that images capture the breadth of actions across the nation. Few people are likely to read more than one thousand individual action reports. But they will watch a digital slideshow on the SIU homepage. Together, a set of images can capture the scope and details of an event or a range of events more efficiently than written text. For a grassroots movement like SIU, images bring together the distinct grassroots groups while maintaining their individuality. The most dramatic photographs (e.g., scuba divers under the water) will be widely disseminated and duplicated by media and supporters alike. For Seattle participants, seeing the polar bear in Puget Sound instantly communicated the urgency for action on climate change. The message's meaning was memorable and easily understood.

4. *Images can communicate complex arguments through representation.* Much has been written about the complexity of climate change and the difficulty of communicating the science to citizens (e.g., Moser 2007; Moser and Dilling 2007; Revkin 2007). An image event like that at Golden Gardens has the possibility of transforming climate science into a tangible experience. Climate change becomes a beloved urban beach marked where it will be covered by water. This local, observable image embodies the impacts of climate change while also implying that action is necessary if people want to protect their local beach.

5. *Image events can transform human-nature relationships.* DeLuca (1999) discussed how some image events work to change the way we see

human-nature relations by putting the human body into more ecocentric forms (e.g., an Earth First! protestor perched in a tree). At the Golden Gardens event, participants not only donned fish and starfish on their heads, they stood together on the shore to enact the forecasted tide line above their heads. In this sense, the participants invited both themselves and possible spectators to identify with the natural world. The polar bear swimming with a sign addressed to humans shifted human-nature relations in a different way. By calling for humans to be concerned about their "habitat," a scientific term conventionally used to mark places animals live, the polar bear challenged humans' unique position of dominance, sending a message of warning that placed humans on the same level (in danger of losing habitat) as the bear. Together, these examples demonstrate how images do important ideological work of realigning values and transforming popular ideas of human-animal relationships—and images can often do so in a way that is not considered preachy.

6. *Image events can also help people see themselves in movement actions.* In the challenges section, we discussed how image events could alienate potential movement supporters. The reverse is also possible. Just as the "mainstream" participant at the Golden Gardens event explained she was participating because it was important to "get people just to notice that there are a lot of normal people who are concerned about the climate," it might be possible that photographs of the "placid" Seattle march can show more conservative viewers that in not advocating for the environment they, in fact, are the ones who might be considered extreme. Likewise, the urban-natural locale of the Golden Gardens event, a city beach, provides a backdrop in a nature that is more familiar to a broader range of viewers than a remote, pristine natural space and may be more inclusive of urban viewers.

Robert Hariman and John Louis Lucaites (2003) argued that daily photojournalistic images define the public through the shared act of viewing events as part of civic life. When images from events appear in the news, these images represent and define the mass public. Through images, people see themselves as part of the public. A similar phenomenon can occur with images from social-movement actions. In the case of SIU, the participation of older people, women, families, and children in events creates a representation of this movement as including diverse groups. Kids rode their bikes in Moscow; older men and women waved signs as they marched in downtown Seattle. This representation enables audiences to see themselves in the representations of the climate-change movement, especially when the movement's images look similar to civic images.

7. *Image events can be experiential, too.* While a potential challenge of the creation of image events is the loss of local participants in favor of outside audiences, the Golden Gardens event offers a different model of an image event—one less focused on changing public discourse through media exposure and more focused on creating meaning and energy among an existing activist group. For instance, while organizers were aware of the photographers, they gave far more attention to participants via socializing and informal speeches in which they made sure participants understood the purpose and meaning of the event. While image events are designed to be widely disseminated, they are also intended to open up possibilities for ways of thinking about the environment and to help mark participants' identities and confirm beliefs. In the best image events, both local participants and broader audiences are struck by the images they have created and feel as though they are part of something bigger than themselves.

SYNTHESIS: PRACTICAL SUGGESTIONS FOR MOVEMENT BUILDERS

The challenges and possibilities demonstrated by these SIU events suggest many reasons why movement organizers might want to use image events. To use them most effectively, organizers need to answer four basic questions.

What Is Our Primary Message?

As the above challenges and possibilities illustrate, some messages work better as images than others. By first identifying the intended message of a particular action or event, organizers can consider whether this message can be effectively conveyed through an image. Going through the possibilities and problems laid out above can help guide an organizer to an answer. For example, images work well to realign traditional values. But images may not be able to articulate the movement's cause and solution in instances where there is no clear enemy.

Who Is Our Primary Audience?

Sometimes an image event targets external publics rather than focusing on event participants. In certain moments, this focus is the best way to build a movement. Other events aim to enrich the commitment of existing supporters. Organizers should decide on the primary audience for a particular action. If the action is geared toward local participants, an experiential image event like the Golden Gardens action may be the most appropriate.

How Far Are We Willing to Go?

The logic of image politics centers on the production of images that are fit for media distribution. As the SIU actions demonstrate, a striking visual image is not always enough to get press coverage. Instead, more extreme actions, including the strategic use of vandalism and violence, are sometimes necessary to attract media attention. Movement leaders must decide whether extreme image events are in line with the larger movement goals. In some cases, staging extreme events can propel movement images onto the public screen, where the mass public is exposed to the movement's message.

How Does This Event Fit Into the Bigger Picture?

Too many image events can compete with each other and minimize the individual impact, as the Seattle actions illustrate. Before committing to the effort of staging an image event, organizers must determine how it will relate to the broader campaign goals and other strategies. Organizers must determine their primary media goals and forgo additional events that may compete for or dilute media attention.

As the nature of society and media change, so do the venues for social movements. The onset of increasingly visual and decentralized media outlets such as YouTube and Facebook clearly indicate that the public sphere of rational debate has turned into the public screen of snapshot images. To respond to this shift, we have highlighted in this chapter the roles, challenges, and possibilities of image events in social-movement processes. As our analysis indicates, image events provide both considerable promise and potential challenges. Action organizers' ability to capture and utilize these possibilities will greatly influence the success of the climate-change social movement in the future.

ENDNOTE

1. The Deception Pass rangers were assigned the task of leading this event by the park's education coordinator, the person who signed up the event on the SIU Web site but was unavailable to lead the action herself. Given their positions as state employees, it is reasonable to presume that promoting the action as a protest event could have endangered their jobs.

REFERENCES

Best, Kirsty. 2005. Rethinking the globalization movement: Toward a cultural theory of contemporary democracy and communication. *Communication and Critical/Cultural Studies* 2:214–237.

DeLuca, Kevin M. 1999. *Image politics: The new rhetoric of environmental activism*. New York: Guilford.

DeLuca, Kevin, and Jennifer Peeples. 2002. From public sphere to public screen: Democracy, activism, and the "violence" of Seattle. *Critical Studies in Media Communication* 19:125–151.

Gamson, William, David Croteau, William Hoynes, and Theodore Sasson. 1992. Media images and the social construction of reality. *Annual Review of Sociology* 18:373–393.

Gregg, Richard. 1971. The ego function of the rhetoric of protest. *Philosophy and Rhetoric* 4:71–91.

Habermas, Jürgen. 1984. *The theory of communicative action, volume one: Reason and the rationalization of society*. Boston: Beacon Press.

——— 1991. *The structural transformation of the public sphere*. Trans. Thomas Burger. Cambridge, MA: MIT Press.

Hariman, Robert, and John Louis Lucaites. 2003. Public identity and collective memory in U.S. iconic photography: The image of "accidental napalm." *Critical Studies in Media Communication* 20:35–66.

Hilding, Tina. 2007. The heat is on: Community discussion on climate change. *Moscow Climate Change*. http://www.moscowclimatechange.com/ (accessed November 30, 2007).

Johnson, Davi. 2007. Martin Luther King Jr.'s 1963 Birmingham campaign as image event. *Rhetoric & Public Affairs* 10:1–26.

Lacitis, Erik. 2007. Placid march seeks action on warming. *Seattle Times*, April 15, 2007. http://seattletimes.nwsource.com/html/localnews/2003667134_stepitup15m.html (accessed August 27, 2009).

Moser, Susanne. 2007. Communication strategies. In *Ignition: What you can do to fight global warming and spark a movement*, ed. Jonathan Isham and Sissel Waage, 11–21. Washington, DC: Island Press.

Moser, Susanne, and Lisa Dilling, eds. 2007. *Creating a climate for change: Communicating climate change and facilitating social change*. Cambridge: Cambridge University Press.

Revkin, Andrew C. 2007. Stuck on coal and stuck for words in a high-tech world. *New York Times*, December 4, 2007. http://www.nytimes.com/2007/12/04/science/earth/04comm.html?_r=1&scp=1&sq=Stuck%20on%20coal%20and%20stuck%20for%20words%20in%20a%20high-tech%20world&st=cse (accessed August 27, 2009).

Sill, Randolph. 2007. Polar bear sited in waters off Myrtle Edwards Park warns of global warming and encourages decrease of carbon emissions! Step It Up. http://events.stepitup2007.org/april/reports/120 (accessed April 12, 2008).

Stein, Linda. 2007. The Marching Forest of Shoreline takes it to the streets. Step It Up. http://events.stepitup2007.org/april/events/show/1334 (accessed April 12, 2008).

Chapter 10

"Smells Like Folk Life"

Participants' Identity Construction at Step It Up

Tracylee Clarke and Trudy Milburn

At a Step It Up (SIU) campaign event in Seattle, Washington, a young onlooker commented to his mother, "Smells like Folk Life." His mother nodded in agreement. The young man's comparison of SIU with Seattle's annual Folk Life Festival illustrates the importance of group identity, or our ability to identify with other "folks," in order to take action together. We all tell stories about why we engage in social actions. Combining our stories with those of others who engage in similar activities—such as attending a global-warming-campaign event together—reveals more not only about why we gather collectively to address social issues, but about what that means for our sense of personal and political identity. One of the central themes in the literature on social movements is the observation that identity is important to movement dynamics. Indeed, understanding identity is core to understanding social movements (Snow and McAdam 2000; Hobson 2003), as the "collective search for identity" plays a prominent role in social movements (Stryker 2000, 23). Researchers agree that social identity is a crucial motivator for commitment to the goals of a movement, participation in movement events, and personal political choices related to a movement (Almanzar, Sullivan-Catlin, and Deane 1998). This has implications for not only recruitment and the development of movement messaging (Snow and Oliver 1995), but also for understanding the social meaning(s) created by political events associated with a movement.

Although identity has been conceptualized in reference to social movements and political activism, there is considerable debate in the social-movement literature about the nature and relevance of identity for social-movement analysis. While some argue that researchers focus too much attention on social identity to the disadvantage of individual identity (Almanzar, Sullivan-Catlin, and Deane 1998), others argue that notions of self and identity are not nuanced enough (Stryker, Owens, and White 2000). Further, merging individual and collective identities may obscure important theoretical and empirical issues relating to both.

Although understanding group identity is indeed critical in an analysis of social movements, we focus in this chapter on the implications of individual identity. In social-movement literature, identity is often treated as a precursor for or prediction of involvement rather than a product of movement participation. Little research addresses the process through which identities are constructed and maintained through social-movement participation, and the implications of those construction processes for movement success. Snow and McAdam (2000) attempted to remedy this as they elaborated on Melucci's (1996) argument that participants in social movements act out their identities through social-movement activities. They contend that identity formation is a dynamic process and are primarily concerned with the ways in which personal identities and collective identities are linked.[1] They call for an expansion of a constructionist perspective and greater theoretical elaboration and empirical work focused on better understanding of identity-work processes in relation to social movements. This chapter is an attempt to answer that plea by exploring how personal political identities were constructed through participation in SIU.

Drawing upon literature that addresses political-identity construction, we seek to better understand how identity was created and negotiated through participation in SIU, a nationwide, environmental-protest-campaign event. We employ Paul Ricoeur's theory of narrated identity to analyze narratives shared by organizers, participants, and researchers of SIU. We examine the way participants used labels for self and others and how those labels relate to constructed meanings of the event. We use the language of the participants themselves in order to localize our understanding of how selves are constructed in and through narratives (Ricoeur 1983, 1989, 1991b, 1992; Venema 2000). We argue that the political identity of participants at SIU was constructed and negotiated through their global-warming narratives, their accounts about their attendance and participation in the campaign event, and their stated intentions and planned actions after the event. Understanding this process of identity construction provides useful information for activists and organizers about ways they can enhance existing opportunities for participants and increase awareness and participation in future campaigns. We begin with a brief review of research related to political-identity construction, narrative, and political activism/movement participation; introduce Ricouer's theory of narrated identity as it contributes to our analytical method; outline important themes that emerged as a result of our narrative analysis; and close with a discussion of the implications of identity construction for social-movement researchers and practitioners.

The Construction of Political Identity in Environmental Activism/Advocacy

Melucci (1994, 1996) argued that participants in social movements act out their identities through social-movement activities as they negotiate a sense of self in relation to movement allies, antagonists, and observers.

As a result, the identity of those involved shapes the actions and tenor of the movement, influencing choices and behaviors of organizers and participants. This idea is also supported by research conducted by Andrews (1991). Interviews with political activists suggest involvement in movements not only reinforces but shapes identity, as individuals define themselves through their activism, continually restating and recreating who they are. Likewise, Teske (1997) argued that identity is the basis for social activism. Through their involvement in social and political movements, activists construct and confirm their personal identities. Actions taken by individuals beget more actions, and as a result, "the individual becomes politicized and the citizen becomes an activist" (Teske 1997, 123).

Research specific to the environmental movement also supports the premise that political identity is shaped through involvement in movement activities. In his study of environmental activists, Lichterman found that activists participate as individuals and "personal agents of social change," confirming a sense of self through their participation and political actions (1996, 34). Likewise, in her study of beach-cleanup activism, Wheaton (2007) discovered that through their participation in events designed to pressure government and big business to stop dumping refuse and sewage on local beaches, activists in England constructed a political identity, which in turn became a strong motivation to continue involvement. Understanding identity construction as it relates to movement participation is crucial, as it directly influences the environmental choices and behaviors of individuals. Identity directly influences whether values of sustainability are adopted and acted upon (see Almanzar, Sullivan-Catlin, and Deane 1998; Pol 2002; Clayton and Opotow 2003). Building on these and other studies, we seek to answer the question, *how were personal political identities constructed through participation in SIU?* For insight, we turn to the role of narrative in the construction of self.

Constructing Political Identity Through Narrative

Self construction occurs within a political social context that is negotiated and renegotiated within communicative interactions, or social political narratives (Brown 1997). Several theorists have recognized narrative's important role in shaping and expressing political identity, perspective, and ideology (e.g., Cornog 2004; Shenhav 2004, 2005, 2006). Examinations of political discourse show that it relies extensively on narrative patterns. Thus, narrative pattern is a crucial tool for the study of political thought; intended action as narratives work to explain and warrant political incidents. "Political narratives do not just spring into being; they are created in the course of political action" (Shenhav 2006, 248). Narratives are not only indicators of political identity but are also justifications of political activities, as narrators position themselves and rationalize their political actions (Briggs 1996; Dauenhauer 1997). The political self, then, is produced and reproduced through narrative (Brodkin 2007; Teske 1997).

Although researchers recognize the role of narrative in the creation of a political or an environmental identity, their analyses rarely explain *how* identity is constructed through narrative. For this, we turn to Paul Ricoeur's theory of narrated identity. Ricoeur argued that not only does one's experience of identity have a narrative structure, but it also is created and recreated, negotiated and renegotiated through narrative. Identity is constructed and reconstructed through a process he calls *emplotment*, a perpetual weaving and reweaving of past and present events into characters, motives, situations, and actions. Emplotment is the "synthesis between events or incidents which are multiple and the story which is unified and complete" (Ricoeur 1991b, 21). The plot thus organizes events into a coherent story. People in one's life become characters in the stories that are told, retold, and revised as their lives unfold. For example, the organizer of the Oakland, California, SIU event shared pictures and stories of his world travels, illustrating the impact of global warming on various communities. The construction of his narrative was fueled by the need for unity and coherence in diverse, fragmented, and unstable life experiences. Narrative then becomes a mechanism to know oneself and make sense of lived experience. Ricoeur (1991a) argued that to have an identity, we conceptualize ourselves as distinct from others and as continuous over time through the narratives we tell. Narratives enable us to construct identity out of *sameness* and *selfhood*. Sameness refers to the fixed parts of one's identity that remain somewhat consistent over time, making an individual recognizable. For example, a participant in Cheney, Washington, indicated attendance was consistent with, or the same as, her identity when she commented, "I have been an environmental activist for eighteen years and have worked with this issue for a long time." Sameness could also describe similarities with others who share the same attributes. Selfhood, on the other hand, is based on differences from others and is created by contrasting oneself against another person. Selfhood takes on an ethical dimension as it holds us accountable to others for our actions. In narrative, selfhood might sound like the sentiment expressed by a Salt Lake City organizer, who explained, "I felt obligated to do something. I felt responsible as a being on this planet." Selfhood distinguishes oneself from others while, at the same time, it recognizes connections. The task of narrative, according to Ricoeur, is to balance sameness and selfhood and to anchor identity in personal and social history and current experience. Personal narratives function to create consistency or recognizable sameness at the same time as they justify change and articulate selfhood. It is through narrative that one continually creates, constructs, changes, and justifies one's identity.

Narrative Analysis

Ricoeur's theory of narrative identity provides an interpretive lens with which to analyze the narratives and discourse shared by SIU participants.[2] We used the concepts of emplotment, sameness, and selfhood to examine political narratives and to further investigate specific narrative instances that communicated self and social identity for SIU participants. We then identified the emotions that participants expressed and interpreted how emotional responses illustrated what participants were feeling towards the event and towards global climate change in general. It is our intent to offer clues to organizers who seek to provide opportunities for their participants to strengthen their existing identification as environmental activists or refocus it toward the climate-change movement.

To understand the identity of the narrator, we must understand who acts in the stories told by SIU participants: who the characters were, and what their perceived relationship to the speaker was. By examining how persons are addressed or referenced, the notion of identity as expressed by characters within a story is illustrated.[3] The characters, however, do not stand alone, but always act within a plot. Emplotment can be identified by examining the actions participants refer to, or, in other words, what the characters in the story are doing. Identifying pronouns and persons, and then linking these characters to their actions, creates an understanding of the narrative that guides our identification of narrative elements that construct identity.

It is not enough, however, to assess who acts within a narrative. The actions in a narrative must be appraised in light of an actor's ability or agency to act. Actions, understood as deliberate, then become "subject to approval or disapproval," and actors and agents become "subject to praise or blame" (Ricoeur 1989, 164). Reagan (1996) contended that narratives always contain evaluations and value judgments. Through our narratives, we "apply to actions a moral judgment of imputation" (Ricoeur 1989, 101) and engage in the act of accusing or attributing blame. This evaluation leads to characterizing self as a change agent or as a victim of the actions of others. Brown argued, "We recognize ourselves in terms of our suffering as well as our agency in the stories we tell, and in the stories we follow" (1997, 115). People's selfhood—or their ethical accountability to self and others—is both enabled and inhibited by the actions of others or social conditions, leading them to characterize themselves as agents of action or as sufferers (Ricoeur 1991a).

Personal narratives provide the communicative basis for connecting agency and suffering with identity, as they enable speakers to interpret themselves as change agents or as sufferers in relation to social others and social conditions. For example, the statement by a Moscow, Idaho, participant, "We are going to be obliterated by this Bush administration…Then there is not going to be anybody left," identifies the U.S. government as the actor, obliteration via poor policy as the action. It also assigns blame to the Bush administration, making local citizens victims and sufferers.

A second critical part of our analysis was to identify the emotions that participants expressed. Following researchers who study the social construction of emotion (Harré 1986; Potter 1988; Lutz 1988; Pearce 1995), we examined how participants used emotive language to describe their involvement in SIU. Harré (1986) describes emotions as functioning to organize and interpret the disparate parts of individuals' lives, understand specific relationships with others, and express an individual sense of self. As a methodology, Harré suggests examining emotive words within a moral and social context, because recognizing the moral and social particulars will help us understand how an emotion might be described, enacted, and displayed in any given situation. When examined as such, we can learn more about a specific emotion as well as the social identity within which that emotion is enacted (Lutz 1988).

Examining emotive language within Ricouer's narrative framework suggests that actions can be evaluated through the categories of praise or blame. If SIU participants feel their actions are hindered by the actions of others or by social conditions, their restricted abilities become a barrier to the attainment of selfhood, ultimately challenging and stifling their political identity. It is through emplotment that the political identity of participants gets constructed, and it is through narrative analysis that researchers and practitioners can begin to understand how identity may have influenced participant involvement in SIU, as well as subsequent intentions of participants in relation to global warming.

ANALYSIS OF IDENTITY THEMES

Participation as Construction and Confirmation of Self

> I am here for myself.
>
> —Seattle, Washington, participant

SIU acted to construct the identities of those who organized, attended, and participated in the event. When asked what SIU was or why they were there, many attendees took the opportunity to talk about their individual identity or sense of self. One participant in Oakland, California, responded by saying, "This is who I am, it's what I do." When asked to elaborate, she repeated, "It is just what I do, it's who

I am." Unable to further articulate, for her, it was enough that she was there. Her attendance confirmed her identity and her sameness or consistency of self.

The organizer of the Oakland event took the SIU campaign event as an opportunity to share his personal experience as one of one thousand volunteer presenters to take Al Gore's message to local communities. Proud of being named a "climate-change ambassador," the local organizer stated time and time again how excited and pleased he was with this important role he had been given. Sharing his narrative personalized the event, illustrating his emplotment and his creation of self as activist and environmentalist.

The confirmation of self through attendance at SIU was also evident for many participants in Salt Lake City, Utah. Local organizers planned a "sun salutation yoga meditation." When the lead organizer was asked why she chose yoga as a format for the campaign, she replied, "It's true to who I am." Those who attended the sun salutation also echoed that for them, SIU was about practicing yoga; they came because yoga was what they did and it spoke to who they were. One participant explained, "It is about change within." Another stated, "Yoga is something that is strong and beautiful, it makes people strong and beautiful. So this is symbolic, a good way to do this event." Participation at the SIU sun salutation was directly associated with "becoming" and the construction of self (see chapter 2 and chapter 4).

Construction of self through participation at SIU was also evident in the labels organizers and participants chose to give themselves. When asked why they had decided to organize an event, a local coordinator in Moscow, speaking on behalf of her and her partner, stated, "We have been long-time activists and have been working this issue for a long time." As a self-proclaimed activist, participation with SIU seemed a sound and obligatory step to take. Another organizer in Kalamazoo, Michigan, made a similar statement: "I'm a liberal/radical person and I saw this as something that needed to be done, and no one else was doing it." When asked why they were at the event, many participants also responded through labeling of self. A participant in Cheney said, "I have been an environmental activist for eighteen years and have worked with this issue for a long time." Another, from Seattle, stated, "I'm a committed radical. In this society, what radical means is to take positions on what we should do and be vocal about it." For these participants, identity as a radical, activist, or environmentalist meant participation in SIU.

Whereas some participants and organizers readily embraced a direct association with radical activism, others disassociated themselves from "extreme" environmentalism. One organizer from San Antonio, Texas, commented that she wanted her event "to be light-hearted and fun. Not boring and dismal, like what people feel when they think 'environmentalism.'" When giving explanations for their involvement, SIU participants often made disclaimers, such as the Salt Lake City participant who explained, "I am not a protester generally," or a Seattle participant who commented, "I'm not really your out-there environmentalist. We're really more mainstream." Another participant from Salt Lake City claimed, "Extremists ruin it for everyone." Disassociating themselves from environmentalists, many participants instead referred to themselves as "concerned citizens."

While some participants used labels to express a sense of self, others articulated a sense of self through description of personal actions. Ricoeur (1983) argued that the identity of individuals is comprehensible by understanding how they recount actions in the narratives they share. Action can be recollected in stories whose function, in turn, provides an identity to the actor. When asked why she was at the SIU event, one participant in Lansing, Michigan, answered,

> I do recycling, I changed my lightbulbs, keep my heat low, and I bought a Honda Civic, which isn't hybrid, but was smaller than the other Hondas. I eat local. I went to the Slow Foods conference last week. I bought the book *World Changing* and also bought it for my son.

Another participant, in Salt Lake City, said, "I bike everywhere. I biked with my daughter to see the Al Gore movie." Actions such as recycling and biking confirmed concern for the environment and provided an identity for these participants. Consider the following extended statement by a Salt Lake City participant:

> My electrical bill is down to $25. I use less than one unit (100 cubic feet) of water or less per month. I've plumbed my roof with gutters and pipes to collect rain water. I distribute the water in July and August to needy plants. My dryer is my clothesline—inside and out. I moved to the basement when gas prices went up. No heat is required. The dirt holds the temperature there at 50–55 degrees Fahrenheit. I just wear a hat and a base layer of clothing along with a T-shirt and an overshirt. I need to keep moving, and it keeps me warm. I sleep in a sleeping bag. I only use one garbage can every four to six weeks. I want to do away with mine and share with my neighbors. We should just bring our own carrying devices to places. For example, a newspaper can be folded up like a cone. One can't have a thought or indulge in air without participating in the universe, without asking what to share. All is taken, otherwise. We can enjoy and be involved in the whole process. We can become in a graceful manner, and Mother Nature will provide. Trees are full of wisdom. Avalanches—I've survived them. We perceive along these dimensions. Things happen with an intention that can't be explained or reproduced. I've back-country skied hills steeper than 45 degrees and joined with the trees, blindfolded.

In this account, the speaker interlaces past actions and events with current actions and events, including his participation with SIU. His narrative demonstrates emplotment and the use of personal pronouns to describe actions the narrator had taken. In the narrative above, the participant shifts from using "I" to using "we" when advocating a course of actions for others. This is one way a person sets her- or himself up as a standard of what it means to be an environmentalist. Through pronoun usage within narratives, participants both confirmed and constructed their sense of self through their articulation of personal action taken. A sense of selfhood is confirmed when narrators can describe themselves "as the owners and authors of their actions" (Ricoeur 1989, 100–101). Articulating their ability to act in relation to global warming strengthened a sense of self, as it confirmed their agency and ability to be accountable to self and others for their environmental actions.

Participants and organizers rarely gave direct answers to the questions of "Why are you here?" or "Why did you organize this rally?" Rather, participants responded by telling how crucially linked their participation was to their sense of sameness. For these participants, attendance at SIU confirmed a sense of self by providing an opportunity to act consistently, further constructing their identity. Furthermore, the organic and flexible format of SIU rallies allowed individual identities to be constructed. Allowing organizers or participants to create their local event on their own terms provided opportunity for individuality to emerge and gave participants a chance to construct their own definition of self, whether as environmentalists, concerned citizens, or something else. It is crucial for event organizers and other practitioners to understand the role and importance of individual identity creation through participation in social events or rallies. If events provide a forum or opportunity for individual identities to emerge, they are more likely to attract a wide range of participants whose individuality and political identity can be self-articulated and confirmed by other participants. This in turn may lead to increased individual action in support of a particular cause.

Balancing Despair and Hope: Emotional Response to SIU and Climate Change

> It is the biggest problem facing us in our lifetime. We are not here to wring our hands in despair about the magnitude of the problem. We are here to celebrate and embrace solutions.
>
> —Seattle, Washington, organizer

A narrative analysis of SIU participant discourse reveals the construction of self as both a victim, constrained by social and political circumstances brought about by government, and an agent of action, empowered by involvement at the local event. If global warming creates a sense of impending doom and despair, SIU provides a sense of hope and empowerment.

Respondents repeatedly articulated a sense of helplessness brought about by global warming. A Seattle participant stated, "I'm trying to be as active as I can be just to hold off despair." When asked why she was at the campaign event, a young mother stated, "From a feeling of helplessness." Phrases such as "we're all gonna die," "it can't wait," "pretty soon, it's gonna be too late," or "we are too far gone to do anything about it" characterized the responses of despair from participants at events across the nation. The sense of imminent ruin left some feeling besieged and terrified. A Salt Lake City participant lamented, "It's frightening to not know the tipping point on this. We must arrest the downward spiral. It's a sobering thought. The process feels overwhelming." Referring to global warming as a "tipping point" implies a tenuous reality where things could quickly go one way or the other, making the very nature of our existence extremely fragile. Plagued by what could happen as a result of global warming, many feel their lives are at stake. A comment from a Moscow participant illustrates this sentiment: "Anyone who cares about the planet, it's, it's, it's our very survival, the web of life, the biosphere, the message. I think it takes over as the most eminent issue of our lives."

The frequent use by multiple participants of emotive words such as "threats," "immoral," "wrong," "despair," "overwhelming," "terrifying," "dangerous," and "scary" implies helplessness. Participants used these terms to describe both the potential effects of climate change and their own reactions to the idea of climate change. For instance, one participant in Raleigh, North Carolina, stated, "What are we doing? This whole thing is *very scary*. I am very concerned about the climate and the environment." Another participant, from Moscow, proclaimed, "Our whole way of life is *threatened* by this." Laden with emotion, participants' narratives constructed a sense of despair and gloom. Indeed, with their survival in jeopardy, a strong sense of impending doom permeated participants' narratives as they characterized themselves as victims and sufferers of global warming.

Understanding participants' emotional response to global warming gives us insight into their construction of self as victims and sufferers. Individuals telling stories of their suffering include accounts of who is to blame for their conditions. Those who attended SIU blamed not only government and corporations for the current environmental problem, but individuals who did not make environmentally conscious choices. One Seattle participant said, "I look at all the choices you can make, and no matter how virtuous you are, you're looking at a sea of SUVs." Participants' use of the word "should" when speaking of what they and others must do, or stop doing, to address global warming alludes to narrations of blame. A Salt Lake City participant's statement that "Congress *should* act more responsibly," and a Seattle participant's self-definition as "a committed radical [who must] take positions on what we *should* do and be vocal about it," illustrate judgments of blame. By blaming Congress, companies, and other individuals, participants vocalized how the social conditions of global climate change were jointly constructed by all facets of society. No one was excluded.

Juxtaposed against the narratives of doom and despair were narratives of hope and empowerment brought about by involvement with SIU. These narratives came from organizers, presenters, and participants alike. When asked about the primary message of SIU, a participant from Spokane answered, "Increased awareness and learning that it does not need to be an absolutely overwhelming thing—how can I help and what is me changing to a compact florescent going to do and how is that going to possibly save the world." Likewise, a participant from Oakland said, "The most important message is that I can personally do something, that I can start my container garden, I can drive less. This gives me things I can do personally." For those attending SIU events, the message was hope as a result of personal agency.

Presenters and organizers repeatedly told participants that SIU was meant to empower them. "There is hope!" encouraged one Oakland event organizer. In Seattle, local organizers created an SIU fact sheet that read, "We are organizing this event to disseminate the message of hope and optimism that we can and will succeed in working towards solutions to global warming. Our goal is to…instill the hope that it is, indeed, possible." At another SIU event, in northwest Washington, a local park ranger repeatedly used the word "empower" when talking about personal actions people could take, further stating that stopping global warming depended on "individual choices and actions." Exercising personal agency or choice thus became a solution. This message was consistent among organizers and presenters across the nation. In Lansing, the mayor told attendees,

> You can sit there and look at the glass half empty and say, "My God, it's over, we can't do anything," or you can make a difference, starting today. And when you see the statistics, you realize how quickly we could make a difference if we choose to.

A Seattle presenter reinforced the claim that resolution is dependent on choice and individual action by claiming, "What's between us and the solution to global warming is not technology, it's not economics, it's the sheer force of human will. Our vision…[is] the only thing that's separating us from a solution today." According to these participants, personal agency tied to action is the answer. Local organizers provided participants the opportunity to take ownership of the problem of global warming, exercise their agency, and take the necessary actions to manage the problem. "I live in an apartment complex, so I have no control. But I got this lightbulb today; it's a start," said a participant in Lansing. Before attendance at the campaign event, this individual described herself as having "no control" over her actions or global warming, but attending the rally created a sense of empowerment, enabling her to identify as an active change agent rather than a victim.

Where participants used negative emotional language such as "threat" or "despair" to describe global warming, they used positive emotional words to describe their involvement with SIU. The term "hope" was used most frequently. Participants also echoed one another in frequently using words like "care," "excited," "pleased," and "happy." These terms, sprinkled throughout participant narratives about their involvement in SIU, illustrated how participation created a sense of optimism and hope for the present and the future.

Through emotionally laden narratives of doom contrasted against equally emotive narratives of hope, rally participants engaged in emplotment or the perpetual weaving of past and present events. Participant narratives about their experiences at SIU provided an opportunity to negotiate tensions within and between participants' fragmented past experiences and their participation at SIU. The two choices that are present in this discourse are self as victim or self as agent. Participation in SIU enabled participants to experience the dialectic (see chapter 5) relationship between self as victim *and* self as agent. When both are present, it isn't a pure dichotomy, but rather a sequence where self as victim precedes self as agent.

For some SIU participants, confronting global warming meant contributing to the survival of self and community; at the same time, the overwhelming notion of global warming led others to feel powerless and paralyzed. One participant from Spokane observed, "I think when you say 'global warming,' people go, 'Wow, it is too big for me.'" Yet these narratives are juxtaposed with narratives of hope, choice, and action brought about by participation in local SIU events. Overall, participating in SIU empowered individuals to take ownership of the problem.

According to the SIU plot, the suffering caused by global warming can be overcome. Participants can become change agents rather than remaining victims of social and political circumstance. This directly addressed the ability to answer for self, speaking to participants' sense of selfhood and contributing directly to their political identity. An e-mail sent by an organizer to his team in Seattle illustrates the power of overcoming suffering through action:

> Every person is a singular opportunity for creation to become aware of itself. But the prerequisite is an open heart that allows itself to be hurt by the world's suffering. What happened across Seattle on Saturday happened only because each of you took the world's suffering—present and future—to the very foundations of your humanity, let yourself be hurt by it, and came back up to take action.

The "world's suffering" as a result of global warming can only be overcome through self-realization brought about by individual agency and action. According to the SIU plot, suffering is a precondition to action.

Participants at various SIU events imbued the issue of global warming with personal meaning through the narratives they shared. This is important for researchers and practitioners to understand, as the messages they offer communities must take into account how those communities view global warming. Further, SIU provided an opportunity to express personal hopes and hear others' stories of hope in the face of frustration. If practitioners such as event planners, environmentalists, and social organizers are to be successful, they must create opportunities for joint storytelling about hope to empower communities and show that individuals can take action against global warming. It is not enough for a few vocal individuals to convince communities that there is a problem. Those who take a leadership role in the environmental movement should seek opportunities to learn more about the hopes and fears that individuals bring to each particular issue and help them articulate a coherent narrative that blends hopeful and fearful emotional responses into opportunities for positive actions—actions that affirm each participant's individual sense of self.

Moral Responsibility and Accountability

> It's important to ground ourselves in the knowledge that this is not a political issue. This is a human issue, it is a moral issue, and we are all called to it.
>
> —Seattle, Washington, presenter

Climate change has been called a "dramatic challenge to our moral consciousness" (Jamieson 2007, 475). Because it is a moral concern, it speaks to our sense of right and wrong, making us responsible for doing something about the current situation. This sense of ethical accountability is directly related to our sense of self. Individuals conceive of themselves in terms of the values they hold, and their commitment to these values becomes the basis for their moral self-definition.

Communicative accountability locates responsibility for actions in public discourse (Mason 2001) or shared narratives. People think, perceive, imagine, and carry out moral decisions using narrative structures

(Faver 2001; Teske 1997). These moral decisions include environmental attitudes and intentions. In many cases, people espouse environmental attitudes not because of a belief that they will be listened to, but in order to define themselves by expressing fundamental values and ethical standards (Thompson and Gonzales 1997).

A narrative analysis of SIU participant discourse revealed a strong sense of moral responsibility. SIU provided participants the opportunity to espouse their environmental beliefs, further shaping their moral political self through their shared narratives. Those in attendance articulated narratives of personal accountability and personal choice. When asked why they were at the event, participants repeatedly cited moral imperative. They answered, "because I feel responsible," "I want to do my part," or "moral obligation."

When asked what the most important message of the event was, many respondents did not articulate the need for Congress to cut carbon emissions 80 percent by 2050. Instead, they spoke of moral accountability and individual responsibility. A participant in Salt Lake City said, "Politics are very important, but you need individuals to get something to change, to get a solution. It doesn't work without individuals." A participant in Lansing commented, "The government isn't taking responsibility. It starts with you. You're responsible and can contribute to everything." Likewise, an event organizer at one of the Salt Lake City events stated,

> People need to realize that it takes individual action to change. If the change is one at a time, so be it. We need to take responsibility. We need to take responsibility for our own actions. We only can take responsibility for our own actions. It can't just be other people. Solutions do not lie in the hands of politicians but in individual choices.

As one Salt Lake City participant claimed, "We must think about our personal choices."

This theme of moral responsibility relates directly to selfhood, which refers to our ethical answerability to self and others. Moral identity is related to our ability to be accountable to self and others in our community

(Haker 2000). This sense of community obligation was widely present in the narratives of SIU participants and organizers. When asked why they put together the campaign event, one Salt Lake City organizer stated,

> I felt obligated to do something. I felt responsible as a being on this planet, part of the community. We're all interconnected and it is important to take positive action. I talk enough. I talk about these things, but it is another thing to take action, political action.

For this organizer, the move from talk to action was realized in presence at the campaign event. Another event organizer from Salt Lake City stated, "Certainly, our event focused on the fact that we are all implicated in this ecological community, and that we must each do what we can to make a difference." A sense of obligation to others, including the ecological community, became a strong motivator for participation and action.

Participants attending SIU events spoke as if they were taking meaningful action by coming to the events. This was the first step that indicated they would continue to make similar choices in the future. "The power is ours to change this," claimed a Salt Lake City participant. Empowerment through action shaped the personal political identity of those at the campaign event as a confirmation of their selfhood; the ability to be held morally accountable was strengthened. Future advocates for global-warming activism will be more successful if they, too, can tap into the individual sense of moral responsibility and create opportunities where participants can construct or confirm their political identity and affirm their sense of moral obligation and responsibility.

Conclusion and Suggestions

The political identities of participants at SIU were constructed and negotiated through their global-warming narratives, their accounts about their attendance and participation in the campaign event, and their stated intentions and planned actions after the event. Through narrative emplotment, organizers and participants confirmed their sense of sameness and selfhood as they negotiated their political identities in relation to climate change. Based on our analysis of how individuals negotiated their political identities through the SIU campaign, we recommend the following suggestions to campaign-event planners and practitioners.

Allow for multiple campaign formats, created locally and supported nationally. Jamieson (2007) argues that the climate-change movement needs a sense of ownership and identity associated with climate change. The relatively loose, organic, unstructured, and dynamic format of SIU events provided a context for participants to articulate a sense of ownership about both their local event and the issue of global warming. Our analysis revealed how participants used labels for themselves and others to define what it meant for them to participate in an expressly environmental event. Although some associated themselves with radical environmentalism, others' discourse explicitly disassociated their political selves from radical environmentalism, as a potentially alienating concept. Narrative analysis shows that SIU participants believe they are just folks trying to live their lives. What we heard in the narratives shared by participants were multiple redefinitions of what it means to be an environmental activist. As one participant from Lansing stated, global warming "is no longer a liberal issue." The unstructured and organic format of the events provided a context for participants to make SIU an event that confirmed their political self either as an "environmentalist" or as just a "concerned citizen." Both narratives were able to coexist and create positive participant identity. SIU became, for many participants, exactly what they chose it to be—a political event on their own terms.

Provide opportunities for participants to share personal stories of despair, hope, and intent. Our analysis revealed that participant discourse was infused with personalized emotive language about global warming. The deeply emotional content demonstrated how integral this issue was to the construction of participants' political identity. Through the sequencing of despair, then hope, participant narratives worked to negotiate what global warming and their own personal political activism meant for them. Whereas thinking about global warming brings about a sense of doom, events such as SIU provide opportunities to act, to take ownership of the issue. This move empowers individuals and gives them a sense of personal and political security that confirms individual feelings of hope for the future. This hopefulness reinforces people's positive sense of selfhood and confirms their ability to answer for their actions, further constructing a strong political identity.

Provide opportunities for participants to share stories of responsibility and accountability. An examination of participant narratives showed that the opportunity to voice and enact moral responsibility is crucial to encouraging people to define themselves as part of the solution to climate change. Jamieson argues that in relation to climate change, it is difficult to identify the victims or agents responsible for climate change, and thus, it is difficult to assign responsibility, respond with what he calls a "virtue of mindfulness" (2007, 481), or act with moral urgency. SIU enabled participants to assign responsibility and create a moral urgency, which they articulated in their narratives. Rather than being subjected to others' messaging, they voiced personal ownership of the issues. Through their narratives, participants readily identified themselves as victims and agents, recognized their own personal responsibility, and both articulated and demonstrated their intent to act. For social-movement campaigns and political events to be successful, organizers must create opportunities for individual participants to vocalize their stories as they take responsibility for creating potential solutions. By paying attention to the narratives participants share, practitioners can better understand what climate change means to potential members of the social movement, and then create messages that speak directly to those meanings. This in turn can increase participant involvement in climate-change activism as a means of demonstrating an ability to act, rather than be acted upon.

Clearly articulate the goal(s) and outcomes of the campaign. While the unstructured format of SIU encouraged participants to take ownership of the issue and event, what we did not hear was recurrent rephrasing of the event's stated goal—to send a message to Congress. Although every event had a banner proclaiming the central goal, many participants did not make the connection between their local event and the national effort to send Congress the message to "cut carbon emissions 80 percent by 2050." Perhaps local leaders feared that by focusing on what Congress could do, the message sent by the national organizers might stifle or limit a sense of agency for individual community members. For whatever reason, many events strayed far afield from the intended message (see chapter 4 for more discussion).

In this chapter, we have used narrative analysis of participant discourse to highlight the importance of identity and constructed meaning of SIU as well as the role of social participation in the construction of political identity. Although the SIU focus on Congress may have created a disconnect between the issue of climate change and individual or local community responsibility (see chapter 5), local organizers and participants at SIU events across the country redefined the events for their own purposes, shifting the focus from sending a message to Congress to making a strong claim about individual action in local communities. As one participant from Lansing stated, "This movement is about everyday people, real people. It is not a political campaign."

The stories people told about their participation in SIU created personal involvement, passion, and action. By engaging in narratives of sameness and selfhood, SIU participants engaged in the process of emplotment to recreate their political identity and thus expand their political choices. The suggestions described above offer ways movement organizers and advocates may build on this understanding to create more meaningful social-movement campaigns and create positive changes in society.

Endnotes

1. Snow and McAdam (2000) outlined three different perspectives on identity formation: (1) the dispositional perspective, which argues that psychological traits or states predispose individuals to make certain choices, such as participate in a social movement; (2) the structuralist perspective, which argues that individual identities are embedded in elements of social structure, and therefore identity is structurally determined; and (3) the constructionist perspective, which argues that identity is constructed and maintained through joint action, negotiation, and interpretive work.
2. The narratives collected were brief responses to interview questions about respondents' participation at SIU. Familiar features of full narrative stories, such as characters, action, settings, and a coherent chronology, were often missing. In everyday talk, we leave out many of these narrative features; thus, we are left with narrative fragments that can be organized by theme or common ideas (Jasinski 2001). Although participant responses lacked full narrative structure, their narrative fragments provided insight into the identity of participants as their own discourse worked to balance fragmented experiences with a cohesive sense of self as they articulated their storyline or plot.
3. We acknowledge the early work of Brown and Gilman (1960) and Brown and Ford (1961), who established a framework for identifying pronouns used as personal reference and address terms.

REFERENCES

Almanzar, Nelson A. P., Heather Sullivan-Catlin, and Glenn Deane. 1998. Is the political personal? Everyday behaviors as forms of environmental movement participation. *Mobilization: An International Journal* 3 (2): 185–205.

Andrews, Molly. 1991. *Lifetimes of commitment: Aging, politics, psychology*. New York: Cambridge University Press.

Briggs, Charles L. 1996. *Disorderly discourse: Narrative, conflict and inequality*. New York: Oxford University Press.

Brodkin, Karen. 2007. *Making democracy matter: Identity and activism in Los Angeles*. New Brunswick, NJ: Rutgers University Press.

Brown, David D. 1997. On narrative and belonging. In *Paul Ricoeur and narrative: Context and contestation*, ed. Joy Morny, 109–120. Calgary, Alberta: University of Calgary Press.

Brown, Roger, and Albert Gilman. 1960. The pronouns of power and solidarity. In *Style in language*, ed. T. Sebeok, 253–277. Cambridge, MA: MIT Press.

Brown, Roger, and Marguerite Ford. 1961. Address in American English. *Journal of Abnormal & Social Psychology* 62:375–385.

Clayton, Susan, and Susan Opotow. 2003. *Identity and the natural environment: The psychological significance of nature*. Cambridge, MA: MIT Press.

Cornog, Evan. 2004. *The power and the story*. New York: Penguin.

Dauenhauer, Bernard P. 1997. Ricoeur and political identity. In *Paul Ricoeur and narrative: Context and contestation*, ed. Joy Morny, 129–140. Calgary, Alberta: University of Calgary Press.

Faver, Catherine A. 2001. Rights, responsibility and relationship: Motivations for women's social activism. *Affilia* 16:314–336.

Foss, Sonja K. 1996. *Rhetorical criticism: Exploration & practice.* 2nd ed. Prospect Heights, IL: Waveland Press.

Haker, Hermann. 2000. Narrative and moral identity in Paul Ricoeur. In *Creating identity*, ed. Hermann Haring, Maureen Junker-Kenny, and Dietmar Mieth, 59–68. London: SCM Press.

Harré, Rom. 1986. An outline of the social constructionist viewpoint. In *The social construction of emotions*, ed. Rom Harré, 2–14. New York: Basil Blackwell.

Hobson, Barbara. 2003. *Recognition struggles and social movements: Contested identities, agency and power.* Cambridge: Cambridge University Press.

Jamieson, Dale. 1992. Ethics, public policy and global warming. *Science Technology Human Values* 17:139–153.

———. 2007. The moral and political challenges of climate change. In *Creating a climate for change: Communicating climate change and facilitating social change*, ed. Susanne C. Moser and Lisa Dilling, 475–482. Cambridge: Cambridge University Press.

Jasinski, James. 2001. *Sourcebook on rhetoric: Key concepts in contemporary rhetorical studies.* Thousand Oaks, CA: Sage.

Lichterman, Paul. 1996. *The search for political community: American activists reinventing commitment.* Cambridge: Cambridge University Press.

Lutz, Catherine. 1988. *Unnatural emotions: Everyday sentiments on a Micronesian atoll and their challenge to western theory.* Chicago: University of Chicago Press.

Mason, Michael. 2001. Transnational environmental obligations: Locating new spaces of accountability in a post-Westphalian global order. *Transactions of the Institute of British Geographers* 26 (4): 407–429.

Melucci, Alberto. 1994. A strange kind of newness: What's "new" in social movements? In *New social movements: From ideology to identity*, ed. Enrique Larana, Hank Johnson, and Joseph R. Gunsfield, 101–130. Philadelphia: Temple University Press.

———. 1996. *Challenging codes: Collective action in the information age*. Cambridge: Cambridge University Press.

Pearce, W. Barnett. 1995. A sailing guide for social constructionists. In *Social approaches to communication*, ed. W. Leeds-Hurwitz, 88–113. New York: Guilford Press.

Pol, Enrique. 2002. The theoretical background of the city-identity-sustainability network. *Environment and Behavior* 34:8–25.

Potter, Sulamith Heins. 1988. The cultural construction of emotion in rural Chinese social life. *Ethos* 16 (2): 181–208.

Regan, Charles E. 1996. *Paul Ricoeur*. Chicago: University of Chicago Press.

Ricoeur, Paul. 1983. *Time and narrative*. Vol. 1. Trans K. McLaughlin and D. Pellauer. Chicago: University of Chicago Press.

———. 1989. Humans as the subject matter of philosophy. In *The narrative path*, ed. T. P. Kemp and D. Rasmussen, 80–101. Cambridge: Cambridge University Press.

———. 1991a. Life in quest of narrative. In *On Paul Ricoeur: Narrative and interpretation*, ed. David Wood, 20–33. New York: Routledge.

———. 1991b. Narrative identity. In *On Paul Ricoeur: Narrative and interpretation*, ed. David Wood, 188–200. New York: Routledge.

———. 1992. *Oneself as another*. Chicago: University of Chicago Press.

Shenhav, Shaul R. 2004. Once upon a time there was a nation: Narrative conceptualization analysis, the concept of "nation" in the discourse of Israeli Likud Party leaders. *Discourse and Society* 15 (1): 81–104.

———. 2005. Concise narratives: A structural analysis of political discourse. *Discourse Studies* 7 (3): 315–335.

———. 2006. Political narratives and political reality. *International Political Science Review* 27:245–262.

Snow, David A., and Doug McAdam. 2000. Identity work processes in the context of social movements: Clarifying the movement/movement nexus. In *Self identity and social movements*, ed. Sheldon Stryker, Timothy J. Owens, and Robert W. White, 41–67. Minneapolis: University of Minnesota Press.

Snow, David A., and Pamela E. Oliver. 1995. Social movements and collective behavior: Social psychological dimensions and considerations. In *Sociological perspectives on social psychology*, ed. Karen S. Cook, Gary Alan Fine, and James S. House, 571–599. Boston: Allyn and Bacon.

Stryker, Sheldon. 2000. Identity competition. In *Self identity and social movements*, ed. Sheldon Stryker, Timothy J. Owens, and Robert W. White, 21–40. Minneapolis: University of Minnesota Press.

Stryker, Sheldon, Timothy J. Owens, and Robert W. White. 2000. Social psychology and social movements: Cloudy past and bright future. In *Self identity and social movements*, ed. Sheldon Stryker, Timothy J. Owens, and Robert W. White, 1–17. Minneapolis: University of Minnesota Press.

Teske, Nathan. 1997. *Political activism in America: The identity construction model of political participation*. Cambridge: Cambridge University Press.

Thompson, Leigh L., and Richard Gonzales. 1997. Environmental disputes: Competition for scarce resources and clashing of values. In *Environment, ethics, and behavior: The psychology of environmental valuation and degradation*, ed. Max Bazerman, David Messick, Anne Tenbrunsel, and Kimberly Wade-Benzoni, 75–104. San Francisco: New Lexington Books.

Venema, Henry Isaac. 2000. *Identifying selfhood, imagination, narrative, and hermeneutics in the thought of Paul Ricoeur*. New York: State University of New York Press.

Wheaton, Belinda. 2007. Identity, politics, and the beach: Environmental activism in Surfers Against Sewage. *Leisure Studies* 26 (3): 279–302.

Chapter 11

Environmentalism 2.0

New Forms of Social Activism

Leah Sprain, Nils Peterson,
Micheal Vickery, and Julie Kalil Schutten

Before going public with Step It Up 2007 (SIU) in the online environmental magazine *Grist*, the national organizers worked to build support for their campaign by sending e-mails to environmental organizers and friends. What started as a "chain e-mail," according to one national organizer, became a National Day of Climate Action with over 1,400 local events. How did an e-mail message lead to the largest political action on climate change in U.S. history? According to the national organizing team, they facilitated an "open-source campaign," and participants did the rest. Putting trust in people across the country, the national team made a "minimal ask": make a banner, take a picture, and send the picture to Congress. From there, local organizers were free to create the events as they saw fit. In the words of one national organizer, it was "like organizing a potluck." Some actions were small gatherings of friends, including a literal potluck organized by Hampshire College students (McKibben et al. 2007), while others looked like political rallies, and some had the look and feel of a carnival. Organizers and participants at local actions responded to the national invitation by speaking about how citizens can rally to address climate change through political and individual actions, drawing on positive messages about solutions, the green economy, new technologies, and energy consumption.

Prior to the National Day of Climate Action, when local organizers encountered problems setting up their actions, they asked the national organizers for support, often by sending an e-mail asking for a press release, banner, or advice. Through these requests and interactions, the national team developed shared resources for organizers to use (e.g., templates for press releases, sample banners, action ideas, etc.) and posted them on the SIU Web site (see chapter 9 for a discussion of these resources as a "cookbook" for organizing). Local organizers used some of the national SIU tools and created some of their own (e.g., SIU stationary, carbon credits to offset attendee transportation). National organizers suggest these creative moments indicate that all participants had ownership of SIU.

This account of how SIU came into being sounds similar to the way in which Eric Raymond wrote about the development of Linux software:

> The most important feature of Linux...was not technical but sociological. Until the Linux development, everyone believed that any software as complex as an operating system had to be developed in a carefully coordinated way by a relatively small, tightly knit group of people. This model was and is typical of both commercial software and the great freeware cathedrals...Linux evolved in a completely different way. From nearly the beginning, it was rather casually hacked on by huge numbers of volunteers coordinating only through the Internet. Quality was maintained not by rigid standards of autocracy but by the naively simple strategy of releasing every week and getting feedback from hundreds of users within days... (1997, 16)

Raymond's view of how Linux developed through an open, accessible process where people had equal opportunity to access, build, alter, and hack software seems to apply in provocative ways to the open-source campaign through which SIU was organized and enacted. The events and actions that ultimately constituted SIU were the result of diverse and unpredictable responses to an open invitation, disseminated through a distributed network of social actors and interactions to organize and act locally. SIU suggests that the localized forms of activism and rhetoric that constituted SIU may be more analogous to the messiness and excitement of an open-air bazaar of environmental activism than to the formal environmentalism practiced in the more monolithic cathedrals of environmental advocacy and policy.

In addition to considering it as an open-source campaign, SIU should also be viewed within the context of the changing rhetoric of environmentalism in the United States. Robert Cox (2006) argued that the environmental movement in the United States has passed from arguments about preserving and conserving natural beauty and resources, to indictment of despoiling influences of business and industry, to a broad political conversation about environmental justice, into a fourth stage that must now confront problems of the global commons and address fundamental questions about sustainability and healthy communities. This final task is made more difficult by what some have seen as the failure of confrontational, apocalyptic, and institutionalized environmental rhetoric to yield fundamental changes in social, political, and economic institutions (Segal 1991; Shellenberger and Nordhaus 2004; Nordhaus and Shellenberger 2007). If the concerns raised by critics about the efficacy of environmental arguments (Segal 1991) and environmental advocacy organizations (Shellenberger and Nordhaus 2004) are warranted, we may ask the question, will new discursive and organizational practices begin to emerge in the public sphere?

The new means of social organization and mobilizing that SIU used to address climate change suggest important opportunities and challenges for those who organize and participate in social movements and for those who wish to understand the exigencies and constraints that may shape environmental activism and rhetoric. A summary comparison of establishment environmentalism to the rhetorics observed at SIU suggests (1) a shift away from environmental rhetoric intended to express outrage, confront bad environmental actors, and enact conventionally understood roles and shared motives within an established environmental movement, and (2) a shift toward forms of environmental activism based on decentralized but coordinated actions by multiple, unaffiliated, even individual environmental actors (table 1). We argue that SIU is an early manifestation of a new environmental rhetoric of individual and local possibilities—what we call *environmentalism 2.0*—from which structural, social, and political change can be initiated and sustained. The new methods for social organizing and new forms of environmental discourse embodied by SIU may reflect the first social movement tailored to this emerging form of environmentalism. By environmentalism 2.0, we mean to differentiate the organizational and discursive phenomena that were observed in SIU from traditional environmental activism in the United States (table 1, items 1 and 2). Of course, these systems share several fundamental strategies for environmental advocacy (Cox 2006; table 1, items 3, 4, and 5).

TABLE 1. Systems of environmental advocacy.

Establishment Environmentalism	Environmentalism 2.0
1. Central organization directs campaigns and goals 2. Draws on rhetorics of apocalypse and limits 3. Campaigns target institutions and decision makers 4. Mobilization to respond to particular environmental problems (situation-driven) 5. Strategies designed on the basis of answers to three fundamental questions: What exactly do environmental leaders and their followers want to accomplish? Which decision makers are the specific targets of campaign actions? What will work to persuade targeted decision makers?	1. Campaign control and goals shared within an open-source network of individuals, local organizations, and the central organization (open-source) 2. Draws on rhetorics of sustainability and hope 3. Campaigns target institutions and decision makers 4. Mobilization to respond to particular environmental problems (situation-driven) 5. Strategies designed on the basis of answers to three fundamental questions: What exactly do environmental leaders and their followers want to accomplish? Which decision makers are the specific targets of campaign actions? What will work to persuade targeted decision makers?

To understand environmentalism 2.0, we focus on the key differences between establishment environmentalism and this emerging mode of social organizing (items 1 and 2 in table 1). We begin by examining the dynamics and implications of thinking about the entire SIU event—not just the national organizing team—as a network organization in which the rhetorical artifacts of the events are manifestations of an open-source campaign of invitation and response. The section *Network Organization for Environmental Action* explains how open-source, networked campaigns bring together a diverse set of interests and participants but do not facilitate a single, consistent message across actions. In the section *From an Apocalyptic Narrative to a Rhetoric of Sustainability*, we demonstrate that SIU organizers and participants overwhelmingly drew on a rhetoric of sustainability to mobilize action on climate change. Having described these two key aspects of environmentalism 2.0, the final section—*Practical and Critical Implications*—analyzes the implications of environmentalism 2.0 for advocates and critics. For advocates, we consider the advantages and limitations of open-source campaigns and the rhetoric of sustainability. For theorists and critics, we discuss how environmentalism 2.0 provides a new critical lens to evaluate movement building and strategies.

Network Organization for Environmental Action

SIU actions varied drastically from location to location. Some actions privileged the collective call to action to reduce emissions by 80 percent; other events focused on local politics, local organizations, and local concerns. Paid staff people for environmental organizations quickly and efficiently organized some actions; single individuals threw together other actions. Some events attracted thousands of people; other events were cancelled at the last minute, moved locations without telling people, or consisted of fewer than ten people. Some participants clearly articulated the goals and purpose of SIU as established by the national organizers; other people at the events said they had not heard of SIU and did not know what it was. This mix could be considered organizational chaos and therefore be deemed a failure, especially when using an establishment-environmentalism frame. The notion of the bazaar, however, helps us make sense of the diversity of local events in an open-source campaign.

At the beginning of this chapter, we introduced the idea that SIU operated like an open-air bazaar, an open mix of people contributing in different ways to the collective project of pushing for political action on climate change. One Austin participant captured this sense by describing SIU as

> a very open-ended proposition nationally. They're like, "Hey, local groups, do whatever you want, the theme is climate change, take a hike somewhere, go out in a canoe, go to a congressman, but do something and tell me that this climate change thing is here, and step it up, man. We got to start doing something."

This comment captures the invitational nature of SIU's organizing rhetoric, encouraging creativity around a central theme. In their book, the national organizers noted that SIU "has been called the first large-scale 'open-source' demonstration, a web-based collaborative effort that allows anyone involved to add their flavor to the project" (McKibben et al. 2007, 110). One key aspect of open-source organizing involves using digital technologies to facilitate transparent and collaborative communication that enables user-generated content.

The very nature of an open-source campaign means that diverse people and organizations participated in SIU (also discussed in chapter 5, chapter 6, and chapter 7). For example, at the Raleigh event, there were booths with representatives from Interfaith Power, Fair Trade Federation, John Edward's presidential campaign, Population Connection, Environmental North Carolina, Code Pink, and even a local university MBA program. Not all of these groups shared a single message, even at an event united for a single cause (also discussed in chapter 4). When asked why he came to a Seattle event, a Greenpeace worker launched into a pitch on the need to stop Japanese whale harpooning. Only after being prompted by an interviewer about global warming did he switch to a pitch on climate policy. Researcher-observers at events in Texas reported not hearing any mention of the national campaign goal (80 percent reduction by 2050) at local SIU actions, even though they were listening for it. Yet from the perspective of a bazaar, the diversity of interests and motives for attending is a potential advantage: participants can contribute to a collective goal without subjugating their primary concerns to those of the campaign. Moreover, in some cases, diverse interests can be connected in interesting ways to climate change, potentially increasing the base of the movement.

Viewing SIU as building a network organization through an open-invitation bazaar of environmental activism explains why all of the events were unique—individual creativity was encouraged. With few external requirements, however, there was also a lack of consistency or collective identity among participants. Unlike the Linux software example, where everyone is refining a single product, the SIU products were intentionally unique, thus making an overall claim about what happened on the National Day of Climate Action highly problematic. Despite these limitations, SIU demonstrated the effectiveness of using online social-networking technologies to quickly organize and coordinate geographically diverse and simultaneous face-to-face actions.

From an Apocalyptic Narrative to a Rhetoric of Sustainability

The presence and predominance of an "environmental rhetoric" predicated on traditional models of persuasion and advocacy campaigns have been noted by many observers (Segal 1991; Killingsworth and Palmer 1991, 1996; Shellenberger and Nordhaus 2004). The modern environmental movement garnered its legitimacy and effectiveness through rhetorical and organizational strategies designed to confront and challenge what Pirages and Ehrlich (1974) labeled as the "dominant social paradigm" of the age. Cox (2006) characterized the environmental rhetoric that emerged during the early stages of the environmental movement as "insurgent discourses" intended to identify, challenge, and assert alternatives to the taken-for-granted values and environmentally destructive practices of this dominant paradigm (e.g., rapacious growth, malignant materialism, blind faith in science and technology, laissez-faire economics, etc.). Segal (1991) argued that this rhetoric rested on conflicted appeals to scientific credibility by sources who represented environmental-advocacy groups and on logically problematic claims that future catastrophes could be extrapolated from current population and pollution measurements. The emotional appeal of this rhetoric was grounded in apocalyptic narratives of ruin and "collapse" (Killingsworth and Palmer 1996; Diamond 2005). Over time, insurgent rhetorics and public protestations were joined by institutionalized and bureaucratized environmental rhetorics of large and well-financed environmental organizations, lobby groups, and newly minted government agencies (Shellenberger and Nordhaus 2004; Nordhaus and Shellenberger 2007). This establishment-environmental rhetoric included apocalyptic, complaint-based discourse that sought to limit human intrusions on nature (Killingsworth and Palmer 1991). A common topic was pollution, and the focus was on stopping, regulating, and preventing environmental contamination (e.g., Carson 1962). In this framework, economic growth itself was part of the problem. Environmentalist activists and academics blamed environmental degradation on laissez-faire capitalism and a treadmill of energy- and resource-intensive processes of industrial manufacturing (Schnaiberg 1980; Czech 2000). Traditional environmentalism defined itself by rejecting anthropocentric tendencies, viewing humans (both their population and consumption) as the primary threat to a delicately balanced nature, and predicting a looming apocalyptic future if human threats to the environment were not stopped (Catton and Dunlap 1978; Dunlap et al. 2000).

Recent calls for refocused and reenergized "environmentalism," in contrast, have emphasized the need for imaginative, hopeful, future-oriented discourse (Nordhaus and Shellenberger 2007; Shellenberger and Nordhaus 2004) and for new envisioning of ways to create positive and sustainable changes for the environment rather than just working to stop negative environmental impacts. In this vision, environmental goals are met by working to create economic prosperity, develop broad coalitions, and support established social values.

Working from these categories, we examined all of the materials that our national resource team collected while attending the SIU actions (including interviews, speech transcripts, and field notes created by researchers). By an estimated ratio of 5:1,[1] participants at SIU events, local organizers, featured speakers, and national organizers more often talked in terms of hopeful discourses focused on sustainability than in terms of established environmentalism. To be sure, there were mentions of doomsday politics. For example, a Duke professor spoke at an action in Raleigh of the damage caused by Hurricane Katrina. But these instances were exceptions to the rule of focusing on sustainable development, positive individual actions to reduce energy consumption, or problem solving through technology. The basic appeals employed by SIU speakers centered on (1) an emphasis on solutions to problems of global warming rather than indictments of bad actors, (2) the strong advantages of transforming to a "green economy," (3) the efficacy of technological solutions to the problem of carbon emissions, and (4) the importance of personal responsibility for energy use and individual efforts to affect global warming.

In exploring the tactics of old establishment environmentalism, we see that when some speakers did use apocalyptic appeals, it did not always work to their advantage. For example, in San Antonio, a state representative engaged the crowd with a series of call-and-response questions:

> Does it make you proud to know Texas is leading the nation in the amount of greenhouse-gas emissions, per capita, in an overall, than any other state? Does that make you proud? (*No!*) Does it make you proud to know that Texas is the leader, number one, in emitting known carcinogens into our air? Does that make you proud? (*No!*) Does it make you proud to know that we are number one in injecting pollution underground? (*No!*) Does it make you proud that we lead the nation in spilling polluted materials into our waters? (*No!*)

These questions typify a focus on pollution as the rallying point of environmentalism. Yet, as one researcher noted in her field notes, proclaiming the negative is not the best way to rally people together:

> Since he had just told us about Texas's bad air quality and how damaging it was to children's health, it seemed that everyone was supposed to respond, "no," and we did. But that doesn't seem like a smart political thing to do. After two or three rounds, a couple of the people standing behind the speakers grimaced every time he asked a question that set people up to shout "no!" They looked in a guarded way at each other also. I think they were thinking this was a bad move.

The problem with this strategy is not simply that it focuses on pollution. Rather, it focuses on the negative. It reminds people that they are the problem. As a call-and-response, it makes people shout out how bad things are over and over and over again before offering a solution.

In comparison, the discourse of sustainability is future-oriented; it is focused on solutions rather than dismal realities. The most common theme across SIU events was that *green economic growth is part of addressing climate change*. Events focused on how economic prosperity and curbing climate change could be mutually reinforcing. While the Texas state representative made the San Antonio crowd wince with negative messages, the mayor of Austin discussed his plan for cutting energy costs by building more efficient buildings. "We are trying to show that you can have a strong economy and a sustainable one," he proclaimed. A participant in Salt Lake City echoed this message, talking about the importance of "being able to integrate economy with ecology." In Lansing, Michigan, the mayor admitted, "Before, I thought I could get economic development. See, I was obsessed with economic development. I was too busy to really consider all this [environmental, sustainability] stuff." But he went on to explain how his outlook had changed: "The reality is, sustainable economic development is where it's at. And, you do the right thing and, and it makes good sense economically…It's the growth model for the future. So, it will pay off. It will pay dividends."

This positive construction of the threat of global warming as an opportunity for economic growth is further developed through claims that *new technologies can solve climate change*. This theme was clearly voiced, for example, in Seattle. One participant, who attended some of the organizing meetings before the day of action, stated an often-heard refrain: "We have the technology to solve these problems. We're here today about solutions." Another suggested, "Americans are innovators. We have the resources to change our energy policy and uses, and we can stop our dependence on harmful forms of energy." These interview comments mirror a Washington state congressman's speech that repeatedly affirmed technology as the source of solutions. He mentioned Washington businesses that are developing new technologies and outlined national policy to support technology development:

> Our research in clean energy to deal with global warming has been slashed by 60 percent. That is the attitude of the ostrich. It is time to increase that by five times and adopt the attitude of the American eagle and create new technologies with our taxpayer dollars.

The promise of technology was likewise sung in Austin, Lansing, San Antonio, Salt Lake City, and Spokane. Electric cars, for example, were on display in Raleigh, Salt Lake City, Bloomington, and Seattle.

In line with the potential economic benefits of global warming, the SIU actions focused on *positive vision for the future*, *celebrating solutions*, and *solving problems*. As the Salt Lake City mayor stated, "It's not only about taking care of the negatives, the challenges that are facing us, it's about embracing the opportunities." In Seattle, the master of ceremonies declared,

> We know what it's going to take to stop global warming. What's between us and the solution to global warming…it's not technology. It's not economics. It is the sheer force of human will. It is our determination. Our vision. The guts that we have to build a better future for ourselves and our kids. That's the only thing separating us right now from the solution. And all across America

> today, we are putting together that sheer force of human will. And it will not stop until we solve global warming.

Here, the SIU actions themselves represented the vision and will necessary to address, even solve, global warming. In Lansing, organizers took it a step further by inviting participants to take the microphone and share what they were doing personally to "step it up," that is, to "live sustainably." Adults and children shared their strategies, from selling SUVs (in an auto state) to riding bikes to improving energy efficiency in appliances.

Together, these themes demonstrate that SIU reflects a shift in environmental discourse from negative to affirmative arguments about environmental problems and an explicitly positive visioning of the future. Notably, these changes were manifested at multiple actions and at multiple levels within the emerging and shifting organization of the events. We recognize that these rhetorical tactics can gloss over crucial material, political, and economic realities. In the following section, we shift to considering the advantages and limitations of environmentalism 2.0.

PRACTICAL AND CRITICAL IMPLICATIONS

By arguing that SIU represents environmentalism 2.0, we are not suggesting that every organizational and discursive detail is new, different, or unlike any previous modes of activism. National and local organizers were clearly informed by the history of environmental and social movements in the United States. The model of local events on a shared day nationally reflects the first Earth Day, for example. The sights and sounds of SIU actions seem similar to other protest events, to the point of adapting popular signs (e.g., "Make love not climate change") and chants (e.g., "Hey hey, ho ho, climate change has got to go") from the global social-justice movement, civil rights movement, and antiwar rallies. The key differences were a shift in how the campaign and its goals were controlled and what rhetorics were used to mobilize participants (table 1, items 1 and 2). These differences have significant practical implications for those hoping to organize movements and theoretical implications for those studying movements.

Implications for Activists and Organizers

Activists and organizers can continue to use the tools, rhetorical appeals, and organizing philosophies that mark environmentalism 2.0 or can combine these approaches with methods from establishment environmentalism. To inform these choices, this section discusses the positive and negative aspects of environmentalism 2.0, particularly the switch to a network mode of organizing and a rhetoric of sustainability.

Network Organizing

Open-source networks are communal, benefiting from the collective intelligence and participation of large groups. It is the most efficient way of problem solving. The concept of open-source has been adopted into other realms, including politics (Hindman 2006) and cultural production (see Harrold 2007). The 2004 Howard Dean presidential campaign and MoveOn.org are examples of open-source politics, where online forms were used to bring together local groups to participate in and further political campaigns. These experiments with open-source models suggest that online technologies can be used to mobilize large groups of people. Nonetheless, having several key traditional players (e.g., established politicians, charismatic leaders, well-known authors) is essential to success (Hindman 2006). Often, a small core of elite professionals is necessary to organize the central framework of a political campaign. Open-source organization depends on the strength of the original architecture and the presence of a charismatic leader to promote the project (Raymond 1997). For SIU, Bill McKibben functioned as a central leader, and individual participants and local organizers (such as the Salt Lake City mayor) attributed their participation directly to McKibben's influence. While all the local actions were products of local creativity, the national campaign goal was determined by a small group of organizers without input from local groups.

The SIU coalition was similar to the global social-justice movement (also called the antiglobalization movement), where environmental, labor, and human-rights activists joined together to support actions and campaigns using social networks and online technologies. Since protests in Seattle, Prague, Quebec, and Barcelona, activists have used e-mail lists, Web pages, and open-editing software to organize actions, share information, and coordinate direct action (Juris 2005). Lance Bennett (2003) has argued that networked activism organized online changes the political game by making political processes more accessible for resource-poor players. This might explain how a group of recent college graduates could plan a national day of action to address climate change. Open-source campaigns provide multiple sources of entry for individuals who might want to get involved through any of the participating organizations or nodes in the network. SIU took this approach one step further by allowing individuals to sign up to organize an event online rather than requiring organizational affiliation as a prerequisite for participation as a leader. By filling out the online form or e-mailing the national organizers, local SIU organizers instantly became part of the campaign. Thus, the campaign organized itself via communication systems as opposed to the communication reflecting existing political organization. Yet, Bennett notes this approach makes it difficult to control or achieve coherent collective-identity frames, which have traditionally been seen as crucial to social-movement mobilization (see Klandermans 1997). In light of this challenge, open-source campaigns may face pressure to form more institutional structures of organization, making the benefits of an open-source approach short-lived.

Open-source organizing encourages and enables broad, easy-to-accomplish "participation." This provides space for diverse motives among movement participants to be expressed and lowers the accessibility costs of movement participation, inviting new people to participate who might not typically see themselves as activists. Rather than making people uncomfortable by forcing a party line, a bazaar thrives on the creativity, energy, and contributions of members. This constant influx of input enables a bazaar to adapt and change to meet the needs of participants. As Philip Brick and R. McGreggor Cawley stated in their argument for environmental advocates to expand their tent to achieve policy objectives, "diversity is widely recognized as a positive force in the life of ecosystems. Our analysis suggests much is the same for political ecosystems" (2008, 216). A bazaar also makes the movement more resistant to institutionalization, as its strength lies in grassroots support rather than central leadership likely to become an institution over time.

The downside of the network organization is that participant members may not share a vision of the movement's purpose or goals (see chapters 4, 5, and 10 for discussions of this same issue). An open-source campaign necessarily gives over some control of the campaign goal even when a central goal is established nationally. This can yield a lack of consistency as participants express different reasons for participating and define the movement itself in different terms, sometimes doing so in ways that actually undermine or contradict the movement's message. Broad coalitions can decenter the focus on key issues, such as taking the environment out of the environmental movement (Blühdorn and Welsh 2007), and undermine core philosophies and values that traditionally mobilized a movement. Not only can this be confusing to outside publics and the media, it also sends an ambiguous message to policy makers that the movement demands are not consistent or universally supported by action participants. Open-source, bazaar-like network organization makes it easy to attract people for one-time actions. But this participation may not lead to sustainable structures of activism or to sharply focused and sustained efforts that may be needed to produce policy change. Indeed, SIU2, the second day of climate-change action in November 2007, inspired only 305 confirmed events, as compared to 850 confirmed events for SIU1.

Rhetoric of Sustainability

By replacing apocalyptic millennialism with a rhetoric of sustainability, discourses in environmentalism 2.0 focus energy and efforts on creating a better future rather than lamenting loss of a better past. This reflects what Stacey Sowards (2006) called the rhetoric of perpetual potential, which creates motivation for change by avoiding cynicism, pessimism, and passivity. Working with dominant social values and structures, such as economic gain through free markets and technological solutions, this discourse is probably more palatable to most Americans because it does not call for sacrifice. Instead, it subtly allows that people can curb climate change without curbing consumption (at least certain types of consumption). This possibility is more comfortable, less scary, and more mainstream. The politics of possibility helps foster a better image of environmentalism. Instead of fighting as antagonists, environmentalists speaking in positive terms are collaborators working together with other interests and organizations for common goals.

Yet, by embracing market solutions, this rhetoric also trivializes the environmental and social destruction wrought by free markets and globalized capitalism (Schnaiberg 1980). If economics provide the solutions, then economic and material desires cannot be fully acknowledged as part of the problem. This weakens motivation to sacrifice in cases when it is actually needed, such as rejecting the American car culture in favor of comprehensive public transportation rather than endorsing green cars that save gas and make money but do little to challenge the entitlement and convenience expected by the single car-driver. By making everything seem possible, positive discourses dilute the urgency and motivation associated with a focus on imminent destruction. In the attempt to make people believe that solving global warming is possible, advocates may fail to acknowledge the scientific realities and the enormity of global warming.

An exclusive focus on positive discourses robs the movement of a key power: the ability to confront and criticize. Without confrontational rhetoric, a movement can struggle to fundamentally change the status quo, instead getting trapped in reformist refrains that do not go far enough to alter the fundamental problem. Not only can these approaches fail to achieve the desired political ends, they can also be easily co-opted by outside interests. For example, many corporations have tried to cash in on the green market by making themselves appear more environmentally friendly through greenwashing, creating marketing campaigns that highlight environmental friendliness but often do little to alter fundamental business practices. Positive rhetorics can feed these corporations by arguing in favor of environmental and economic cooperation while staying quiet about questionable alliances in order to remain positive.

The potential benefits of environmentalism 2.0 depend on whom a movement wants to attract. Mass audiences who do not identify as environmentalists might be attracted to positive, hopeful discourses, whereas existing environmentalists might find them empty of meaning, a sellout to globalized capitalism. This is the issue of framing that gets discussed in detail in chapter 4. The rhetorical choices of a movement have implications beyond attracting or alienating supporters. Rhetoric helps shape the expectations, motivations, and actions of movement participants. Discourses of sustainability enable and encourage people to address global warming through technology and green businesses while reminding themselves that curbing global warming is possible. If this positive thinking has no basis in environmental realities, people may fail to make substantive behavior changes, call for policy solutions, or sacrifice to the point of personal discomfort. Environmental discourse will likely fail to identify problems, mobilize action, and promote sustainable positive change if it does not balance unrelenting, empirically sound analysis of the environmental impact of taken-for-granted life practices with unapologetically positive framing of a future in which more sustainable life practices have become the norm.

To best manage this tension, activists should be mindful of how both positive and negative messages fit into their demands for social change. Fundamental problems need to be acknowledged and addressed. For example, global warming will have negative impacts on ecosystems and the people who rely on them. But focusing exclusively on the problem might not be the best way to motivate people toward actions to curb climate change. Instead, acknowledging problems can be used as leverage for a move to sustainable economic systems, greener technologies, less consumptive lifestyles, and new definitions for the quality of life—and rhetorical strategies can focus on these outcomes and how to realize them. Positive messages must not mean forgoing real change, however.

Positive discourses do not require pandering to the lowest common interest, agreeing to work with everyone at any cost. Activists can and should criticize those who attempt to co-opt their visions.

Critical and Theoretical Implications

There is a necessary relationship between the nature of global environmental problems and the forms of human organization and communication that are likely to successfully address those problems. It is important for theorists, critics, and practitioners to pay close attention to the consonance or dissonance between the processes of communication that are intended to organize responses to interdependent and rapidly changing problems and the processes of communication that tend to make collaboration and change more difficult. One lesson that could be drawn from SIU is that the bazaar of activism resulting from open-source processes may look disorganized and short-lived but may, in fact, cultivate an adaptive mentality that facilitates evolution into new campaigns and efforts.

SIU indicates ways that rhetorical appeals for political mobilization and action may become fundamentally intertwined with and, in some ways, indistinguishable from an ongoing process of networked grassroots organizing. The "new" form of environmental organization may be the fundamental communicative action of "organizing" itself. The most powerful rhetoric at work may simply be the always-open invitation to participate and cooperate. Howard Rheingold coined the phrase "smart mobs" to describe the potential that exists at any time among a diversity of digitally networked communicators to cross some threshold of participatory engagement and foster "outbreaks of cooperation" (2002, 174). Rheingold's analysis of the smart-mob phenomenon makes clear how "sudden epidemics of cooperation" (ibid.) can emerge, but he also pointed out that these are largely ad hoc social structures and are always morphing into new forms, with new thresholds of cooperation, as members join and leave the network.

The ad hoc and constantly evolving "organization" that results may just as readily reflect as alter the socioeconomic system in the ways that

Robert Brulle and Craig Jenkins (2006) suggested are desirable. From this point of view, SIU could simply be seen as a proof that ad hoc environmental advocacy to address global warming may more effectively reinforce conventional models of economic (albeit "green") growth than fundamentally challenge the structural and cultural impediments to change in national energy and environmental policy. This perspective, however, may not reflect the emerging and profound influence of social-networking technologies on grassroots responses to a constantly present, always-open invitation to continually organize and act to produce change. Evidence of this new mode of constantly evolving environmental organization and advocacy is reflected in an e-mail distributed by the SIU team on the anniversary of SIU to celebrate the accomplishments of the 2007 event, recognize the collaboration of the SIU organizers with "lots of other grassroots organizers across the country to form 1Sky," and announce a new venture called 350.org, which will serve as an international invitation to organize actions that will affect international negotiations and cooperation on global warming (McKibben 2008). The diversity of rhetoric, organizations, and actions suggested in the existence of a global communication system that can, at any time, foster a sudden epidemic of cooperation is a variable not only to be utilized by social-movement organizers but to be understood by those who would attempt to describe and understand the nature and scope of environmental "movement" that is developing at the grassroots level.

Another important theoretical implication of SIU is that it is impossible to account for the process of organizing and sustaining social movements without recognizing the effect of social-networking technology and the resulting structures of organization and collective action that very quickly shift in membership, focus, and duration. Many scholars have noted the importance of networked interactivity in an increasingly participatory web of networked communication technologies (Stein and Sinha 2006; Rheingold 2002; Frederick 1993). Laura Stein and Nikhil Sinha argued that current scholarship clearly documents the impact of computer networks on "growing trans-border cooperation between various types of citizen groups" on a broad range of issues that include human-rights, peace, and environmental activism (2006, 431). It is not possible to account for the effectiveness of environmental rhetoric without locating the symbolic resources of social influence within the shifting organizational patterns of interaction facilitated through new media. Movements, rhetoric, and communication technology are inextricably connected. This factor is one reason that SIU has to be seen as a manifestation of environmentalism 2.0 that—while it clearly had a moment of materiality as an SIU organization, events, and actions—ultimately receded into constantly shifting organizational structures of new environmentalism.

Endnote

1. We reached this ratio by coding all of the electronic documents collected by the research teams (speech transcripts, interview transcripts, and researcher field notes) along the categories of "establishment environmentalism" and "beyond environmentalism," as defined by Nordhaus and Shellenberger (2007). The ratio refers to both the number of instances and length of these mentions of the rhetoric of sustainability.

REFERENCES

Bennett, W. Lance. 2003. Communicating global activism: Strengths and vulnerabilities of networked politics. *Information, Communication, & Society* 6:143–168.

Blühdorn, Ingolfur, and Ian Welsh. 2007. Eco-politics beyond the paradigm of sustainability: A conceptual framework and research agenda. *Environmental Politics* 16:185–205.

Brick, Philip, and R. McGreggor Cawley. 2008. Producing political climate change: The hidden life of US environmentalism. *Environmental Politics* 17:200–218.

Brulle, Robert J., and J. Craig Jenkins. 2006. Spinning our way to sustainability? *Organization & Environment* 19:82–87.

Carson, Rachel. 1962. *Silent spring*. Boston: Houghton Mifflin.

Catton, William R., and Riley E. Dunlap. 1978. Paradigms, theories, and primacy of hep-nep distinction. *American Sociologist* 13:256–259.

Cox, Robert. 2006. *Environmental communication and the public sphere*. Thousand Oaks: Sage.

Czech, Brian. 2000. *Shoveling fuel for a runaway train: Errant economists, shameful spenders, and a plan to stop them all*. Berkeley: University of California Press.

Diamond, Jared M. 2005. *Collapse: How societies choose to fail or succeed*. New York: Penguin.

Dunlap, Riley E., Kent D. Van Liere, Angela G. Mertig, and Robert E. Jones. 2000. Measuring endorsement of the new ecological paradigm: A revised NEP scale. *Journal of Social Issues* 56:425–442.

Frederick, Howard. 1993. Networks and the emergence of global civil society. In *Global networks: Computers and international communication*, ed. Linda M. Harasim, 283–295. Cambridge, MA: MIT Press.

Harrold, Christine. 2007. *Ourspace: Resisting the corporate control of culture*. Minneapolis: University of Minnesota Press.

Hindman, Matthew. 2006. "Open-source politics" reconsidered: Emerging patterns in online political participation. In *Governance and information technology*, ed. Viktor Mayer-Schonberger and David Lazer, 183–208. Cambridge, MA: MIT Press.

Juris, Jeffery. 2005. The new digital media and activist networking within anti-corporate globalization movements. *The ANNALS of the American Academy of Political and Social Science* 597: 189–208.

Killingsworth, M. Jimmie, and Jacqueline S. Palmer. 1991. *Ecospeak: Rhetoric and environmental politics in America.* Carbondale: Southern Illinois University Press.

———. 1996. Millennial ecology: The apocalyptic narrative from *Silent Spring* to global warming. In *Green culture: Environmental rhetoric in contemporary America*, ed. Carl G. Herndl and Stuart C. Brown, 221–245. Madison: University of Wisconsin Press.

Klandermans, Bart. 1997. *The social psychology of protest*. Oxford: Blackwell.

McKibben, Bill. 2008. A note from Bill McKibben. http://www.350.org (accessed April 14, 2008).

McKibben, Bill, Phil Aroneanu, Will Bates, May Boeve, Jamie Henn, Jeremy Osborn, and Jon Warnow. 2007. *Fight global warming now: The handbook for taking action in your community*. New York: Holt.

Nordhaus, Ted, and Michael Shellenberger. 2007. *Break through: From the death of environmentalism to the politics of possibility*. New York: Houghton Mifflin.

Pirages, Paul R., and Dennis Ehrlich. 1974. *Ark II: Social response to environmental imperatives*. San Francisco: W. H. Freeman.

Raymond, Eric. 1997. *The cathedral and the bazaar: Musing on Linux and open source by an accidental revolutionary*. Sebastopol, CA: O'Reilly and Associates.

Rheingold, Howard. 2002. *Smart mobs: The next social revolution*. Cambridge, MA: Perseus.

Schnaiberg, A. 1980. *The environment*. New York: Oxford University Press.

Segal, Judy Z. 1991. The structure of advocacy: A study of environmental rhetoric. *Canadian Journal of Communication* 16:409–415.

Shellenberger, Michael, and Ted Nordhaus. 2004. The death of environmentalism: Global warming politics in a post-environmental world. The Breakthrough Institute. http://www.thebreakthrough.org/PDF/Death_of_Environmentalism.pdf (accessed April 17, 2008).

Sowards, Stacey K. 2006. Rhetoric of perpetual potential: A case study of the environmentalist movement to protect orangutans. In vol. 3 of *The Environmental Communication Yearbook*, ed. Steven P. Depoe, 115–136. Mahwah, NJ: Lawrence Erlbaum Associates.

Stein, Laura, and Nikhil Sinha. 2006. New global media and the role of the state. In *The handbook of new media*, ed. Leah A. Lievrouw and Sonia Livingstone, 415–442. Thousand Oaks, CA: Sage.

Chapter 12

An Ecologist's Response to the SIU Campaign

The Role of Natural Scientists in Environmental Movements

Markus J. Peterson

It is not true that all science is equal, or that we cannot justly compare the effectiveness of scientists by any method other than a mutual-recommendation system. The man to watch, the man to put your money on, is not the man who wants to make "a survey" or a "more detailed study" but the man with the notebook, the man with the alternative hypotheses and the crucial experiments, the man who knows how to answer your Question of disproof and is already working on it.

—John R. Platt (1964, 352)

> The ecological field-worker acquires a deep-seated respect, or even veneration, for ways and forms of life. He reaches an understanding from within, a kind of understanding that others reserve for fellow men and for a narrow section of ways and forms of life. To the ecological field-worker, the equal right to live and blossom is an intuitively clear and obvious value axiom.
>
> —Arne Naess (1973, 95–96)

The Step It Up (SIU) organizers made use of ecological and climatological data, but established natural scientists[1] did not drive SIU at the national or local level as activists. This situation should come as no surprise. After all, environmental activists typically use data and data interpretations provided by natural scientists, but ecologists[2] and other environmental scientists rarely lead environmental activism (Peterson and Peterson 1996). It may seem paradoxical that ecologists, given their holistic understanding of ecosystems and their passion for nature, seem to avoid leading or even actively participating in environmental activism. I demonstrate, below, that this paradox largely can be explained by the philosophical underpinnings of how ecologists and other natural-science researchers are trained.

To maximize their effectiveness, environmental activists must understand why it is that ecologists, conservation biologists, and other environmental scientists can care passionately about biodiversity conservation, climate change, and other environmental issues, yet logically (from their perspectives) shun public participation in environmental activism. This knowledge is required if environmental activists are to successfully negotiate this paradox and thus obtain unfettered access to natural-science data and, more importantly, knowledgeable interpretation of these data to support their claims in the court of public opinion and policy.

This chapter explores the role of natural science in the SIU campaign and suggests why environmental scientists are reluctant participants in environmental activism. I first briefly review how SIU organizers obtained and utilized scientific knowledge and the role these data played in the campaign. I then address how the philosophy of science, thence scientific training, limits the likelihood that established natural-science researchers will become environmental activists. Given this situation, I address challenges and opportunities environmental activists face as they make use of natural scientists, not only for access to their data, but also for their interpretation of these data. Finally, I close with suggestions for accessing environmental scientists while organizing and advocating environmental action.

SIU and Natural Science

Readers may have noticed that not a single chapter to this point has referenced a natural scientist or specific data produced by such individuals, although the authors of chapter 6, a speaker (chapter 7), and a rally organizer (interlude between chapter 7 and chapter 8) cited the report by the Intergovernmental Panel on Climate Change (IPCC 2007). Does this minimal mention of natural science imply that natural science had little or no influence on SIU? Certainly not. Instead, natural scientists and their data were essentially taken as givens by the national SIU organizers. Local organizers followed suit, although a few rallies included presentations by climatologists and ecologists. My objective here is to make explicit the implicit role of natural science in the "cookbook" on how to organize a rally and other materials made available online by national and local SIU organizers.[3]

The SIU call to action, "Step It Up, Congress: cut carbon 80 percent by 2050," serves as a useful microcosm for evaluating the role of natural science in the SIU campaign, as it summarizes precisely what the national organizers were trying to accomplish. Local organizers were asked to construct a banner featuring this call to action. They then were instructed as follows: "at the peak of your event, gather the largest group you can and take a picture of your group, and **be sure to have your banner front and center** displaying our Call to Action" (Step It Up 2007b; emphasis original). These photographs, preferably taken in iconic places, were then placed on the SIU Web site and sent to Washington, DC, in order "to send the most powerful message to Congress." Essentially, the call to action demands that Congress pass one or more statutes requiring regulatory agencies, such as the Environmental Protection Agency, to make and implement rules that can be expected to result in an 80 percent reduction in carbon (primarily CO_2)[4] emissions in the United States by 2050.

The national SIU organizers did not choose an 80 percent reduction in carbon emissions by 2050 arbitrarily. Instead, they worked with climate-change author Bill McKibben to come up with the magnitude of the decrease in carbon emissions and the timeline required. McKibben, in turn, had developed close working relationships with several leading climate scientists over the years, including James Hansen, director of the NASA Goddard Institute for Space Studies and a leading climatologist. The 80 percent reduction in carbon emissions by 2050 was a best estimate based on available climatological and ecological data. Moreover, under "Why are we asking for 80% carbon cuts by 2050?" the national SIU organizers used Hansen to justify their decision:

> America's foremost climatologist, NASA scientist James Hansen, has said that we have just a few years to start reducing carbon emissions, and he's endorsed our goal of 80% by 2050. That won't prevent global warming—it's already too late for that—but it may be enough to stave off the most catastrophic effects. (Step It Up 2007a)

Thus, what many might have considered draconian reductions in carbon emissions were not only validated by "America's foremost climatologist," but were actually insufficient to prevent any but the most "catastrophic" changes in climate.

Data from natural scientists were required to justify SIU's call for an 80 percent reduction in carbon emissions by 2050, even for many rally leaders. For example, the primary organizer of SIU in Bloomington, Indiana, stated:

> When I first read that Bill McKibben and his team were calling for 80 percent cuts in carbon emissions by 2050, I thought that was pretty extreme. Actually, I assumed they were purposely setting the goal really high, in hopes of getting *something*. But then

> the IPCC reports started coming out and it became clear that 80 by 50 is what's needed, not an exaggeration. (interlude between chapter 7 and chapter 8)

In this case, the carefully considered evaluation of natural-science and economic data compiled by the IPCC (2007) was required to convince this organizer that reductions in carbon emissions of this magnitude were required over a relatively short time frame in order to avoid catastrophic climate change. Similarly, it is unlikely that both Hillary Clinton and Barack Obama would have repeatedly echoed the call for 80 percent reduction in carbon emissions by 2050 unless key staff members had at least some confidence in scientific data supporting this objective.

There are several additional examples of the implicit role of natural-science data in the SIU campaign. For example, the idea of reducing the carbon footprint of individuals, households, municipalities, or states was prominent at several local SIU rallies, including Salt Lake City, San Antonio, and Seattle. This idea was also discussed in thirteen SIU-related blogs and thirty-four Web-based summaries of local rallies linked to stepitup2007.org. Although natural-science data were not referenced, the context makes it clear that writers felt data supported their conclusion that our carbon footprint was too large and that something approaching a carbon-neutral footprint was needed. For example, the Empowerment Institute, in a blog posted by the SIU organizing team, encouraged participants in their "Global Warming Cafés" to take a personal CO_2 reduction pledge at the end of each workshop. Organizers planned to use these pledges collectively as follows when they delivered the SIU message to Congress: "We, the people, have collectively pledged to reduce our own individual carbon footprints by ____million pounds, but as individuals we can only go so far. Now, it's your turn. Step It Up, Congress: Cut Carbon 80% by 2050." This example implies there is some scientifically supported, yet attainable, level of carbon emission below which the most deleterious anthropogenic climate changes will not occur if prompt action is taken. In sum, natural scientists and their data were essential components of nearly all aspects of the SIU campaign at both the national level and at local events, even though scientists and their data were infrequently mentioned explicitly.

A PRIMER ON THE PHILOSOPHY OF NATURAL SCIENCE

The reluctance of natural scientists, including ecologists and others who directly study the environment, to participate in political controversy has long been recognized. Arne Naess, in his keynote address at the Second International Conference of Conservation Biology, directly addressed this issue. Naess stated, "The high-level experts I asked to comment on the eight points of a so-called deep-ecology platform answered in a way favorable to a remarkably strong conservation policy" (1986, 512). He concluded, however, that

> These experts and most others do not propagate their strong views on conservation in public. Why? Here are some possible reasons:
>
> 1. Time taken away from professional work
> 2. Consequent adverse effects of this on promotion and status
> 3. Feeling of insufficient competence outside their "expertise"
> 4. Lack of training in the use of mass media and in facing non-academic audiences
> 5. Negative attitude towards expressing "subjective" opinions and valuations, or violating norms of "objectivity"; reluctance to enter controversial issues
> 6. Fear that colleagues or bosses think that they dabble in irrelevant, controversial fields, and that their going public is due to vainglory and publicity seeking
> 7. Fear of fellow researchers, institution personnel or administrations; fear of the stigma "unscientific" (513)

Naess clearly was aware of the paradox environmental scientists find themselves in: caring passionately about the environment, yet finding it inappropriate to even comment publicly, let alone lead environmental activism. Despite the fact that the proceedings of this conference (Soulé, 1986) essentially became the founding document for the discipline of conservation biology, Naess' critique is just as salient today as it was more than two decades ago. I argue, below, that Naess' reasons for lack of public involvement by environmental experts revolve around how natural scientists have been trained since at least the early 1960s. This includes most natural scientists with active research programs today.

My objective in this section is to provide a primer on the ontological, epistemological, and axiological perspectives inculcated by graduate training in the natural sciences. A basic understanding of this philosophy is required for environmental activists and those studying environmental social movements to understand (1) why natural scientists are so reluctant to become publicly involved in environmental activism, (2) why natural scientists almost never seem certain about anything, and (3) how activists can negotiate the behavioral paradox embodied by today's environmental scientists.

The Nature of Reality

If asked to define reality, most contemporary natural scientists probably would find the question somewhat silly. After all, to them, reality is the state of the material universe around us. Since Aristotle, the empiricist tradition of philosophy has held that material reality was indeed largely independent of human thought and best understood through experience. Natural science is still informed to a large degree through this empiricist perception. Rationalism, however, has an equally long tradition in philosophy. Rationalists such as Pythagoras, Socrates, and Plato argued that the ideal, grounded in reason, was in many ways more "real" than the material. From this perspective, the criterion for reality was not sensory experience, but instead was intellectual and deductive. Certain aspects of the rationalist perspective still are an integral part of modern natural science, such as the importance of inductive, deductive, and retroductive reasoning. It is probably fair to say, however, that most natural scientists are unaware that many contemporary philosophers, social scientists, and humanists maintain that reality is ultimately a social construction (Berger and Luckmann 1966; Lincoln and Guba 1985; Jasinoff et al. 1995), even though philosophers voiced arguments consistent with social constructionism as far back as Heraclitus. This perspective on reality means that natural scientists deal comfortably with things they can measure or address using logic and/or mathematical models, but find the socially constructed realities embodied by social movements and related political controversy not only uncomfortable but even disorienting.

HOW WE COME TO KNOW

In Plato's 369 BC dialogue, *Theaetetus* (1973), Socrates concluded that knowledge was justified true belief. This definition survived without serious challenge for thousands of years until Edmund L. Gettier (1963), in a remarkably brief paper (fewer than three pages), provided what he maintained were examples of beliefs that were both true and justified, but that should not be considered knowledge. Philosophers have been wrestling with the "Gettier problem" since, and have yet to agree on a single definition of knowledge. A second problem with Plato's definition relates to reality. If reality is socially constructed, then the truth regarding this reality must also be a social construct, whether justified true belief or not.

Beginning with the age of Enlightenment, the empiricist tradition of inquiry exhibited new vigor. Important thinkers associated with the maturation of empiricism include Francis Bacon, John Locke, David Hume, and John Stuart Mill. From the empiricist perspective, scientists acquire knowledge only through experience, particularly as gained by observations of the natural world and carefully designed experiments. Also during this period, philosophers informed by the rationalist tradition were busily honing their epistemological perspective. René Descartes, Baruch Spinoza, and Gottfried Leibniz are often associated with this tradition and were responsible for integrating mathematics into philosophy. For rationalists, reason takes precedence over experience for acquiring knowledge, and, in principle, all knowledge can be acquired through reason alone. In practice, however, rationalists realized this was unlikely except in mathematics.

Philosophers during the classical era probably would not have recognized any crisp distinction between empiricism and rationalism.

The seventeenth-century debate between Robert Boyle and Thomas Hobbes regarding Boyle's air-pump experiments and the existence of vacuums fleshed out this division (Shapin and Schaffer 1985). Hobbes argued that only self-evident truths independent of the biophysical could form knowledge, while Boyle promoted experimental verification, where knowledge was reliably produced in a laboratory and independent of the researcher (Latour 1993). Even in the seventeenth century, many rationalists found empirical science important, and some empiricists were closer to Descartes methodologically and theoretically than were certain rationalists (e.g., Spinoza and Leibniz). Further, Immanuel Kant began as a rationalist, then studied Hume and developed an influential blend of rationalist and empiricist traditions. At least two important combinations of empiricism and certain aspects of rationalism followed that are key to modern natural science.

The first of these syntheses, pragmatism, originated with Charles Saunders Peirce in the early 1870s and was further developed and popularized by William James, John Dewey, George H. Mead, and others. Their perspectives on pragmatism were influenced by Kant, Mill, and Georg W. F. Hegel, respectively (Haack and Lane 2006, 10), although other thinkers such as Bacon and Hume undoubtedly were influential as well. Philosophers who consider themselves pragmatists still generally hold that truth, knowledge, and theory are inexorably connected with practical consequences, or real effects. This is key to understanding contemporary science, because, as I detail below, most scientists are trained as pragmatists, whether they know it or not.

The other important philosophical blend of empiricism and rationalism, logical positivism, emerged during the 1920s and 1930s from the work of Moritz Schlick and his Vienna Circle, and Hans Reichenbach and his Berlin Circle (Rosenberg 2000). Logical positivists maintain that a statement is meaningful only if it is (1) analytical (e.g., mathematical equations) or (2) can reasonably be verified empirically. To logical positivists, ethics and aesthetics, for example, are metaphysical and thus scientifically meaningless because one cannot evaluate such arguments analytically or empirically. A common, often implicit assumption of those informed by logical positivism is that given sufficient ingenuity, technology, and time, scientists can ultimately come to understand material reality in all its complexity. Similarly, the notion that researchers should work down to the ultimate elements of the system of interest, then build the causal relationships back up to eventually develop a complete explanation of the universe in question, characterizes logical positivism. The recently completed mapping of the human genome and promised medical breakthroughs related to this genomic map illustrate this tendency.

Evolution of Postpositivism

The publication of Karl R. Popper's *Logik der Forschung* by the Vienna Circle in 1934 (given a 1935 imprint) called into question the sufficiency of logical positivism; after World War II, Popper translated the book into English and published it as *The Logic of Scientific Discovery* (1959). Popper's perspectives were further developed in *Conjectures and Refutations: The Growth of Scientific Knowledge* (1962). Unlike most positivists, Popper was not concerned with distinguishing meaningful from meaningless statements or verification, but rather with distinguishing scientific from metaphysical statements using falsification. For him, metaphysical statements were unfalsifiable, while scientific statements could potentially be falsified. On this basis, scientists should ignore metaphysical contentions; instead, they should deductively derive tests for hypotheses that could lead to falsification. This approach often is called the hypothetico-deductive model of science. Popper argued that hypotheses that did not withstand a rigorous test should immediately be rejected and researchers should then move on to more productive alternatives. He acknowledged, however, that metaphysical statements in one era could become scientific later if they became falsifiable (e.g., due to changes in technology[5]). Under Popper's model of science, while material reality probably exists, the best scientists can do is determine what it is not, by systematically falsifying hypotheses related to the topic of interest. Thus, for Popperians, knowledge regarding an issue is approximated by the explanatory hypothesis that has best survived substantive experimental challenges to date. From this perspective, often called postpositivism, knowledge ultimately is conjectural and can be modified based on further investigation. Under this model, scientific findings on climate change, for example, are seen as conjectural because the exact numbers of temperature increase are subject to further investigation, even if the underlying assumption is that human-caused climate change is a verifiable phenomenon.

Thomas Kuhn, in *The Structure of Scientific Revolutions* (1962), also argued that because science contains a social dimension, it does not operate under the simple logical framework outlined by the logical positivists. Kuhn's model of science includes "normal science," or periods where there is general consensus within a scientific community regarding theory, methods, terminology, and types of experiments likely to contribute useful insights. He argued that although advances occur during normal science, they typically are incremental in nature. Normal science at some point is interrupted by "revolutionary science," where a shift in paradigm occurs, followed by a new version of normal science, and eventually another paradigm shift, and so on. Kuhn argued that transition from an old to a new paradigm is neither rapid nor seamless, largely because the two paradigms are incommensurable. That is, a paradigm shift is not just about transformation of theory, but includes fundamental changes in terminology, how scientists perceive their field, and, perhaps most importantly, what questions are deemed valid and what decision rules and methodological approaches are determined to be appropriate for evaluating scientific concepts. Thence, new paradigms are not simply extensions of the old, but are radically new worldviews, or, as Kuhn put it, "scientific revolutions." Despite the importance of societal influences, Kuhn's model of science resonated with scientists because it provided a workable explanation for the obvious revolutionary changes observed historically in science (e.g., ecology from the Linnaean versus Darwinian perspective; Worster 1994).

Imre Lakatos attempted to resolve the perceived conflict between Popper's falsification and Kuhn's revolutionary models of science in "Falsification and the Methodology of Scientific Research Programmes" (1970). He held that groups of scientists involved in a research program shielded the theoretical core of their efforts with a "protective belt" of hypotheses related to their central area of inquiry. Hypotheses within this belt could be found inadequate while the core theoretical construct remained protected from falsification. This approach protected the core ideas from premature rejection due to anomalies or other problems, something many viewed as a shortcoming of Popper's model of science. Under Lakatos' model, the question is not whether a given hypothesis is false, but whether a research program is progressive, marked by growth and discovery and potentially leading to a shift in paradigm, or degenerative, marked by lack of growth and novel facts and leading to oblivion.

Pragmatism: Knowledge Acquisition Today

Even if we ignore the serious deficiencies Kuhn (1962), Lakatos (1970), Feyerabend (1975, 1978, 1987), and other philosophers of science found in Popper's model of science, and argue that hypothesis falsification defines natural science, there still is a major disconnect between this ideal and what respected natural-science researchers actually do. For example, if the array of study designs commonly employed by ecologists (e.g., Morrison et al. 2008) is evaluated, only a few can be construed as clearly Popperian. This does not imply the remaining designs are not useful. Instead, while Popper's postpositivist model of science sometimes is useful, it often is insufficient for the scope of natural science.

Susan Haack (2003) developed a model of scientific evidence that makes perfect sense in light of the array of study designs actually employed by natural scientists. She argued that natural science is a pragmatic undertaking. Her retro-classical version of pragmatism places science firmly within the empiricist sphere of epistemology as well, due to the criticality of experience. She maintained that natural-science research programs are conducted in much the way one completes a crossword puzzle, with warranted scientific claims anchored by experiential evidence (analogous to clues) and enmeshed in reasons (analogous to the matrix of completed entries). As she put it,

> How reasonable a crossword entry is depends not only on how well it fits with the clue and any already-completed intersecting

> entries, but also on how plausible those other entries are, independent of the entry in question, and how much of the crossword has been completed. Analogously, the degree of warrant of a [scientific] claim for a person at a time depends not only on how supportive his evidence is, but also on how comprehensive it is, and how secure his reasons are, independent of the claim itself. (2003, 67)

Following the crossword analogy, a group of researchers, each with twenty years of experience working with a system of interest, should be able to solve a scientific puzzle more easily than a first-semester graduate student. While many traditional philosophers of science find the social nature of science problematic, Haack maintained it is beneficial. Different investigators employing different study designs and different methodologies might solve different portions of the puzzle. Further, just as someone completing a crossword puzzle might make inappropriate entries and be forced to rethink his or her approach, scientists are fallible as well. In fact, learning from mistaken results, concepts, or theories and having to begin certain aspects of a research program repeatedly characterizes natural science (Hafner and Presswood 1965; Haack 2003). In other words, as pragmatists, natural scientists use mistakes to make progress. For example, mistakes incorporated into early climate-change models predicted that climate change would occur at a slower rate than it is actually occurring. Rather than viewing these mistakes as fatal errors, the IPCC used them, along with new technologies, to produce much more accurate models.

Others noted the puzzle-like nature of natural science prior to Haack. For example, Albert Einstein wrote that

> the liberty of choice [of scientific concepts and theories], however, is of a special kind; it is not in any way similar to the liberty of a writer of fiction. Rather, it is similar to that of a man engaged in solving a well designed word puzzle. He may, it is true, propose any word as the solution; but, there is only *one* word which really solves the puzzle in all its forms. It is an outcome of faith that nature—as she is perceptible to our five senses—takes the

> character of such a well formulated puzzle. The successes reaped up to now by science do, it is true, give a certain encouragement to this faith. (1936, 353–354)

Haack added that despite the fact that our knowledge of the world is to some degree a social construction, "there is a real world knowable to some extent by creatures with sensory and intellectual powers such as ours" (2003, 125).

In sum, Haack provided a pragmatic model for knowledge acquisition that does more than explain why natural scientists commonly employ numerous study designs incongruent with Popper's (1959, 1962) falsification approach. Her pragmatic model of science allowed for material reality on Earth before (and possibly after) human existence. It allows the social aspects of science to be explicitly included within the enterprise. It also permits any study design that can provide reliable solutions to the scientific puzzle, including various types of descriptive research, impact assessment, information-theoretic approaches using model selection, replicated manipulative experiments attempting to falsify retroductively derived research hypotheses, and qualitative designs, to name just a few examples. Finally, according to Haack, truth, knowledge, and theory are inexorably connected with practical consequences, or real effects. Thus, the best way to evaluate a scientifically derived climate-change model would be to see how well it predicts global climate in the future. Similarly, the metric of success for climate-change activists would not be the number of statutes passed by Congress, but whether these laws led to desired environmental outcomes. This pragmatic perspective undoubtedly resonates with environmental activists, ecologists, climatologists, and conservation biologists, for whom practical conservation consequences are the ultimate metric of success.

Values in Science

As discussed above, for logical positivists and postpositivists, ethics and aesthetics are metaphysical and thus unscientific by definition. This does not imply that natural scientists such as ecologists do not perceive aesthetics. A field trip with any ecologist I know would quickly dispel any such notion, a fact well captured by the second epigraph at the beginning of the chapter. Instead, natural scientists are trained to keep aesthetic judgments in their personal domain so they do not impinge on their scientific "objectivity." While activists may want scientists to make ethical and aesthetic judgments, scientists would likely prefer to keep these judgments to themselves for the reasons Naess (1986) suggested.

There are at least three aspects of "value" that natural scientists are trained to address: (1) researcher ethics, (2) personal values researchers bring to science, and (3) how scientists determine the quality of research. Ethical concerns in research run the gambit from humane and appropriate treatment of animal or human subjects to honesty in recording, evaluating, and reporting data to who receives credit for research conducted (e.g., who is listed as first author on a multiauthored article). Ethical concerns were sufficiently important that the U.S. National Academy of Sciences formed the Committee on the Conduct of Science to provide ethical guidelines primarily for students beginning careers in scientific research (Committee on the Conduct of Science 1989), and most research methods courses in the natural sciences now include a module on ethics.

The personal values researchers bring to natural science have long been perceived as problematic by scientists. For example, Thomas Chrowder Chamberlin (1890) argued that scientists should make use of multiple working hypotheses to help protect themselves from their own biases and to ensure they did not develop tunnel vision. John R. Platt (1964) rediscovered Chamberlin's contention and presented it to a new generation of scientists. As science philosophers such as Kuhn, Lakatos, and Feyerabend maintained, however, there inevitably are cultural aspects of natural science regardless of scientists' attempts to be "objective" and "unbiased."

Finally, how do natural scientists determine whether knowledge they are developing matters? To whom does it matter (e.g., themselves, colleagues, some constituency, society)? How do scientists and the public determine whether science is of high quality? These questions defy clear answers. To complicate matters further, they must be answered at multiple scales. One person might think that science that actually changes things on the ground to directly benefit biodiversity conservation, for example, is the most valuable sort of scientific inquiry. If this person is an academician, however, she or he cannot safely ignore what colleagues working for funding agencies, peer-reviewed journals, or tenure and promotion committees perceive to be valuable work. I could make the same sort of argument for scientists who maintain theoretical breakthroughs are the ultimate metric of quality in science. Moreover, society is influenced by, and influences, these value judgments. Society ultimately controls the purse strings for governmental, industrial, and nongovernmental organizations (NGOs) and thus, indirectly, controls scientific funding. In sum, the quality of scientific knowledge is important to scientists and nonscientists alike, and social influence on the scientific process—at whatever scale—is axiomatic.

ENVIRONMENTAL ACTIVISTS AND NATURAL SCIENTISTS

A Crib Sheet on Natural Scientists

What are the primary points activists should take from the primer on natural-science philosophy presented in the previous section? In general,

1. Natural scientists assume there is a material reality independent of human thought. For this reason, natural scientists assume one should indeed be concerned about human impacts on the world around us, because our environment is *not* entirely a social construction that can safely be ignored.

2. Natural scientists believe they *should* utilize the postpositivist, hypothetico-deductive model of science initially promulgated by Popper (1959, 1962) and reiterated by Platt (1964), Romesburg (1981), and others.

3. For this reason, natural scientists perceive knowledge regarding an issue to be approximated by the explanatory hypothesis that has best survived substantive experimental challenges to date. They assume this

knowledge will be modified based on further investigation. This is why natural scientists typically appear uncertain about most topics. After all, certainty only comes when scientific theories attain the stature of laws, such as the laws of thermodynamics.

4. For this same reason, natural scientists typically avoid voicing opinions regarding even their own area of expertise in public forums. Such "subjective" behavior could be perceived by many of their peers as violating norms of "objectivity," as evidence of hubris, and as being "unscientific"—the ultimate taboo. There is a real and not unreasonable fear that such behavior could have adverse effects on one's professional life (e.g., funding, tenure, promotion, ability to publish, and ultimately, one's research program).

5. Despite lip service to hypothesis falsification, most researchers, particularly in the more "soft" natural sciences such as ecology, conservation biology, and climatology, only rarely employ controlled, manipulative experiments designed to falsify deductively or retroductively derived research hypotheses. Instead, they rely on mensurative experiments, modeling, impact analysis, and descriptive research, to name just a few alternatives (see Morrison et al. 2008).

6. Thus, study designs actually employed by most natural scientists fit the pragmatic model proposed by Haack (2003) far better than Popper's postpositivist model of science (1959, 1962). The pragmatic model also is congruent with the revolutionary nature of natural science (Kuhn 1962) and measuring the contribution of research programs based on how progressive they are (e.g., how much the program is characterized by growth and discovery; Lakatos 1970). An important caveat, however, is that most natural scientists are not familiar with Haack's work and would not necessarily think of their approach to science as pragmatic.

7. Natural scientists, particularly those working in environmental disciplines such as ecology, conservation biology, and climatology, typically are passionate about the natural world around them. As Naess (1986) argued, they generally agree, at least in principle, with even rather radical contentions propagated by environmental activists, such as the eight

planks of the deep-ecology movement. They are unlikely, however, to voice these views in public for reasons already stated.

8. The rationale for the paradoxical behavior of environmental scientists is clearly Darwinian. Despite caring passionately about the environment, many natural scientists fear they will be destroyed professionally and thus personally (e.g., financially) if they join environmental activists and speak in public forums about their environmental concerns. Just as importantly, they realize that if their professional career is destroyed or curtailed, they will not be able to complete and publish further work relevant to their program of research (Lakatos 1970) that could potentially benefit environmental causes in the future.

The World of Activism Versus the World of Natural Science

Niklas Luhmann's (1989) social theory provided an explanation for contextualizing how activists and natural scientists perceive the world. He argued that human society structures itself by communication within and between function systems, and that all communication is constrained by these subsystems. He identified the economy, law, science, politics, religion, and education as the most important function systems. Most environmental activists' objectives involve convincing governmental bodies to pass laws they would like to see enacted or to enforce laws already on the books in ways more favorable to the environment. For example, SIU national organizers called for the U.S. Congress to pass one or more statutes that, when appropriately implemented by regulatory agencies, would lead to an 80 percent reduction in carbon emissions by 2050. Similarly, many influential environmental nongovernmental organizations, such as the Sierra Club, often rely on litigation to reach their objectives. Although environmental activists certainly raise funds, try to help educate others, and use scientific data in various ways, their overriding objective is to alter what happens in the political and legal function systems to benefit the environment.

Conversely, natural scientists obtain grants, help educate students, and deal with political and legal issues primarily to benefit their scientific-research programs. This difference is more important than it might seem at first glance, as most natural scientists are rather poorly prepared to enter the political arena or work in courts of law. After working in the political and legal quagmire associated with the Interagency Spotted Owl Scientific Committee (ISC) convened by the U.S. Congress in 1989, wildlife scientists Dennis D. Murphy and Barry D. Noon aptly summarized the problems they faced as natural scientists working in the political and legal arena:

> Our experience on the ISC suggests that lawyers do not understand how science is done. They perceive that we do science like criminals build successful alibis. Alibis are built of evidence that must stand in total, akin to a structure built from so many metaphorical bricks. The structure, it is often argued, is no stronger than the weakest brick used in its construction. The lawyer's job is then to find that weak brick and, in doing so, bring down the structure...What biologists actually do...is quite different. Biologists do not construct conclusions from data; they construct hypotheses that are tested with data. A conservation plan is not built from bricks of hard data, but is loosely constructed with pertinent information on distributions, abundances, natural history observations, and habitat associations. The resultant structure is tested with statistical analysis of empirical data, predictions from ecological theory and population models, and inferences drawn from studies of related species. In essence, we fling bricks of data at a structure—our plan, our hypothesis—as a means of identifying its weaknesses. Drawing inference from the results of these tests, we adjust and reshape the structure to strengthen it. The testing process continues until relevant data, models, and inferences are exhausted. The conservation plan so produced is not as weak as the weakest data or softest test of its integrity—it is as strong as the strongest data or toughest test. (1991, 775–776)

As illustrated by this quotation, the world of natural science is indeed quite different than the world of politics and law, where activists operate. Regardless, resonance among these function systems is required if substantive social change is to occur (Luhmann 1989). Thus, activists must learn to bridge this void.

Activist Access to Key Natural Scientists

Natural-science data can be critical to political and legal efforts to promote effective climate policy. While challenges to activist access to such data and its interpretation should now be obvious, there also are opportunities. These opportunities involve pathways for negotiating the behavioral paradox embodied by environmental scientists and fall into three broad categories: (1) the aesthetic regard ecologists, conservation biologists, climatologists, and other environmental scientists have for Earth, (2) the fact that most natural scientists are pragmatists, whether they know it or not, and (3) the fact that because natural scientists do not perceive the universe around us to be a social construction existing primarily in our collective minds, they feel we must be concerned about negative human impacts on a to-some-degree material Earth.

So, how can environmental activists use these pathways? From the onset, activists must recognize natural scientists of the caliber they are looking for already have full plates and are not looking for things to do. They are unlikely to show up at a rally, offering to make coffee or do whatever else is needed for the cause. Activists instead must seek out these individuals and form personal relationships over time, characterized by mutual trust. This is essentially the same way leading science journalists and authors make and maintain their contacts. I have overheard several climate-change activists note with awe the apparent information pipeline science journalist Ross Gelbspan and author Bill McKibben seem to have with world-renowned climate scientists such as James Hansen. Rest assured that these critical relationships were neither built overnight nor built without substantial effort. Similarly, such writers undoubtedly individualize their interactions with different scientists in light of each person's comfort zone. For example, the career of a first-year assistant professor at a state university is much more vulnerable to peer- and institutional pressure than is that of a famous, endowed professor or the director of the NASA Goddard Institute for Space Studies.

Perhaps a personal example of how a skilled environmental activist went about developing relationships with natural scientists who initially were strangers might be illustrative. In 2005 I was contacted via e-mail by Bill, a colleague who had recently retired from a federal environmental regulatory agency (names are pseudonyms). In his former capacity, he had funded some of my research, and we had served together on review panels and task forces since the late 1980s. He mentioned that he was now consulting with a major environmental NGO that needed a review of a wildlife disease. He felt this task was outside his area of expertise. He ended the e-mail by stating that he had given Leroy, the president of the NGO, my contact information, as he thought I was better situated to accomplish what the NGO wanted than he. A few hours later, an e-mail from Leroy arrived, asking if I would be willing to speak with him on the telephone at my convenience sometime during the next two to three days. Before explaining his request, Leroy briefly described a recent wilderness pack trip he had completed in the region of concern. The way he told the story made it evident that he had taken the time to learn that I was a native of the region and regularly spent time in the same wilderness. Leroy then pointed out that the environmental NGO he led believed certain wildlife-management practices employed in the region would exacerbate the threat of the disease, which would be detrimental to wildlife populations in the region. He asked if I would be willing to produce a review, even though available funds were insufficient to cover the time required to complete the task. After I expressed tentative willingness, he offered to have a staff member send me details so I could make an informed decision.

After the report was completed, Jodi, an NGO attorney, reviewed the report. She was careful to communicate that her suggestions and questions were not addressed at report content but were an attempt to ensure that other attorneys would be unable to misconstrue my message. She left the ultimate decision regarding report wording to me. Shortly thereafter, Leroy invited me to a gathering at a small resort in the region. The primary objectives for the meeting were to ensure that NGO staff and leaders of other NGOs with similar interests were up-to-date on the ecology of the disease in question, to complete policy-oriented brainstorming, and to provide space for informal interactions among ecologists and NGO personnel. All participants were asked to read my report prior the meeting. Although four invited ecologists made brief formal presentations on assigned topics, most of the meeting was spent in discussion. Questions and answers flew all directions (e.g., ecologists to NGO members, NGO members to each other, ecologists to each other, NGO members to ecologists). The ecologists answered questions just as bluntly as they would have at an informal gathering with colleagues. Importantly, meeting organizers also made certain there was plenty of time for informal interactions among participants, including time spent soaking in a hot spring, listening to wolves, and enjoying well-prepared, leisurely meals. Leroy told me afterwards the meeting accomplished exactly what he had hoped.

Since then, Leroy has kept in touch by personally forwarding relevant news articles and press releases from NGO or regulatory agencies, along with brief messages. Similarly, I have shared scientific articles that might be of interest to the NGO. Finally, Leroy called me during late May 2008 and informed me that his NGO had filed a federal lawsuit that relied on several quotations from my 2005 report. He asked if I would be willing to take questions from the press. I informed him that my employer would not want me listed on the NGO's Web site as a spokesperson, but that he was welcome to pass my contact materials to key reporters. A few days later, he gave me the name of a *New York Times* science reporter who would be contacting me. A remarkably accurate article appeared in the *Times* the day after the reporter spoke with me.

Leroy's interactions with me and the other three ecologists illustrate how activists can negotiate the paradox embodied by natural scientists who are passionate about the environment but feel it is somehow unscientific to be involved in activism. He worked though a common friend to expedite initial contact, then immediately *accessed not only my keen interest in environmental protection, but my sense of place* (Tuan 1977; Cresswell 2004). Leroy convinced all four invited ecologists that wholehearted *participation during the meeting would help assure that practical, real improvements in environmental policy and practice were made*. With Leroy's expert facilitation, I moved from having no role in environmental activism, to writing a report used to directly inform activism, to making personal and lasting connections with several activists, to acting as a spokesperson for NGO goals regarding this issue.

Summary and Suggestions

Those organizing and advocating environmental action often need access to natural scientists' data and, particularly, to their interpretation of these data. This requires open access to key natural scientists. Because most active natural-science researchers, including ecologists, conservation biologists, and climatologists, were trained in the postpositivist tradition, they cannot help but feel they should not be involved in activism, as such activity is "subjective" and "unscientific." This is despite the fact that most care passionately about Earth. To gain access to these scientists, *activists must negotiate this paradox by appealing to environmental scientists' values and beliefs*. For example, most ecologists, conservation biologists, and climatologists' aesthetic perspectives toward Earth are congruent with those of environmental activists. These natural scientists assume there are material aspects of the world around us despite the fact that our perceptions of reality are to some degree socially constructed; thus, action to protect the environment is warranted. Most environmental scientists are pragmatists epistemologically, whether they recognize this fact or not, so changes in environmental policy resulting in practical consequences, or real environmental effects, appeal to natural scientists.

To take advantage of these openings, *activists must seek out individual natural scientists and take the time to form personal relationships that are characterized by mutual trust*. These interactions must be individualized based on each person's situation and allowed to evolve over time. Working through common friends or colleagues is an effective way to make initial contact. As activists utilize environmental scientists' worldview to facilitate relationship building, they must continually ensure this is accomplished without jeopardizing the scientist's professional reputation among her or his colleagues and thence his or her research program. Typically, the natural scientist's comfort zone will expand with time if she or he is working with skilled and empathetic activists. Finally, such relationships inevitably develop into multiway interactions, resulting in a rich atmosphere of discovery for everyone who is involved.

ENDNOTES

1. "Natural science" refers to the physical and life sciences.
2. In this chapter, "ecologist" refers to natural scientists trained in one of the ecological disciplines rather than environmental activists or others interested in the environment.
3. Jodi M. Minion et al. (chapter 8) graciously gave me access to archived texts, including working pages, links, and pictures provided on stepitup2007.org from January 4, 2007 (the start of SIU), through May 14, 2007 (one month after the April 14 events). See chapter 8 for details regarding the role of new media in SIU.
4. Probably the most important source of anthropogenic carbon emissions other than CO_2 is methane (CH_4, the principle component of natural gas). An emission of CH_4 has approximately seventy-two and twenty-five times the impact on the earth's temperature over a twenty- and one-hundred-year period, respectively, than does a CO_2 emission of the same mass emitted at the same time (this standard comparison is called the "global warming potential"). Thus, CH_4 is a far more potent greenhouse gas than is CO_2, particularly during the first two decades after emission, although it occurs at lower concentrations. Primary anthropogenic sources of CH_4 include livestock (ruminants), natural gas facilities and equipment (e.g., leaking pipelines, machinery, and appliances, etc.), landfill venting, biomass burning, and fecal-waste-treatment facilities.
5. In 1895 Swedish physical chemist Svante A. Arrhenius argued that anthropogenic influences on atmospheric CO_2 concentrations would lead to increased global temperature. He used available empirical data and logic (including mathematics) to estimate these influences and introduced the hothouse analogy that dominates climate-change discussions today. One could argue that, from a Popperian perspective, Arrhenius' statements were metaphysical until computer and other new communication technologies (e.g., the Internet) enabled scientists to efficiently collate temperature data taken the same way and at the same times from across the earth. Even though thermometers were available in 1895, it was logistically difficult to empirically test Arrhenius' hypothesized relationship between atmospheric gas concentrations and global temperature. Just as Arrhenius predicted, both CO_2 concentrations and temperatures increased from 1900 to 2000. However, atmospheric CO_2 increased much more rapidly than Arrhenius predicted, and the earth has not warmed more slowly than he thought it would.

References

Berger, Peter L., and Thomas Luckmann. 1966. *The social construction of reality: A treatise in the sociology of knowledge*. Garden City, NY: Doubleday.

Chamberlin, Thomas Chrowder. 1890. The method of multiple working hypotheses. *Science* 15 (366): 92–96.

Committee on the Conduct of Science. 1989. On being a scientist. *Proceedings of the National Academy of Sciences of the United States of America* 86: 9053–9074.

Cresswell, Tim. 2004. *Place: A short introduction*. Oxford: Blackwell Publishing.

Einstein, A. 1936. Physics and reality. *Journal of the Franklin Institute* 221 (3): 349–382.

Feyerabend, Paul. 1975. *Against method: Outline of an anarchistic theory of knowledge*. London: NLB.

———. 1978. *Science in a free society*. London: NLB.

———. 1987. *Farewell to reason*. London: Verso.

Gettier, Edmund L. 1963. Is justified true belief knowledge? *Analysis* 23 (6): 121–123.

Haack, Susan. 2003. *Defending science—within reason: Between scientism and cynicism*. Amherst, NY: Prometheus Books.

Haack, Susan, and Robert Lane, eds. 2006. *Pragmatism, old and new: Selected writings*. Amherst, NY: Prometheus Books.

Hafner, E. M., and Susan Presswood. 1965. Strong inference and weak interactions. *Science* 149 (3683): 503–510.

Intergovernmental Panel on Climate Change (IPCC). 2007. *Climate change 2007: The physical science basis: Contribution of Working*

Group I to the Fourth Assessment Report of the IPCC. Cambridge: Cambridge University Press.

Jasinoff, Sheila, Gerald E. Markle, James C. Petersen, and Trevor Pinch, eds. 1995. *Handbook of science and technology studies*. Thousand Oaks, CA: Sage.

Kuhn, Thomas S. 1962. *The structure of scientific revolutions*. Chicago: University of Chicago Press.

Lakatos, Imre. 1970. Falsification and the methodology of scientific research programmes. In *Criticism and the growth of knowledge*, ed. I. Lakatos and A. Musgrave, 91–195. Cambridge: Cambridge University Press.

Latour, Bruno. 1993. *We have never been modern*. Trans. C. Porter. Cambridge, MA: Harvard University Press.

Lincoln, Yvonna S., and Egon G. Guba. 1985. *Naturalistic inquiry*. Newbury Park, CA: Sage.

Luhmann, Niklas. 1989. *Ecological communication*. Trans. J. Bednarz Jr. Chicago: University of Chicago Press.

Morrison, Michael L., William M. Block, M. Dale Strickland, Bret A. Collier, and Markus J. Peterson. 2008. *Wildlife study design*. 2nd ed. New York: Springer.

Murphy, Dennis D., and Barry D. Noon. 1991. Coping with uncertainty in wildlife biology. *Journal of Wildlife Management* 55 (4): 773–782.

Naess, Arne. 1973. Shallow and deep, long-range ecology movement: A summary. *Inquiry—An Interdisciplinary Journal of Philosophy* 16 (1): 95–100.

———. 1986. Intrinsic value: Will the defenders of nature please rise? In *Conservation biology: The science of scarcity and diversity*, ed. M. E. Soulé, 504–515. Sunderland, MA: Sinauer Associates.

Peterson, Markus J., and Tarla Rai Peterson. 1996. Ecology: scientific, deep and feminist. *Environmental Values* 5 (2): 123–146.

Plato. 1973. *Theaetetus*. Trans. J. McDowell. Oxford: Clarendon Press.

Platt, John R. 1964. Strong inference: Certain systematic methods of scientific thinking may produce much more rapid progress than others. *Science* 146 (3642): 347–353.

Popper, Karl Raimund. 1934. *Logik der forschung: Zur erkenntnistheorie der modernen naturwissenschaft*. Wien, Österreich: Springer.

———. 1959. *The logic of scientific discovery*. London: Hutchinson.

———. 1962. *Conjectures and refutations: The growth of scientific knowledge*. New York: Basic Books.

Romesburg, H. Charles. 1981. Wildlife science: Gaining reliable knowledge. *Journal of Wildlife Management* 45 (2): 293–313.

Rosenberg, Alex. 2000. *Philosophy of science: A contemporary introduction*. London: Routledge.

Shapin, Steven, and Simon Schaffer. 1985. *Leviathan and the air-pump: Hobbes, Boyle, and the experimental life*. Princeton, NJ: Princeton University Press.

Soulé, Michael E., ed. 1986. *Conservation biology: The science of scarcity and diversity*. Sunderland, MA: Sinauer Associates.

Step It Up. 2007a. Organizer FAQs. http://april.stepitup2007.org/article.php?id=184 (accessed June 13, 2008).

———. 2007b. Photographing your action. http://april.stepitup2007.org/article.php?id=184 (accessed June 13, 2008).

Tuan, Yi-Fu. 1977. *Space and place: The perspectives of experience*. Minneapolis: University of Minnesota Press.

Worster, Donald. 1994. *Nature's economy: A history of ecological ideas*. 2nd ed. Cambridge: Cambridge University Press.

Chapter Interlude

Excerpt from SIU San Antonio Rally Transcript

Transcribed by Jodi Minion

Background Note: What follows is part of a transcription of the Step It Up (SIU) event in San Antonio, Texas. This action took place at the Alamo from noon to 2 p.m. on April 14, 2007. James Andrews and Heather DeGrella were part of the San Antonio SIU organizing team.

Crowd: Step It

Crowd: Up!

Crowd: Step It

Crowd: Up!

James Andrews, rally speaker: Well done. That's pretty good. OK, now we are going to do another one. OK, this time it's gonna be, How much? 80 percent! Are we ready?

Crowd: How much?

Crowd: 80 percent!

Crowd: How much?

CROWD: 80 percent!

CROWD: How much?

CROWD: 80 percent!

ANDREWS: Well done!

CROWD: How much?

CROWD: 80 percent!

ANDREWS: Right, thank you very much. OK, last one: When do we want it? 2050! Here we go…

CROWD: When do we want it?

CROWD: 2050!

CROWD: When do we want it?

CROWD: 2050!

CROWD: When do we want it?

CROWD: Now! 2050!

[later]

HEATHER DEGRELLA, RALLY SPEAKER: [*muffled*]…I wanted to say, San Antonio, thank you so much for the turnout today. Higher, can you hear me now? Is this better? All right. Thank you so much for this turnout today. This is an historic event and I am so proud that San Antonio is part of it. Your enthusiasm and your support is unprecedented, and that is great. Today, all across the United States, men and women, young and old, are joining in the biggest nationwide rally for action on global warming that the U.S. has ever seen. You're part of something historic and something quite amazing. People standing up to protect the places they love, to protect the rest of creation, and to protect our future. That's why we're here today. So, I think we should give everyone a round of applause.

CROWD: [*clapping*]

DEGRELLA: Today, we join more than 1,300 other rallies in all fifty states, twenty-nine of which are taking place right here in Texas. The political world that we need has begun together. Dozens of senators and representatives are joining the rallies today. And we know that all of you will make sure our politicians hear this message in the weeks ahead. If you have wanted to know what you personally can do to make a difference, we have petitions and form letters ready for you to sign. And we will be taking a group photo shortly to compile with other events from around the country to demonstrate the large turnout we're experiencing. Particular changes that Americans are calling for today are for Congress to cut carbon emissions 80 percent by the year 2050. It is neither radical nor is it impossible. It will require a steady 2 percent drop in carbon emissions as we learn to conserve, and we find new sources of clean energy. We have the technical expertise and innovation to do that. The question is, do we have the political will? We are also taking this opportunity today to urge our 80th Texas State Legislature to take positive action on several bills that address global warming. Texas leads the nation in energy and electric consumption. It's imperative that we enact change.

[later]

DeGrella: A quick word about the petitions, if you've signed just one, you haven't signed them all. We have, I think, four or five of them. One is the Step It Up petition, which is asking for the 80 percent reduction in carbon for the year 2050. That will go to national Congress. We have another one that goes to the Texas state legislature for a lot of bills that we are—that the Sierra Club has determined are very good bills for supporting this effort. If you want to read more about those bills, we have some more information on that so that you know exactly what you are signing; it's always good to know that. And we also have a petition going around; the city of San Antonio, we understand, is considering a good green-building resolution for city-owned buildings. And we want to write our support on that as well.

Crowd: [*cheers and clapping*]

DeGrella: We have a table over there so you can learn about carbon offsetting, and I just want to mention a few things: the photo is actually really important. We're gonna take that around 1:10, right after the speakers are finished, and it will go on the Web site with everyone else across the nation. With the Alamo in the background, I think will be a fantastic photo, so stick around for that.

Chapter 13

SIU and Beyond

Challenges of Scale and the Strategic

Robert Cox

The climate has begun to change, that is, the political climate: in the past five years, public awareness of the signs of global warming has grown, with news of melting glaciers, endangered polar bears, extreme weather, and the impacts of Hurricane Katrina. And the elections of a new Congress and U.S. president also have begun to prompt government action on energy and the reduction of carbon emissions. At the same time, growing public concern has spurred action at the state and local levels: more than half of the states have adopted renewable portfolio standards mandating greater efficiencies and renewable sources of energy. Wind farms now generate electricity from "red" states like Kansas and Texas as well as "blue" states like California and New York. Meanwhile, ordinary citizens are crowding into public hearings to oppose new coal-fired power plants, and over one thousand U.S. mayors have pledged to reduce carbon emissions in their cities.

An inspiring moment during this period occurred on April 14, 2007, at 1,400 locations across the United States, as thousands of citizens called on Congress to "Step It Up" by reducing carbon emissions by 80 percent by 2050. The contributors to this volume, therefore, have asked what we might learn from this effort. What role did the rallies, concerts, speeches, banners, and nature walks play in raising awareness or influencing solutions? And, perhaps more importantly, what are the prospects for similar initiatives in the future to influence action on climate change in the United States?

In answering these questions, it has been impossible to ignore the larger challenges posed by the sheer scope and complexity of the changes required to reduce emissions by 80 percent by midcentury, as many scientists have urged. As a result, our attempts to evaluate SIU's effectiveness have had to recognize not only its heuristic value but also its basic incompleteness, its *what-is-not-yet-done* quality. It might be fairer, then, to ask, what are Web-based, grassroots initiatives such as SIU *capable* of accomplishing? What limits this potential? And, in this light, what can we learn from the particular efforts of SIU on April 14, 2007?

In what follows, I argue that unresolved tensions in SIU's design, and the strategic requirements called forth by the scale and complexity of change, may have limited the influence of this otherwise innovative initiative. I also suggest that SIU held much broader, strategic potential: *the ability of a loosely coordinated network of agents, opportunities, and practices at multiple local sites of decision to "signal" in critically important ways within our energy system*. I develop this analysis in three sections: (1) the challenge of assessing "effect" in social movements, particularly in contexts of larger scale and complexity; (2) the implications of a failure to resolve key tensions in the April 14 events; and (3) a rethinking of SIU's design to suggest a greater, *strategic* ability to impact decisions about energy. My hope is that by examining both the promise and limitations of the SIU approach, we might begin to think in fresh ways about what constitutes a social movement in the twenty-first century and how a consideration of scale invites a rethinking of its practices, modalities, and the demands placed on the role of strategy.

Assessing "Effect" and Scale

By any measure, the SIU events of April 14, 2007, were innovative, inspiring, and empowering for a new generation of activists. And, lest we forget, SIU was launched by only one professor and his six students on a small New England campus. Yet, April 14 witnessed an outpouring of diverse and imaginative actions, including:

- speakers at the Alamo in San Antonio calling for a new energy future;
- neighborhood activists lining the shore of Seattle's Golden Gardens with bamboo poles to mark the levels forecast for oceans to rise;
- scuba divers giving underwater demonstrations near endangered coral reefs;
- the faith-based group Interfaith Power and Light handing out materials on climate change at a Raleigh, North Carolina, SIU event; and
- sun salutation yoga performed in Salt Lake City as a collective intention to cut CO_2 emissions.

These and the other SIU events have spurred further efforts to organize thousands of others to influence the debates over energy policy in the U.S. Congress. As Bill McKibben said in his foreword to this volume, "We upped the stakes, moved the debate to the left, and helped create a bunch of great organizers."

These are, in themselves, significant achievements, particularly if SIU "moved the debate" on climate change in the United States. But did it? What does it mean to claim that the April 14 events "upped the stakes" or "helped create a bunch of great organizers"? Asking such questions is hardly innocent; contributors to this volume and SIU participants themselves do not always agree on what the claims of "effect" mean or what roles, if any, the inspiring day's activities played in achieving these or other outcomes.

Several chapters in this volume grappled with question of efficacy or the "success" of SIU actions—their influence on media, policies, and wider movement building. In some cases, this effort was wide-ranging. For example, Endres, Callister, Garrison, Senda-Cook, and Schutten (chapter 4) took up this question by examining the "disconnects or ruptures in communication" between SIU organizers and participants in the April 14 events. They argued that such "campaigns can easily be crippled by communicative disconnects" when their objectives are framed in a way that fails to reach multiple audiences and motivate them to act consistently with these goals. Others such as Sprain, Norton, and Milstein (chapter 9) looked more specifically at the effects of "image events" like nature walks, protests, and marches in the Pacific Northwest. Finally, Kinsella, Temple, and Shields (chapter 7) raised the question of success directly: "Is SIU truly a new social movement? If so, is it a successful one?" More importantly, they asked, "*What counts as success, for social movements generally and for this self-proclaimed movement in particular?*"

Kinsella et al. leaned toward affirming both SIU's "newness" and "success," insofar as it exhibits many of the characteristics that Stewart, Smith, and Denton (2007) argued that successful movements possess—organized collective action, broad geographic reach, use of persuasive tactics, and so forth. I am not persuaded, however, that Stewart et al.'s criteria exhaust "what counts as success" in the face of the scale and complexity posed by climate change. Let me briefly illustrate some of the difficulties, then, in assessing "effect," before turning to the challenges of scale itself.

By their own account, McKibben and his students set out ultimately to achieve two goals: (1) to launch a new movement that would (2) persuade the U.S. Congress to reduce greenhouse-gas emissions by 80 percent by 2050. By these criteria, SIU was obviously an unfinished effort. Hence, any attempt to evaluate its "success" will need to look more closely at the outcomes of April 14 themselves, as well as the expectations for these and the lines of influence they were intended to initiate. One of the more important, expected outcomes was media attention for its core message. (SIU planners provided tips for attracting such media coverage in their "cookbook.") Let's look at this briefly.

Several authors used the benchmark of media interest in assessing the SIU activities. Sprain et al. (chapter 9) raised this explicitly in arguing that "the logic of the public screen suggests that image events are effective *when they attract media attention and receive wide dissemination*" (emphasis added). Did, in fact, the SIU rallies, marches, and image events in Idaho and Washington state "attract media attention and receive wide dissemination"? (The authors report that the participants themselves were disappointed by the media coverage.) On the other hand, some progressive blogs and online media reported enthusiastically on the diverse events on April 14, 2007. *Grist* reported,

> On April 14, 2007, gaggles of enthusiastic Americans gathered at more than 1,400 spots around the U.S. to demand action against climate change, part of a coordinated Step It Up campaign…At all the rallies, marches, parties, and other hullabaloos, the message was the same: Step it up, Congress! Enact immediate cuts in greenhouse-gas emissions, and pledge an 80 percent reduction by 2050. (2007)

News reports in the mainstream media were markedly less enthusiastic. Noting the disparaging news coverage of SIU events by the *New York Times* and other media in Boston and nationally, Prelli (chapter 3) observed these did "not sufficiently *manifest* the magnitude of support alleged or make the movement *visibly present* through conventional news sources." He contrasted media attention to SIU with earlier, greater media coverage of successful one-day events, such as the 1963 march on Washington, suggesting that coverage of it "embodied" the much larger movement and "made its existence *visibly present*, and did so indisputably."

In all fairness, however, the success of the 1963 rally at the Lincoln Memorial in Washington, DC, came on the shoulders of bloody marches and sit-ins in the American South and a growing media interest in the unfolding drama of the struggle for civil rights. As with Gandhi's "salt march" in India, these activists had been *building a movement* for years. A fairer measure for assessing the SIU events of April 14, therefore, would be to take seriously this wider SIU objective. Kinsella et al. (chapter 7) pointedly observed, "One notable aspect of SIU is its self-conscious self-identification: the organizers are setting out, deliberately and strategically, to *build a movement*" (emphasis added).

But, this may not be a simple proposition. Is a single campaign a "movement," or is "movement" itself any longer an adequate category to explain the dynamics of social change in the context of global energy systems? I believe, therefore, that it is important to return to McKibben and the SIU team's impulse, to "do more than change lightbulbs," and to ask, what *was supposed to happen* as a result of the April 14 events? If a new movement was expected, or if the April 14 events were to launch new lines of influence, then two questions become important: What is the *scale* of change required to achieve an 80 percent reduction of carbon emissions by midcentury? More specifically, how do the scale and complexity of these changes alter the ways we think about the practices, modalities, and strategies of a movement capable of achieving of this goal?

Challenges of Scale and Complexity

The scope of the changes needed to measurably slow and then reverse anthropogenic climate change can be hard to conceive. This is true not only of systemwide economic changes but also of the ideological and political transformations required. These resemble what Althusser has described as a "complex whole"—a social structure on three levels: economic, political, and ideological. The relations among these make up a "complex whole…a structure articulated in dominance" (1970, 202). In the present case, neoliberal, carbon-based economies include global systems of production, pricing, security, and transportation and distribution of oil, coal, and natural gas. Such energy supplies, in turn, undergird modern life—the generation of electricity, manufacture and transportation of consumer goods, food, and industrial material, heating and cooling of buildings, as well as our everyday activities from work commutes and leisure travel to the Dasani bottled water in the front seat.

This economy, in turn, is sustained and rationalized by ideological constructions of "economic growth," "property," and "free markets," as well as the political and communication apparatuses that reproduce and disseminate these constructs: media, corporate advocacy campaigns, contributions to political parties, subsidies to oil and gas companies, the politicizing of climate science by governmental officials, and more. As a consequence, the economic, political, and ideological alterations required to effect a comprehensive transformation of this "structure articulated in dominance" exceed any one solution or capacities of one campaign, project, or even one movement.

How, then, to begin to challenge this "complex whole" is the pivotal question of a movement that would be part of altering the structures and norms that sustain anthropogenic global climate change. While it may be important—on one level—to "get the message out" or attract news media, the point, ultimately, is not the dissemination of a message but the transformation of a system "articulated in dominance" and resistant to the usual pressures for change.

In evaluating SIU's effectiveness, then, contributors to this volume faced a high bar in providing an account of the articulations among communication practices, modalities, strategies, and the lines of influence needed to leverage change at the scale required. Most authors acknowledged the enthusiasm of SIU participants and the creativity of the diverse actions, locations, image events, and appeals. On the other hand, many questioned what power, influence, or sustained outcomes these events achieved and what obstacles may have lain in the way of more visible outcomes. In the following, then, I want to review some of these obstacles and also identify some of the ways the basic design and goals of SIU may have overshadowed the wider, strategic influence of this innovative project.

Step It Up: Unresolved Tensions

In a very real sense, the "openness" of SIU's approach created an interesting, even heuristic dilemma—a seeming trade-off between grassroots flexibility and a larger, more coherent focus required to effect change at the scale required. Clearly, the minimal requirements for staging the April 14 events helped to explain the responsiveness of so many local organizers and the number of SIU events. On the other hand, the absence of more explicit direction may have fostered a lack of clarity in the purpose, communication style, message, and overall strategy of these events.

To begin, several authors lauded the SIU team for its "open-source" approach to organizing: a compelling but largely unspecified request to "step it up," a minimum "ask" (a message of 80 percent reduction in emissions by 2050, a banner, etc.), and an online "tool kit" to support local organizers. The advantage of such an approach lay in enabling local initiative and control, as well as empowering many first-time activists. (See chapter 7, chapter 8, and chapter 9.) In using this approach, McKibben and his students joined a growing trend in using Web-based social networks and resources for activists in organizing on social and environmental issues.

The very openness of SIU's design, however, posed a difficult dilemma. While explaining the success of numerous SIU events, it also perpetuated a number of tensions in the April 14 activities: Was the purpose of SIU to create an *awareness* of the dangerous consequences of climate change or to encourage specific *actions*? Did either of these goals require a consistent "message," or could local organizers interpret the language and meaning of the April 14 events more freely? Was a moral "confrontation" with "villains" important, or were speakers expected to identify practical or incremental steps individuals could take? And, if the latter, how were individuals expected to relate their *personal* experiences or actions to the wider *collective action* needed to address climate change? As the SIU rallies, speeches, concerts, and nature walks unfolded, many of these tensions remained unresolved. A relevant question, then, becomes, did this failure weaken the potential effectiveness of the local actions to function *at the scale required*, and if so, how?

Let me briefly describe four tensions fostered by the dilemma of SIU's choice of approach, tensions that were implicitly or explicitly identified by other contributors as well.[1] I do this not so much to suggest how they might be resolved, but to illustrate some of the ways in which the *failure* to resolve them may have sacrificed opportunities for more coherent or focused action. Indeed, I believe this failure led to a basic ambivalence about strategy in SIU's use of local rallies and attempts at movement building. In the end, this eclipse of more explicitly strategic considerations may be the most important lesson for future Web-based initiatives on climate change.

Education Versus Action

Any evaluation of the success of the SIU events raises a prior question about the objectives for the April 14 events: Were they intended principally to educate those who were previously unaware of the seriousness of climate change? Or, was the purpose to mobilize those who attended the rallies, nature walks, or exhibits to "do something"? If "something," was it strategically chosen or left to the discretion of local SIU speakers and organizers? For example, Hall, Bernacchi, Milstein, and Peterson (chapter 2) reported that climate-change activists intended to use community art to "reconfigure perceptions of climate change" among local residents (awareness), while speakers at SIU rallies in Austin, Boston, and other cities issued calls for signing petitions, supporting local legislation, and so on. A number of authors in this volume puzzled over this tension between "awareness" and action, or between "education" and a kind of "moral confrontation." (See chapter 4, chapter 5, and chapter 3.) Endres et al. (chapter 4) noted this tension occurring in SIU speakers' framing of events as they alternated between urging "awareness" (an effort to educate and "potentially gather more supporters") and encouraging "protest." Similarly, Norton and Paveglio (chapter 5) observed speakers' mention of *competing* objectives which, they argued, indicated a tension in the purpose of the events. These included "raising of awareness" and "making voices heard," implying "a symbolic struggle of competing discourses."

Going further, Prelli (chapter 3) insisted that the purpose of any movement—and SIU in particular—must be to *confront* existing power with a moral demand for change: "A dramatic situation so urgent [as global climate change] calls for heroes who must prevail over the villains who resist needed change." He observed that although speakers at the Boston Common depicted heroes who are "opposed by the villainous coal and oil industries," their warnings seemed to lack "urgency." Their calls for action—sign a postcard, recycle, reduce garbage, attend another rally—seemed "a rather weak denouement to the drama they had characterized, literally, in earth-threatening terms."

The point Prelli is making is that the rhetoric at the Boston Common SIU rally didn't match what speakers repeatedly described as a problem of "earth-threatening" proportions ("awareness"). Speakers, he suggested, missed an opportunity to tell those at the rally what they should "do next," an omission with wider strategic implications. More specific calls for action were "lost behind a welter of varied suggestions about how to alter one's own life practices and abstractions about the movement's growing power." (I will return to this below.)

Controlled Versus Diffuse Message(s)

At the core of SIU was an emphasis on the local, decentralized nature of the new movement. Its open-source model offered flexibility for grassroots initiatives and local control of the April 14 message itself. Indeed, several contributors echoed Kinsella et al.'s (chapter 7) praise of the "diffuse and decentered characteristics of [this] Web-based communication...which provide[d] some protection against the risk of creeping institutionalization" of the movement. Minion, Kinsella, O'Neil, and Peterson (chapter 8) similarly celebrated the capacity of SIU's open-source mode of organizing to "provide simple and clear instructions [and]...empower leaders at the local level." And, Hall et al. (chapter 2) noted that SIU organizers encouraged local organizers "to interpret the SIU policy message of cutting national emissions of CO_2 '80 percent by

2050' in whatever manner they deemed most appropriate" (Barringer 2007). Finally, Sprain et al. (chapter 11) argued the contrasts between the local activism of SIU and the more established environmental groups suggested that a new "environmentalism 2.0" was in the making. This version stressed "decentralized but coordinated actions by multiple, unaffiliated, even individual environmental actors," a shift that opened "individual and local possibilities" through a variety of media such as art (chapter 2), speeches, concerts, yoga, and rallies at state capitals.

Despite the claims of environmentalism 2.0 at work, limitations also became apparent. One limitation was the failure to ensure a diversity of participants—especially minorities, low-income, and environmental-justice groups—in the April 14 events. While the idea of an open-source campaign was meant to invite new audiences, Endres, Clarke, Garrison, and Peterson (chapter 6) found that the digital divide may have made this difficult for some. As they observed in their discussion of SIU and climate justice,

> Understanding the digital divide can help to explain why heavily relying on the use of the Internet for movement participation (i.e., online sign-ups to host events, dissemination of information about the campaign through the SIU Web site, blogs, and *Grist*, etc.) can result in a lack of participation of marginalized people, who may suffer the most from climate change.

A second limitation may have occurred in the relationship between local messaging and wider effects (or absence of effects) of the SIU actions. Perhaps this was inevitable once the national SIU planners took a hands-off approach in ensuring that local messages were consistently delivered or linked to specific actions. This hands-off attitude was clearly in tune with the decentered and diffuse nature of the "local possibilities" fostered by SIU but may have elided the strategic focus necessary to coordinate at larger scale. Sprain et al. (chapter 9) touched upon this in discussing the use of "image events" by SIU organizers in the Pacific Northwest. They argued such images "are powerful because they…invite readers and audiences to construct *their own meaning of the message*" (emphasis added). Similarly, Minion et al. (chapter 8) noted that "SIU used the Internet to mediate a shift *away from the federal government* to the local grassroots level," a "shift" that, at the very least, pulls attention from SIU's intended message, telling the U.S. Congress to "step it up."

SIU events in Plano, Texas, illustrated this tension when the local organizer offered her own interpretation of the day's message:

> The Plano rally seemed less about SIU and more about resources and activities offered by Live Green in Plano. Throughout the event, people came and went based on individual activities, and none described themselves as attending an SIU event...The brochure on the concert/film series did identify April 14 as a "National Day of Climate Action," but SIU was not emphasized. As with all scheduled activities, the organizer focused on more environmentally friendly living, *while minimizing the connection to climate change.* (Feldpausch-Parker, Hall, Horton, Minion, Munoz, Parker, and Peterson, chapter 1; emphasis added)

Understandably, for many conservative areas, calls for "environmentally friendly" actions, including renewable energy, succeed only when they do *not* reference climate change or SIU's goal of 80 percent reduction in emissions by 2050. The Pew Center on Global Climate Change discovered this shift away from climate change to other rationales when states "anticipate significant economic development benefits from promoting renewables, particularly given the promise of developing homegrown energy sources that could lead to instate job creation" (Rabe 2006). Feldpausch-Parker et al. (chapter 1) observed this shift at the Austin SIU rally when a speaker stressed the "new economic opportunities and the importance of retaining the competitive edge during the next energy boom."

Even allowing for the flexibility needed in some regions to play down the issue of climate change, considerations of "effect" still require attention to the strategic; that is, how do the "diffuse" (or absent) interpretations of SIU's basic message contribute to a sustained influence at scale? These tensions between the disciplined and diffuse, between action and effect, may have drawn attention from accounts of how SIU envisioned the movement would gain power to affect actual reductions in CO_2 emissions.

Apocalypse Versus Hope

Ambiguities in overall purpose and message inevitably impact decisions about communication style and the motives presented for action. This is apparent in both the rhetorical choices of SIU speakers and the analyses of some contributors to this volume. For some speakers, climate change promises catastrophic consequences for the planet, suggesting equally dramatic warnings. I noted earlier that Prelli (chapter 3) insisted the "earth-shattering" dangers of climate change meant that SIU speakers needed to *confront* existing power rhetorically with a moral demand for change. Other speakers, however, focused on sustainable lifestyle choices and a vision of hope. Kinsella et al. (chapter 7) observed a similar tension in the statements by McKibben and others on the SIU Web site conveying "a sense of moral import," while other, more pragmatic messages circulated at rallies on April 14. These differing visions, they suggested, mirrored "an ongoing debate within the environmental movement regarding radical versus incremental change."

Such sharp rhetorical differences led Sprain et al. (chapter 11) to contrast two very different systems of advocacy in their reports of SIU events on April 14: the more "establishment" rhetorics of "apocalypse and limits" versus newer environmentalism 2.0 "rhetorics of sustainability and hope." The latter, they argued, were the majority—speakers who talked in terms of "hopeful discourses focused on sustainability," emphasizing "technological solutions" and "the importance of personal responsibility for energy use and individual efforts to affect global warming."

Ultimately, the heart of the debate about the "right" approach, rhetorically, is an assumption of the efficacy of an information-driven movement or public-education campaign; i.e., with the *right* framing or message, individuals' behaviors can be changed. But this assumption runs headlong into an enduring conundrum of social change—the so-called "attitude-behavior" gap. This is the assumption that new beliefs or attitudes will be followed by new behaviors. This clearly was an assumption of some who drew on a more apocalyptic style at the SIU events. For example, a participant in the waterfront image event in Seattle explained that performances that visually illustrated rising ocean levels would "help people *see* what is happening [and]...*more and more awareness will lead to action*" (emphasis added). Yet, the available evidence suggests this is not necessarily true. In other words, environmentally appropriate behavior does not always follow from new information, even when individuals are aware of the dangers of climate change.[2]

This conundrum is related to a final tension in SIU's design and local practice. Even if individuals respond to "rhetorics of sustainability and hope" and are willing to alter their own behaviors (riding a bike, purchasing CFL bulbs, etc.), a further question remains: what do SIU planners and local organizers assume is the relationship *between individual lifestyle changes and the wider, coordinated actions* needed to challenge the systems sustaining the "complex whole" of our carbon economies?

Individual Versus Collective Action

Quite a few contributors to this volume picked up on the challenge posed by the ambiguous articulation in SIU speeches between individual and collective actions. That is, how do individual steps or lifestyles help to advance a wider strategy? Endres et al. (chapter 4) took up this "*collective-action/individual-activity* tension" at length in exploring the problem of "framing": did the participants at SIU events understand themselves to be "acting in coordination with others toward collective social change," or did they not see their individual actions as linked to a larger movement or campaign? Norton and Paveglio (chapter 5) sensed this tension in the comments of SIU participants. While a Michigan resident stressed his personal choices—"I ride my bike and walk...I'm not an eco-Nazi or anything"—a participant in Utah pointed out, "These kinds of concerns will only be solved through a populist *movement*."

Still, it's not clear, even if individual actions such as "green consumerism" could be linked to broader impacts, that these actions are available to everyone. As Endres et al. (chapter 6) pointedly remind us in their discussion of climate justice, "the ability to engage in green behaviors is a function of personal resources (i.e., money) and opportunity." As a result, by "emphasizing green consumerism as a primary part of the solution to the problem of global warming," we effectively exclude "a large portion of the population from being a part of the 'solution.'" To be clear, organizers usually do recognize the importance of individuals' sense of agency and identity in engaging in activism; encouraging this can be an important outcome of an event. And, campaigns often organize on the assumption that even a small act (replacing a lightbulb with a CFL bulb) may serve as an individual's "first step" in activism, that is, *changing a lightbulb today, urging the Congress to act tomorrow*. Pezzullo's interview (interlude between chapter 7 and chapter 8) in Bloomington, Indiana, is illustrative: there, the local SIU organizer confessed that her experience "has made me much more willing to take on projects...I guess I'm thinking on a bigger scale than I ever would have before." Further, Sprain et al. (chapter 9) identified a closely related outcome—a normalizing of protest or oppositional practices, particularly when individuals "see themselves" as being like other participants, "especially when the movement's images look similar to civic images."

On the other hand, McKibben's call to "do something" other than change a lightbulb brings into relief a larger, strategic challenge. Endres et al. (chapter 4) argue that tensions in a movement arise when individuals' lifestyle choices appear to satisfy their desire to "do something about climate change," while absolving them of the need to see this as necessarily part of a collective solution to the problem. Clarke and Milburn (chapter 10) captured this tension in observing that while every SIU event had a banner with the message of "80 percent by 2050," local organizers often avoided a connection between their activities and the avowedly political act of "sending a message" to Congress. Perhaps, they explained, "local leaders feared that *by focusing on what Congress could do, the message sent by the national organizers might stifle or limit a sense of agency* for individual community members" (emphasis added).

Confusion about the relations between individual agency and collective action, unfortunately, ignores the strategic work needed to influence structures or practices beyond the individual. As Norton and Paveglio (chapter 5) noted, "symbolic acts and individual gestures of opposition are interesting but largely impotent because the basic structure of a situation may be unaltered…In this way, SIU challenges us to think about not just the symbolism of resistance but also *the possibilities of changing material conditions*" (emphasis added).

How, then, to articulate such *collective* actions in coordination with individual responsibility and calls for "everyone to do their part"? Did the SIU events enable individuals to "do their part"? What was supposed to happen after some agreed to ride his or her bike or walk instead of driving his or her car? After he or she changed a lightbulb or sent an e-mail? What were the SIU organizers' explanations for the link between these acts and the larger goal of persuading Congress to reduce emissions by 80 percent by 2050? Despite calls to sign a petition, send e-mail, and so forth, the dilemma at the heart of SIU's "open-source" approach and its local implementation remained.

RETHINKING THE STRATEGIC IN SIU AND BEYOND

Ultimately, the ambiguous linkages of education/action, coordinated/diffuse, apocalypse/hope, and individual/collective in the SIU events appeared to overshadow or deflect any account of the *strategic*. *How, exactly, does the new movement gain power to achieve the ends it seeks?* SIU organizers remained quite vague on how the 1,400 events on April 14, 2007—rallies in Seattle, Raleigh, and Boston, speeches at the Alamo, scuba diving off coral reefs, a nature walk on Whidbey Island, and sun salutation yoga in Salt Lake City—contributed to the wider influence needed to move closely to their goal of 80 percent reduction in greenhouse-gas emissions by 2050. What "leverage" or wider lines of influence did the organizers hope to create by sending e-mails, raising banners, and posting photos on the SIU Web site, and how did these hoped-for lines of influence relate to the scale of change sought?

"The Politicians Can't Ignore It."

Inevitably, the SIU speakers and organizers believed their efforts did make some difference—that their speaking, making banners, circulating petitions, and so forth were relevant to wider outcomes and effects. As McKibben said in the foreword of this volume, "We offered a strategy with a reasonable chance of affecting larger politics." People did turn out for the rallies; they believed they were sending a "message." Yet, surprisingly, more concrete explanations of the strategic appeared to be an afterthought, not only in the "tool kit" but also in local planning for the SIU events.

Several contributors to this volume picked up on this limited account of strategy. As I noted above, although rally speakers spoke of media attention as one measure of effectiveness, it remained unclear—even had there been more coverage in the press—how such media representations of SIU activities actually "worked" or were articulated to subsequent effects. I believe it was this inattention to the *strategic* that Prelli (chapter 3) challenged: "The many rallies…did not sufficiently *manifest* the magnitude of support alleged or make the movement *visibly present* through conventional news sources…[It] was not *brought into being* for many who were not already participants or like-minded supporters."

Prelli identified this as a "rhetorical deficit" in the SIU campaign. That is, a (strategic) requirement of movement rhetoric is to manifest or exhibit "something *previously unseen or concealed*," making the new movement *present* to a wider audience. Sprain et al. (chapter 9) put the matter more pointedly when they asked of image events in Idaho and Washington state, "How does this event fit into the larger picture?'

Strategy need not be obscure, although many campaigns often tend to ignore or understate its role. Traditionally, movement strategy has been conceived as a set of actions or "identification of the *specific* steps" to bring about a desired outcome (Cox 2006, 258). (I'll return below to the challenges that "scale" brings to this traditional idea.) Yet, such specification is often missing in self-accounts of movements. Orr has compared this absence to the *American Scientist* cartoon of scientists at a blackboard: puzzling over how to balance an algebraic equation, one scientist scribbled in the middle of the equation, "*then a miracle occurs*." Orr observed, "Most strategies of social change have a similar dependence on the miraculous" (1992, 61). That is, many movements' ideas about strategy tend to lack any account of the intervening lines of influence required for their desired outcome or effects. Torgerson has similarly observed that dependence on the "miraculous" is often true in environmental strategies, the gauzy outlook that "problems are sure to get worse…and when they do, more and more people will be moved to join the green cause, thus enhancing…its chance of making a real difference" (1999, 22).

A similar assumption about the miraculous seemed to underlie the expectations of participants interviewed at SIU events. Implicitly, some assumed their attendance at the rally, speaking up, or marching somehow would be sufficient to persuade those in power. As one person at the Boston Common rally explained, "*Just having all these people here means that the politicians can't ignore it*" (Prelli, chapter 3; emphasis added). In other cases, Prelli reported, those attending the rallies spoke in confused generalities, for example, "I just hope the people, like the local politicians and the national politicians, actually do something." Another confessed, "I don't know. We should do something about the environment. I don't know how that happens, but I guess through everything—politics and art and all that. Eventually, people will take action and do something about it."

I believe the SIU events did offer "a strategy with a reasonable chance," and (some) people did respond. Nevertheless, as I have argued, the SIU approach too often appeared to be handicapped by unresolved tensions in the rallies themselves. Thus, while Prelli's conclusion may seem harsh, it is understandable: despite the excitement among SIU participants, many left the next steps of the movement "up to the politicians, or were confused about what they could do specifically…There was no logic of demonstrative protest rhetoric that compelled or constrained opponents' or participants' subsequent discourse and actions."

I do not believe this outcome was inevitable. The potential for a wider effectiveness remains quite strong within the SIU model—its openness of design, ability to coordinate activism across multiple sites, and the genuine enthusiasm of those involved. I believe implicit in the diverse SIU speeches, rallies, image events, and invitations to "become involved" are the makings of a wider strategic contribution.

Strategic Elements in SIU and Beyond

What would it mean to resolve the strategic ambiguity of the SIU initiative? More specifically, how might we go about rethinking—that is, placing within a more strategic context—some of the key elements of this innovative effort? I believe three elements are particularly relevant: (1) the *local* nature of its focus, (2) interest in what a new *movement* could be in a connected, online age, and (3) the intent to "send a message." What might we learn if we were to rethink these aspects, both in themselves and in relation to each other? Can we rethink the strategic potential of the local, the idea of "movement," and "signaling" (sending a message) in ways that contribute to leveraging change at the scale required?

In what follows, I want to explore the nature of the strategic in the face of the wider challenges posed by the scale and complexities of changes needed to reduce carbon emissions by 2050. For a movement can no longer simply plot "specific steps," but must consider more broadly how it can achieve power to *leverage* or take advantage of wider openings or decisions to set in motion other effects or lines of influence that contribute to change. Because this is a little abstract, I want to illustrate what I mean here.

Specifically, I want to rethink "local," not for its utility for attracting media, but as the *site of decisions* about the production and/or end-uses of energy; to view "movement" as a *multi-dimensional network of agents, opportunities, practices, and entities* with access to these sites; and redefine "message" as *altering the signals* in the "complex whole" of the energy system itself. If we can rethink these three aspects of the SIU initiative, I believe we may uncover several strategic opportunities for a movement to alter the momentum of forces now sustaining the complex relations of our carbon economy.

Rethinking "Local" as a Site of Action

The SIU Web site urged organizers from the beginning to choose locally appropriate places to stage their events. Indeed, SIU activists and contributors to this volume alike acknowledged the importance of location (see especially chapter 1, chapter 4, and chapter 2). These locations enabled rally speakers to connect *global* climate change to its impacts in specific places and regions. The SIU events in Salt Lake City and San Antonio, for example, did this by linking climate change to declining snowpacks in the Wasatch Range near Salt Lake City and lowering aquifers in Texas and the Southwest generally.

The advice to emphasize the local is not surprising. Leiserowitz similarly identified a paradox in recent efforts to communicate the risks of global warming to American audiences. While large majorities believe that global warming is real, he explained, they nevertheless assign it "a low priority" and generally feel a "lack of urgency" about it (2007, 53). He found that the highlighting of "potential local and regional climate-change impacts" was a communication strategy that had the potential to change this.[3]

Still, these uses of the local remained unclear *strategically*. SIU speakers' appeals to the event participants often were framed as individual or voluntaristic actions, apart from the "outside" systems or larger structures. Such an eclipse of wider contexts continued to obscure the relationships among places (locales), decisions, and forces of the wider energy economy—forces that often flow through local and regional sites.

This is not to suggest that it is inappropriate to tell individuals what they can do locally, in their communities. Instead, by viewing the local as one site within *multiple sites within a wider energy system*, one might ask, can the aggregation or the cumulative influence of actions at these (local) sites leverage more systemic change? In other words, are there consequential relationships between specific local actions—blocking a permit for a coal-burning power plant in state X, persuading city council Y to require municipal facilities be powered by solar or wind energy, and so forth—and wider signals being sent to state and federal policy makers and investors in the energy sector? That is, is there a synergy among actions in Salt Lake City, Seattle, Moscow, New York City, Raleigh, Boston, and Austin which can help to leverage wider political and economic pressure? This is a question not of what constitutes a "movement," but of a *strategy*, that is, whether an effort to *coordinate* actions at multiple, local sites contributes to, leverages, or enhances a larger effect.

Ironically, SIU organizers' emphasis on a simple, single "message" may have overshadowed this more complex relationship and made the task of articulating the relationships between local and national actions more difficult. Norton and Paveglio (chapter 5) noted that several SIU speakers chose to reduce the complexity of political change by focusing solely on Congress instead of the opportunities for individuals to act not only within their own communities but in a way that built upon similar efforts across the country. There is, therefore, some danger in reifying the "solution" for climate change as a (single) law that calls for emissions to be reduced by 80 percent by 2050. A comprehensive response to climate change, in fact, may flow from multiple sources, for example, cities enacting green-building codes, financing light-rail public transportation, making shifts of capital to renewable technologies, or adopting renewable portfolio standards, regional compacts for CO_2 emission trading, or local initiatives to preserve forests for carbon sequestration, and so on.

In rethinking the "local" in this manner, we are already assuming other relevant or coordinated actions and the influence of a wider network of agents at other sites where decisions about energy are being made.

Rethinking "Movement" as Network

Earlier, I questioned the idea of "movement" as an analytical category that is able to explain the dynamics of change at the scale required to transform a carbon economy. Yet, it is clear that McKibben and SIU activists viewed their efforts as launching a new "movement." Kinsella et al. (chapter 7) noted,

> From its inception, Bill McKibben and his associates...envisioned the Step It Up (SIU) project as a new social movement.

> McKibben (2008) recounted that upon returning from a trip to India in 2006, "[F]or some reason, the idea of [a] movement had an almost visceral appeal. I'd always loved the story of Gandhi's march to the sea to make salt. It seemed to me that the time had come to march here too, odd as it seemed in an Internet age."

SIU activists also spoke of their involvement the same way. As one enthusiastically told a reporter in Boston, "This country hasn't seen a movement this size since the civil rights movement!" (in Prelli, chapter 3).

Some contributors to this volume resist this claim, noting that SIU activities never achieved a level of visibility that would constitute a powerful political movement. Others questioned whether SIU met basic "functions" of a movement (see, for example, Kinsella et al., chapter 7). But asking whether the SIU project constituted a discrete social *movement*—in either sense—may be posing the wrong question.

How would our assessment of SIU be different if we asked, *how did the activities of multiple agents (activists, audiences, planners, decision makers, etc.) at multiple local sites help SIU to gain power or leverage change?* Norton and Paveglio (chapter 5) posed a somewhat similar question in focusing on the role that "resistance" plays in different movements. Leaving aside whether e-mail messages or signatures on petitions can be considered as *resistance*, the authors raised an important question of strategy when they asked if SIU can be "transformative," that is, can it change material practices or policies?

Norton and Paveglio's purpose was to reframe the SIU efforts—not as a discrete "movement," but as the kind of organization that can "create *networks of groups* that are linked by larger purposes, but operate in different locations and on various issues" (emphasis added). Such a framing puts the emphasis on a network's *function* rather than its "form," that is, it asks, what work does such a network do? In this context, then, we can ask the question, did (or how could) the 1,400 local SIU events—and the network of agents, practices, and organizations linked by these events—contribute to a transformative politics?

By using "network," rather than "movement," we may be able to locate the SIU efforts within the wider context of climate activism and the actions being taken by multiple subjects and entities, such as online initiatives, local governments, environmental organizations, labor, religious communities, and others. Indeed, Kinsella et al. (chapter 7) regarded SIU less as an autonomous *movement* and "more as part of a broader, *rapidly evolving network* of organizations, groups, and individuals seeking action on global climate change." This broader "network" includes conferences such as Power Shift; larger environmental groups' advocacy campaigns; local community, business, and other civic groups; Web-based climate initiatives like Focus the Nation (www.focusthenation.org), It's Getting Hot in Here (www.itsgettinghotinhere.org), Climate Challenge (www.climatechallenge.org), and Sierra Student Coalition (http://www.ssc.org); and the loose coalitions emerging among climate activists and faith-based organizations, as well as the Blue-Green Alliance of labor and environmental groups (www.bluegreenalliance.org) and environmental-justice groups who are organizing around the issue of a "green economy"—sustainable jobs, a revival of the construction sector, and jobs for low-income communities (e.g., wind turbines and solar panels). Other possibilities also exist. Peterson (chapter 12), for example, pointed to the growing collaboration between climate-change activists and scientists.

Many of these entities, in turn, are locating themselves in wider networks of other organizations and allied actions. Kinsella et al. (chapter 7) observed that by January 2008, the SIU Web site provided links to 103 groups. Similarly, Climate Challenge (2008) announced on its own Web site that it "unites 42 organizations and over 617 local groups in 56 states and provinces…[who] have worked to pass 100s of local and regional climate policies." More recently, SIU archived its own site, and the original team migrated to http://www.350.org/ and its global links to other online climate activists and groups.

Nevertheless, there are challenges that remain in the building of such networks. As Endres et al. (chapter 6) pointed out, the mainstream environmental movement has largely failed to involve certain interests and populations in its organizing. For example, despite SIU planners' intentions to network with environmental-justice groups, this proved difficult in the April 14 events. In their discussion of climate justice, Endres et al. identified several reasons for this, including a failure to "tap into local communities and social networks" that can "mobilize environmentally burdened groups on the ground." They observed that one reason for the lack of participation by underresourced people "may be related to the SIU campaign's reliance on the Internet as its main form of outreach and organizing." As a result, they suggested closer links with *faith-based* groups who may serve as "bridges" between mainstream and environmental-justice groups in building a climate-change network.

Focusing on such a loose network of agents, organizations, initiatives, and activities linked by a "larger purpose" allows us to begin to think about their aggregate influence or power. Kinsella et al. appear to do this when they observe that "the new media principles of aggregation, coproduction, and mobilization are key parts of this communication strategy." Still, while praising the "diffuse and decentered characteristics of Web-based communication," they remain unclear about its larger potential, noting only that it is a "rapidly evolving and highly variegated" movement. But this may confuse "movement" and "network," where the latter invites us to foreground communicative *functions* rather than "form," that is, how the strategic work of "highly variegated" voices, entities, and practices may actually leverage wider effects.

If we take seriously the idea of the communicative functions of networks, then two possibilities immediately present themselves. First, the aggregation of voices and practices of the SIU events of April 14 may, in fact, have functioned as a kind of new "register" of public notice or urgency, an amplification of other initiatives at that time. Second, such amplification is not inconsistent with the leverage that the wider climate movement recently has begun to exercise: *an altering of the signals that help sustain the larger energy economy*. It is this second function of "sending a message" that I turn to now.

Rethinking "Message": Altering Signals

From the beginning, national SIU organizers made clear that an important goal of the April 14 events was to "send a message" to the U.S.

Congress. Speakers and banners urged those attending the rallies and concerts to proclaim, "Step It Up, Congress: cut carbon 80 percent by 2050." Yet, if this "message" is understood in terms of traditional strategy—if enough people "speak up," the politicians will respond—then Congress, so far, hasn't gotten the message. But is this the only meaning of "message"? To put it another way, how might the actions at local sites produce alternative meanings or senses of "reality," which, if taken together, might begin to challenge the "complex whole" of our energy society? Can these dispersed efforts at sense-making be taken collectively as a coherent *signal*? And if so, a signal about what, and to whom?

One possibility presents itself: over the past two years, growing numbers of citizens, community groups, activists, and larger environmental organizations have been successful at the local level in blocking or delaying permits for coal-fired power plants, one of the principal sources of CO_2 emissions in the United States. While the U.S. Department of Energy reported that 151 new coal-fired plants had been proposed at the start of 2007, during that year alone, "59 of those proposed plants were either refused licenses by state governments or quietly abandoned. Almost 50 more are being contested in the courts," and opponents now are challenging the other plants as they come up for approval (UK Reuters 2008; see also Brown 2008). At the same time, local and state governments are mandating that a percentage of energy in these localities come from renewable sources or greater efficiencies. Cumulatively, these decisions have begun to "signal" quite clearly to the sources of investment—capital markets—that underwrite our energy economy. I am also suggesting that *an amplification of these signals* has a wider strategic relevance for a climate movement.

It is not simply that movement strategists *hope* that the halting of coal-fired plants and other accomplishments at local sites might have some meaning. Benford and Snow reminded us that movement actors are "signifying agents actively engaged in the production and maintenance of meaning for constituents, antagonists, and bystanders" (2000, 613; quoted in Kinsella et al., chapter 7). I am arguing that such signifying also occurs as a consequence of a consciously initiated set of practices or decisions. Indeed, the growing effectiveness of the grassroots movement in affecting decisions for coal-generated electricity has had consequences in the U.S. energy economy. These decisions have now begun to signify or "signal" increased risk for capital markets and equity firms that finance construction in the energy sector, and this, in turn, has the potential to shift momentum in investments toward other, renewable energy sources.

The U.S. Department of Energy cautioned in early 2007 that "proposals to build new power plants are often speculative...based upon the ever changing economic climate of power-generating markets" (National Energy Technology Lab 2007). Hence, the slowing or cancellation of permits to build coal-fired plants has particular meaning in this business sector. In 2008 three of the nation's largest investment firms—Citigroup, J. P. Morgan Chase & Co., and Morgan Stanley—announced they were imposing new requirements for financing construction that "will make it harder for companies to build coal-fired power plants in the US" (Ball 2008, 1). Such reluctance is, itself, "sending *a potent signal to the energy sector* that it views dirty coal as shaky financial prospects and that the smart money is heading toward cleaner, more sustainable energy options" (Beinecke 2008, 1). And this, in turn, has the potential, *amplified by continuing efforts at multiple local sites of decision*, to build an agenda for a more comprehensive political action on energy in the United States.

Admittedly, attention to local fights against coal power plants or local energy usage was not the target of most SIU speakers on April 14, although clearly this potential was there. For example, Feldpausch-Parker et al. (chapter 1) reported that a local Sierra Club speaker at the Austin rally challenged listeners to bring their own local municipalities into the group of 442 (at that time) cities and towns that had crafted "Cool Cities" agreements[4] to lower their carbon footprint. Such actions—linked by "larger purposes, but operating in different locations," and sustained by wider networks of actors and practices—have strategic promise that is only now beginning to be realized.

Still, the efforts of local activists, networks, advocacy campaigns, and lobbying by environmental groups remain in a race against time and climate "tipping points" that surely will erase further opportunities. In the weeks before the world's energy ministers met in Bali, Indonesia, in December 2007 to discuss a post-Kyoto treaty to reduce greenhouse-gas emissions, the director of the UN Intergovernmental Panel on Climate Change, Rajendra Pachauri, cautioned, "If there's no action before 2012, that's too late. What we do in the next two to three years will determine our future. This is the defining moment" (Rosenthal 2007). A new movement clearly has begun in the United States and globally, attempting to influence this future. Its success, ultimately, depends not simply on inspiring rallies, marches, or Web sites, but on a network able to act at local sites of relevant decisions and capable of affecting the "complex whole" of energy systems. Collectively, these actions have the very real promise of functioning strategically, not only to shift the momentum in capital markets, but also to build an agenda for change in Washington, DC, as well as Salt Lake City, Bloomington, Seattle, or San Antonio.

ENDNOTES

1. Norton and Paveglio (chapter 5) similarly argued that there is a core tension in the SIU approach, but while they suggested the "centralized control" of SIU undercut the "bottom-up nature of the movement," I suggest there is a very different danger.
2. Leiserowitz reported evidence from a national study on risk perceptions and global warming in stating, "Large majorities of Americans believe that global warming is real and consider it a serious problem, yet global warming remains a low priority" (2007, 44). More recently, Crompton, summarizing recent research, argued, "We should not expect information campaigns to create behavioural change" (2008, 6), and instead highlights the need to reexamine the more basic values and motivations that lie behind behavior itself.
3. Leiserowitz identified five variables that are likely to aid efforts to communicate the dangers of climate change to the general public: (1) "Highlight potential local and regional climate-change impacts," (2) "Climate change is happening now," (3) "Highlight the potential impacts of climate change on human health and extreme weather events," (4) "Talk openly about remaining uncertainties," and (5) "Tailor messages and messengers for particular interpretive communities" (2007, 53–61).
4. "Cool Cities" is a nationwide, grassroots campaign of the Sierra Club, encouraging cities to pledge to reduce their carbon footprints. Some 1,084 cities had become "Cool Cities" as of this writing. See http://www.coolcities.us.

References

Althusser, Louis. 1970. *For Marx*. Trans. Ben Brewster. New York: Random House.

Ball, Jeffrey. 2008. Wall Street shows skepticism over coal: Banks push utilities to plan for impact of emissions caps. *Wall Street Journal*, February 4, 2008. http://online.wsj.com/article_email/SB120209079624339759-lMyQjAxMDI4MDAyNDAwOTQwWj.html (accessed July 26, 2008).

Barringer, Felicity. 2007. Renewing a call to act against climate change. *New York Times*, March 14, 2007.

Beinecke, Frances. 2008. The twilight of dirty coal. *NRDC Switchboard*, February 11, 2008. http://switchboard.nrcd.org/fbeinecke/the_twilight_of_dirty_coal.html (accessed July 26, 2008).

Benford, Robert D., and David A. Snow. 2000. Framing processes and social movements: An overview and assessment. *Annual Review of Sociology* 26:611–639.

Brown, Lester R. 2008. U.S. moving toward ban on new coal-fired power plants. Earth Policy Institute. http://www.earth-policy.org/Updates/2008/Update70.htm (accessed August 13, 2008).

Climate Challenge. 2008. www.climatechallenge.org (accessed August 5, 2008).

Cox, Robert. 2006. *Environmental communication and the public sphere*. Thousand Oaks, CA: Sage.

Crompton, Tom. 2008. *Weathercocks and signposts: The environmental movement at a crossroads*. WWF-UK. http://www.wwf.org.uk/filelibrary/pdf/ weathercocks_report.pdf (accessed August 3, 2008).

Grist. 2007. One giant step for greenkind. *Grist Environmental News and Commentary*, June 19, 2007. www.grist.org/news/maindish/2007/04/19/stepitup (accessed June 25, 2008).

Leiserowitz, Anthony. 2007. Communicating the risks of global warming: American risk perceptions, affective images, and interpretive communities. In *Creating a climate for change: Communicating climate change and facilitating social change*, ed. Susanne C. Moser and Lisa Dilling, 44–63. Cambridge: Cambridge University Press.

National Energy Technology Lab. 2007. Tracking new coal-fired power plants. http://cmnow.org/NETL%20New%20Coal%205.2007.pdf.

Orr, David W. 1992. *Ecological literacy: Education and the transition to a postmodern world.* Albany: State University of New York.

Rabe, Barry G. 2006. Race to the top: The expanding role of U.S. state renewable portfolio standards. Pew Center on Global Climate Change. http://www.pewclimate.org/global-warming-in-depth/all_reports/race_to_the_top (accessed August 3, 2008).

Rosenthal, Elizabeth. 2007. U.N. report describes risks of inaction on climate change. *New York Times*, November 17, 2008. http://www.nytimes.com/2007/11/17/science/earth/17cnd-climate.html?pagewanted=print (accessed August 16, 2008).

Stewart, Charles J., Craig Allen Smith, and Robert E. Denton. 2007. *Persuasion and social movements*. Long Grove, IL: Waveland.

Torgerson, Doug. 1999. *The promise of green politics: Environmentalism and the public sphere*. Durham, NC: Duke University Press.

UK Reuters. 2008. Time is up for coal. UK Reuters, February 14, 2008. http://uk.reuters.com/article/environmentNews/idUKN1445731320080214?pageNumber=2&virtualBrandChannel=0 (accessed August 13, 2008).

Appendix A

Principles of Environmental Justice (First National People of Color Environmental Leadership Summit, 1991)

Delegates to the First National People of Color Environmental Leadership Summit held on October 24-27, 1991, in Washington DC, drafted and adopted 17 principles of Environmental Justice. Since then, *The Principles* have served as a defining document for the growing grassroots movement for environmental justice.

Preamble

WE, THE PEOPLE OF COLOR, gathered together at this multinational People of Color Environmental Leadership Summit, to begin to build a national and international movement of all peoples of color to fight the destruction and taking of our lands and communities, do hereby re-establish our spiritual interdependence to the sacredness of our Mother Earth; to respect and celebrate each of our cultures, languages and beliefs about the natural world and our roles in healing ourselves; to insure environmental justice; to promote economic alternatives which would contribute to the development of environmentally safe livelihoods; and, to secure our political, economic and cultural liberation that has been denied for over 500 years of colonization and oppression, resulting in the poisoning of our communities and land and the genocide of our peoples, do affirm and adopt these Principles of Environmental Justice:

1) **Environmental Justice** affirms the sacredness of Mother Earth, ecological unity and the interdependence of all species, and the right to be free from ecological destruction.

2) **Environmental Justice** demands that public policy be based on mutual respect and justice for all peoples, free from any form of discrimination or bias.

3) **Environmental Justice** mandates the right to ethical, balanced and responsible uses of land and renewable resources in the interest of a sustainable planet for humans and other living things.

4) **Environmental Justice** calls for universal protection from nuclear testing, extraction, production and disposal of toxic/hazardous wastes and poisons and nuclear testing that threaten the fundamental right to clean air, land, water, and food.

5) **Environmental Justice** affirms the fundamental right to political, economic, cultural and environmental self-determination of all peoples.

6) **Environmental Justice** demands the cessation of the production of all toxins, hazardous wastes, and radioactive materials, and that all past and current producers be held strictly accountable to the people for detoxification and the containment at the point of production.

7) **Environmental Justice** demands the right to participate as equal partners at every level of decision-making, including needs assessment, planning, implementation, enforcement and evaluation.

8) **Environmental Justice** affirms the right of all workers to a safe and healthy work environment without being forced to choose between an unsafe livelihood and unemployment. It also affirms the right of those who work at home to be free from environmental hazards.

9) **Environmental Justice** protects the right of victims of environmental injustice to receive full compensation and reparations for damages as well as quality health care.

10) **Environmental Justice** considers governmental acts of environmental injustice a violation of international law, the Universal

Declaration On Human Rights, and the United Nations Convention on Genocide.

11) **Environmental Justice** must recognize a special legal and natural relationship of Native Peoples to the U.S. government through treaties, agreements, compacts, and covenants affirming sovereignty and self-determination.

12) **Environmental Justice** affirms the need for urban and rural ecological policies to clean up and rebuild our cities and rural areas in balance with nature, honoring the cultural integrity of all our communities, and provided fair access for all to the full range of resources.

13) **Environmental Justice** calls for the strict enforcement of principles of informed consent, and a halt to the testing of experimental reproductive and medical procedures and vaccinations on people of color.

14) **Environmental Justice** opposes the destructive operations of multi-national corporations.

15) **Environmental Justice** opposes military occupation, repression and exploitation of lands, peoples and cultures, and other life forms.

16) **Environmental Justice** calls for the education of present and future generations which emphasizes social and environmental issues, based on our experience and an appreciation of our diverse cultural perspectives.

17) **Environmental Justice** requires that we, as individuals, make personal and consumer choices to consume as little of Mother Earth's resources and to produce as little waste as possible; and make the conscious decision to challenge and reprioritize our lifestyles to insure the health of the natural world for present and future generations.

Appendix B

SIU Bloomington Schedule

April 14, 2007

9:00–10:00	People gather, check out tables, marchers arrive
9:50–10:00	Director of local youth and music center, Brad Wilhelm [master of ceremonies]
10:00–10:15	Mayor Mark Kruzan
10:15–10:20	Pianist Julian Bransby and fusion band Para Luxe perform "Cool It"
10:20–10:35	Monroe County Commissioner Iris Kiesling and State Representative Matt Pierce
10:40–11:15	Kaia, an a capella vocal group
11:15–11:35	Author and Professor Scott Russell Sanders reads "A Letter to My Children"
11:35–11:50	U.S. Congressman Baron Hill
11:55–12:20	Para Luxe
12:20–12:25	Buff Brown, BTOP [Bloomington Transportation Options for People]
12:30–12:45	Singer/songwriter Carrie Newcomer
12:45	Wrap Up
1:30	Tree-planting ceremony next to the Monroe County Public Library

Appendix C

SIU Bloomington Partner Organizations

Step It Up, Bloomington! was presented in cooperation with the Bloomington Environmental Commission, the Bloomington Commission on Sustainability, and the City of Bloomington Department of Parks and Recreation. In addition to those three government entities, the following is a list of organizations that hosted tables at the event:

BTOP [Bloomington Transportation Options for People]
Center for Sustainable Living
Local Growers' Guild
INPIRG [Indiana Public Interest Research Group]
Hoosier Environmental Council
Indiana Forest Alliance
Campus Climate Challenge
Caldwell Center for Culture and Ecology
Citizens for Appropriate Rural Roads
SPEA: Earth Week [Indiana University School of Public and Environmental Affairs]
Democracy for Monroe County
Hoosier Hills Food Bank

Bibliography

1Sky. "The 1Sky Campaign." http://www.1sky.org/wp-content/uploads/pdfs/ 1sky_background.pdf.

———. "About 1Sky." http://www.1sky.org/about.

570 KNRS. "Community Calendar, April 14, 2007." http://www.570-knews.com/slc-common/community_calendar/.

ABC News. *Good Morning America* transcript, April 14, 2007. Lexis-Nexis *Academic* database. http://www.lexisnexis.com.

ABC News. *World News Saturday* transcript, April 14, 2007. LexisNexis *Academic* database. http://www.lexisnexis.com.

Abram, David. *The Spell of the Sensuous*. New York: Vintage, 1996.

Abramson, Mark. "Locals Fight Global Warming: Palo Alto Joins Cities for National Day of Climate Action." KRTBN *Knight-Ridder Tribune Business News—Palo Alto Daily News*, April 15, 2007. LexisNexis *Academic* database. http://www.lexisnexis.com.

Agarwal, A., S. Narain, and A. Sharma. "The Global Commons and Environmental Justice—Climate Change." In *Environmental Justice: Discourses in International Political Economy*, edited by J. Byrne, L. Glover, and Cecilia Martinez, 171–199. New Brunswick, NJ: Transition Publishers, 2002.

Agyeman, Julian. *Sustainable Communities and the Challenge of Environmental Justice.* New York: NYU Press, 2005.

Agyeman, Julian, R. Bullard, and T. Evans. "Exploring the Nexus: Bringing Together Sustainability, Environmental Justice and Equity." *Space and Polity* 6 (2002): 70–90.

Agyeman, Julian, and T. Evans. "Towards Just Sustainability in Urban Communities: Building Equity Rights with Sustainable Solutions."

Annals of American Academy of Political and Social Science 590 (2003): 35–53.

Allen, Kim, Vinci Daro, and Dorothy C. Holland. "Becoming an Environmental Justice Activist." In *Environmental Justice and Environmentalism: The Social Justice Challenge to the Environmental Movement*, edited by Ronald Sandler and Phaedra Pezzullo, 1–24. Cambridge, MA: MIT Press, 2007.

Allsopp, Michelle, Adam Walters, David Santillo, and Paul Johnston. "Plastic Debris in the World's Oceans." United Nations Environment Programme. http://www.unep.org/regionalseas/marinelitter/publications/docs/ plastic_ocean_report.pdf.

Almanzar Nelson, A. P., Heather Sullivan-Catlin, and Glenn Deane. "Is the Political Personal? Everyday Behaviors as Forms of Environmental Movement Participation." *Mobilization: An International Journal* 3, no. 2 (1998): 185–205.

Althusser, Louis. *Essays on Ideology*. London: Verso, 1984.

———. *For Marx*. Translated by Ben Brewster. New York: Random House, 1970.

Andrews, James R. "Confrontation at Columbia: A Case Study in Coercive Rhetoric." *Quarterly Journal of Speech* 55 (1969): 37–46.

Andrews, Molly. *Lifetimes of Commitment: Aging, Politics, Psychology*. New York: Cambridge University Press, 1991.

Anonymous. "Mainers Rally to Curb Carbon Emissions." *Associated Press State and Local Wire*, April 14, 2007. LexisNexis *Academic* database. http://www.lexisnexis.com.

Appell, David. "Toxic Bulbs." *Scientific American* 297 (2007): 30–32.

Aronowitz, Stanley. *Politics of Identity: Class, Culture, Social Movements*. New York: Routledge, 1992.

Arthos, John Jr. "The Shaman-Trickster's Art of Misdirection: The Rhetoric of Farrakhan and the Million Men." *Quarterly Journal of Speech* 87 (2001): 41–60.

Asen, Robert. "Seeking the 'Counter' in Counterpublics." *Communication Theory* 10 (2000): 424–446.

Asen, Robert, and Daniel C. Brouwer. "Introduction: Reconfigurations of the Public Sphere." In *Counterpublics and the State*, edited by Robert Asen and Daniel C. Brouwer, 1–32. Albany: SUNY Press, 2001.

Ashcraft, Karen Lee. "Resistance through Consent?" *Management Communication Quarterly* 19, no. 1 (2005): 67–90.

Ashford, Kate, Carolyn Bigda, Tara Kalwarski, Sarah Max, Donna Rosato, Asa Fitch, Michelle Kalkhoff, Shelly Ng, and Ingrid Tharasook. "Best Places to Live." *Money* (August 2006): 95–108.

Bakhtin, Mikhail. *The Dialogic Imagination: Four Essays by M. M. Bakhtin*. Austin: University of Texas Press, 1981.

Bakken, Peter W., Joan Gigg Engel, and Ronald J. Engel. *Ecology, Justice and Christian Faith: A Critical Guide to the Literature.* Westport, CT: Greenwood Press, 1995.

Ball, Jeffrey. "Wall Street Shows Skepticism Over Coal: Banks Push Utilities to Plan for Impact of Emissions Caps." *Wall Street Journal*, February 4, 2008. http://online.wsj.com/article_email/SB120209079624339759-lMyQjAxMDI4MDAyNDAwOTQwWj.html.

Barber, Benjamin. *Strong Democracy: Participatory Politics for a New Age*. Berkeley: University of California Press, 1984.

Barra, Allen. "Trying for the Truth About the Alamo: A Report from the Field on the Attempt to Make a Historically Accurate Movie About an Event That Has Always Been Both Highly Controversial and Wildly Romanticized." *American Heritage* 54, no. 6 (2003): 50–55.

Barringer, Felicity. "Renewing a Call to Act Against Climate Change." *New York Times*, March 14, 2007.

Bauers, Sandy. "Rallying the Masses in Support of the Earth; Big Buzz Builds for Step It Up; A Nationwide Global-Warming Event

Tomorrow." *Philadelphia Inquirer*, April 13, 2007. LexisNexis *Academic* database. http://www.lexisnexis.com.

Bauman, Joe. "Guv Commits Utah to Climate Accord." *Deseret Morning News*, May 22, 2007. http://deseretnews.com/dn/view/0,1249,660222-802,00.html.

Baxter, Leslie A. "A Dialectical Perspective on Communication Strategies in Relationship Development." In *Handbook of Personal Relationships*, edited by Steve Duck, 257–274. London: Wiley, 1988.

Beck, Ulrich. *Ecological Enlightenment: Essays on the Politics of the Risk Society*. Atlantic Highlands, NJ: Humanities Press, 1995.

———. *Ecological Politics in an Age of Risk*. London: Polity, 1995.

———. *Risk Society: Towards a New Modernity*. Thousand Oaks, CA: Sage, 1992.

Begley, Sharon. "The Truth About Denial." *Newsweek*, August 13, 2007.

Beinecke, Frances. "The Twilight of Dirty Coal." *NRDC Switchboard*, February 11, 2008. http://switchboard.nrcd.org/fbeinecke/the_twilight_of_dirty_coal.html (accessed July 26, 2008).

Bekkers, V., H. V. Duivenboden, and M. Thaens. *Information and Communication Technology and Public Innovation: Assessing the ICT-Driven Modernization of Public Administration*. Fairfax, VA: IOS Press, 2006.

Benford, Robert D., and David A. Snow. "Framing Processes and Social Movements: An Overview and Assessment." *Annual Review of Sociology* 26 (2000): 611–639.

Bennett, Lance. "The Uncivic Culture: Communication, Identity, and the Rise of Lifestyle Politics." *Political Science and Politics* 31 (1998): 740–761.

Bennett, W. Lance. "Communicating Global Activism: Strengths and Vulnerabilities of Networked Politics." *Information, Communication, & Society* 6 (2003): 143–168.

Benson, Thomas W., and Bonnie Johnson. "The Rhetoric of Resistance: Confrontation with the Warmakers, Washington, D.C., October, 1967." *Today's Speech* 16 (1968): 35–42.

Berg, David. "Rhetoric, Reality, and Mass Media." *Quarterly Journal of Speech* 58 (1972): 255–263.

Berger, Peter L., and Thomas Luckmann. *The Social Construction of Reality: A Treatise in the Sociology of Knowledge*. Garden City, NY: Doubleday, 1966.

Best, Kirsty. "Rethinking the Globalization Movement: Toward a Cultural Theory of Contemporary Democracy and Communication." *Communication and Critical/Cultural Studies* 2 (2005): 214–237.

Bhattacharya, C. B., and Kimberly D. Elsbach. "Us Versus Them: The Roles of Organizational Identification and Disidentification in Social Marketing Initiatives." *Journal of Public Policy & Marketing* 21 (2002): 26–36.

Bitzer, Lloyd F. "The Rhetorical Situation." *Philosophy and Rhetoric* 1, no. 3 (1968): 1–14.

Blitefield, Jerry. "It's Showtime! Staging Public Demonstrations, Alinsky Style." In *Rhetorics of Display*, edited by Lawrence J. Prelli, 255–272. Columbia: University of South Carolina Press, 2006.

Blühdorn, Ingolfur, and Ian Welsh. "Eco-Politics Beyond the Paradigm of Sustainability: A Conceptual Framework and Research Agenda." *Environmental Politics* 16 (2007): 185–205.

Bobel, Chris. "'I'm Not an Activist, Though I've Done a Lot of It': Doing Activism, Being an Activist and the 'Perfect Standard' in a Contemporary Movement." *Social Movement Studies* 6, no. 2 (2007): 147–159.

Bowers, John W., Donovan J. Ochs, and Richard J. Jensen. *The Rhetoric of Agitation and Control*. Prospect Heights, IL: Waveland Press, 1993.

Brick, Philip, and R. McGreggor Cawley. "Producing Political Climate Change: The Hidden Life of US Environmentalism." *Environmental Politics* 17 (2007): 200–218.

Briggs, Charles L. *Disorderly Discourse: Narrative, Conflict and Inequality*. New York: Oxford University Press, 1996.

Brodkin, Karen. *Making Democracy Matter: Identity and Activism in Los Angeles*. New Brunswick, NJ: Rutgers University Press, 1997.

Brown, David D. "On Narrative and Belonging." In *Paul Ricoeur and Narrative: Context and Contestation*, edited by Joy Morny, 109–120. Calgary, Alberta: University of Calgary Press, 1997.

Brown, Lester R. "U.S. Moving Toward Ban on New Coal-Fired Power Plants." Earth Policy Institute. http://www.earth-policy.org/Updates/2008/Update70.htm.

Brown, Roger, and Albert Gilman. "The Pronouns of Power and Solidarity." In *Style in Language*, edited by T. Sebeok, 253–277. Cambridge, MA: MIT Press, 1960.

Brown, Roger, and Marguerite Ford. "Address in American English." *Journal of Abnormal & Social Psychology* 62 (1961): 375–385.

Brulle, Robert J., and J. Craig Jenkins. "Spinning Our Way to Sustainability?" *Organization & Environment* 19 (2006): 82–87.

Bryner, Gary. "Failure and Opportunity: Environmental Groups in US Climate Change Policy. " *Environmental Politics* 17 (2008): 319–336.

Buechler, Steven M. *Social Movements in Advanced Capitalism: The Political Economy and Cultural Construction of Social Activism*. New York: Oxford University Press, 2000.

Bullard, Robert D. *The Quest for Environmental Justice: Human Rights and the Politics of Pollution*. San Francisco: Sierra Club Books, 2005.

Bullard, Robert D., G. S. Johnson, and A. O. Torres. "Addressing Global Poverty, Pollution, and Human Rights." In *The Quest for*

Environmental Justice: Human Rights and the Politics of Pollution, edited by R. D. Bullard, 279–297. San Francisco: Sierra Club Books, 2005.

Burke, Kenneth. "Dramatism as Ontology or Epistemology: A Symposium." With Bernard L. Brock, Parke G. Burgess, and Herbert W. Simons. *Communication Quarterly* 33 (1985): 17–33.

———. *Language as Symbolic Action*. Berkeley: University of California Press, 1966.

———. *The Philosophy of Literary Form: Studies in Symbolic Action*. 3rd ed. Berkeley: University of California Press, 1973.

———. *A Rhetoric of Motives*. Berkeley: University of California Press, 1969.

Bush, George W. "Address to a Joint Session of Congress and the American People." http://www.whitehouse.gov/news/releases/2001/09/20010920-8.html.

Campbell, Karlyn K. "The Rhetoric of Women's Liberation: An Oxymoron." *Quarterly Journal of Speech* 59 (1973): 74–86.

Canham, Matt. "Utah Less Mormon Than Ever." *Salt Lake Tribune*, November 18, 2007.

Carroll, Joseph. "Americans Assess What They Can Do to Reduce Global Warming." *Gallup News Service*, April 24, 2007.

Carson, Rachel. *Silent Spring*. New York: Houghton Mifflin, 1962.

Cathcart, Robert S. "Movements: Confrontation as Rhetorical Form." *Southern Speech Communication Journal* 43 (1978): 233–247.

Catton, William R., and Riley E. Dunlap. "Paradigms, Theories, and Primacy of Hep-Nep Distinction." *American Sociologist* 13 (1978): 256–259.

CBS News. "Republicans Stall Climate Change Bill." CBSNews.com, June 6, 2008. http://www.cbsnews.com/stories/06/06/national/printable 4158834.shtml.

Chamberlin, Thomas Chrowder. "The Method of Multiple Working Hypotheses." *Science* 15, no. 366 (1890): 92–96.

Champagne, Anthony, and Edward J. Harpham, eds. *Texas at the Crossroads: People, Politics, and Policy*. College Station: Texas A&M University Press, 1987.

Chesters, Graeme, and Ian Welch. *Complexity and Social Movements: Multitudes at the Edge of Chaos*. New York: Routledge, 2006.

Christiansen, Adrienne E., and Jeremy J. Hanson. "Comedy as Cure for Tragedy: Act Up and the Rhetoric of AIDS." *Quarterly Journal of Speech* 82 (1996): 157–170.

City of Boston. "History of Climate Action in Boston." http://www.cityofboston.gov/ climate/history.asp.

City of Plano. "The City of Plano." http://www.plano.gov.

———. "Live Green in Plano." http://www.plano.gov/Departments/Environmental+Services/ GreenLiving/.

Clayton, Susan, and Susan Opotow. *Identity and the Natural Environment: The Psychological Significance of Nature*. Cambridge, MA: MIT Press, 2003.

Clegg, Stewart. "Organization and Control." *Administrative Science Quarterly* 26 (1981): 545–562.

Climate Challenge. www.climatechallenge.org.

Cloud, Dana. "Doing Away With Suharto—And the Twin Myths of Globalization and New Social Movements." In *Counterpublics and the State*, edited by Robert Asen and Daniel C. Brouwer, 235–263. Albany: SUNY Press, 2001.

Cole, Luke, and S. Foster. *From the Ground Up: Environmental Racism and the Rise of the Environmental Justice Movement*. New York: NYU Press, 2001.

Collinson, David. *Managing the Shopfloor: Subjectivity, Masculinity and Workplace Culture*. New York: Walter de Gruyter, 1992.

Committee on the Conduct of Science. "On Being a Scientist." *Proceedings of the National Academy of Sciences of the United States of America* 86 (1989): 9053–9074.

Cornog, Evan. *The power and the Story*. New York: Penguin, 2004.

Cox, Robert. "The Die is Cast: Topical and Ontological Dimensions of the Locus of the Irreparable." *Quarterly Journal of Speech* 68 (1982): 227–239.

———. *Environmental Communication and the Public Sphere*. Thousand Oaks, CA: Sage, 2006.

Cresswell, Tim. *Place: A Short Introduction*. Oxford: Blackwell Publishing, 2004.

Crompton, Tom. *Weathercocks and Signposts: The Environmental Movement at a Crossroads*. WWF-UK. http://www.wwf.org.uk/filelibrary/pdf/weathercocks_report.pdf.

Curran, John. "Global Warming 'Day of Action' Rallies Planned Nationwide Saturday." *Associated Press State & Local Wire*, April 13, 2007. LexisNexis *Academic* database. http://www.lexisnexis.com.

Curtin, P. "Reevaluating Public Relations Information Subsidies: Market-Driven Journalism and Agenda-Building Theory and Practice." *Journal of Public Relations Research* 11, no. 1 (1999): 53–90.

Curtin, P., and E. Rhodenbaugh. "Building the News Agenda on the Environment: A Comparison of Public Relations and Journalistic Sources." *Public Relations Review* 27 (2001): 179–195.

Czech, Brian. *Shoveling Fuel for a Runaway Train: Errant Economists, Shameful Spenders, and a Plan to Stop Them All*. Berkeley: University of California Press, 2000.

Daley, Beth. "N.E. Chorus Against Global Warming." *Boston Globe*, April 12, 2007. LexisNexis *Academic* database. http://www.lexisnexis.com.

Dauenhauer, Bernard P. "Ricoeur and Political Identity." In *Paul Ricoeur and Narrative: Context and Contestation*, edited by Joy Morny, 129–140. Calgary, Alberta: University of Calgary Press, 1997.

Davis, David Howard. *American Environmental Politics*. Chicago: Nelson-Hall, 1998.

Deetz, Stanley. "Critical Cultural Research: New Sensibilities and New Realities." *Journal of Management* 11, no. 2 (1985): 121–136.

Deetz, Stanley A. *Democracy in an Age of Corporate Colonization*. Albany: State University of New York Press, 1992.

Deetz, Stanley, and Dennis K. Mumby. "Power, Discourse, and the Workplace: Reclaiming the Critical Tradition." In Vol. 13, *Communication Yearbook*, edited by James Anderson, 18–47. Newbury Park, CA: Sage, 1990.

Delicath, John W. "Art and Advocacy: Citizen Participation Through Cultural Activism." In *Communication and Public Participation in Environmental Decision Making*, edited by Stephen P. DePoe, John W. Delicath, and Marie-France Aepli Elsenbeer, 255–266. Albany: State University of New York Press, 2004.

Delicath, John W., and Kevin Michael DeLuca. "Image Events, the Public Sphere, and Argumentative Practice: The Case of Radical Environmental Groups." *Argumentation* 17 (2003): 315–333.

Della Porta, Donatella, and Mario Diani. *Social Movements: An Introduction.* Oxford: Blackwell, 1999.

DeLuca, Kevin, and Jennifer Peeples. "From Public Sphere to Public Screen: Democracy, Activism, and the 'Violence' of Seattle." *Critical Studies in Media Communication* 19 (2002): 125–151.

DeLuca, Kevin M. *Image Politics: The New Rhetoric of Environmental Activism*. New York: Guilford, 1999.

———. "Unruly Arguments: The Body Rhetoric of Earth First!, Act Up, and Queer Nation." *Argumentation and Advocacy* 36 (1999): 9–21.

Democracy Now! "Step It Up: Thousands Gather This Weekend for Largest Ever Rally Against Global Warming." Democracy Now! transcript, April 13, 2007. http://www.democracynow.org/2007/4/13/step_it_up_thousands_gather_this.

Denzin, Norman K. *Interpretative Interactionism.* Newbury Park, CA: Sage, 1989.

Diamond, Jared M. *Collapse: How Societies Choose to Fail or Succeed.* New York: Penguin, 2005.

Donnelly, John. "Early Critic of Warming Steps Up Activist Role." *Boston Globe*, March 19, 2007. LexisNexis *Academic* database. http://www.lexisnexis.com.

Douglas, Susan. "Art + Activism = ATSA." *Women & Environments International Magazine* 72/73 (2006): 10–12.

Dunlap, Riley E., Kent D. Van Liere, Angela G. Mertig, and Robert E. Jones. "Measuring Endorsement of the New Ecological Paradigm: A Revised NEP Scale." *Journal of Social Issues* 56 (2000): 425–442.

Eder, Klaus. "The Institutionalization of Environmentalism: Ecological Discourse and the Second Transformation of the Public Sphere." In *Risk, Environment, and Modernity: Towards a New Ecology*, edited by Scott Lash, B. Szerszynski, and B. Wynne, 203–223. Thousand Oaks, CA: Sage, 1996.

Einstein, A. "Physics and Reality." *Journal of the Franklin Institute* 221, no. 3 (1936): 349–382.

Eisenberg, Eric M. "Ambiguity as Strategy in Organizational Communication." *Communication Monographs* 51 (1984): 227–242.

Entman, Robert M. "Framing: Toward Clarification of a Fractured Paradigm." *Journal of Communication* 43 (1993): 51–59.

Ezzamel, Mahmoud, Hugh Willmott, and Frank Worthington. “Power, Control and Resistance in the Factory That Time Forgot.” *Journal of Management Studies* 38 (2001): 1053–1078.

Faber, Daniel. “A More ‘Productive’ Environmental Justice Politics: Movement Alliances in Massachusetts for Clean Production and Regional Equity.” In *Environmental Justice and Environmentalism: The Social Justice Challenge to the Environmental Movement*, edited by Ronald Sandler and Phaedra Pezzullo, 309–320. Cambridge, MA: MIT Press, 2007.

Fairclough, Norman. “Critical Discourse Analysis and the Marketization of Public Discourse: The Universities.” *Discourse and Society* 4 (1993): 133–168.

———. *Language and Power*. London: Longman, 1989.

Fairhurst, Gail, and Linda Putnam. “Organizations as Discursive Constructions.” *Communication Theory* 14 (2004): 5–26.

Faver, Catherine A. “Rights, Responsibility and Relationship: Motivations for Women’s Social Activism.” *Affilia* 16 (2001): 314–336.

Fehrenbacher, Kati. “T. Boone Pickens: Kicking Off the World’s Largest Wind Farm.” Earth2Tech. http://earth2tech.com/2008/04/14/t-boone-pickens-kicking-off-the-worlds-largest-wind-farm/.

Felski, Rita. *Beyond Feminist Aesthetics: Feminist Literature and Social Change*. Cambridge, MA: Harvard University Press, 1989.

Fernandez, Manny. “For the Environment, Rallies Great and Small (and Unusual Attire).” *New York Times*, April 15, 2007. LexisNexis *Academic* database. http://www.lexisnexis.com.

Feyerabend, Paul. *Against Method: Outline of an Anarchistic Theory of Knowledge*. London: NLB, 1975.

———. *Farewell to Reason*. London: Verso, 1987.

———. *Science in a Free Society*. London: NLB, 1978.

Fink, Marc. "Toward a Sunbelt Urban Design Manifesto." *Journal of the American Planning Association* 59 (1993): 320–333.

First National People of Color Environmental Leadership Summit. "Principles of Environmental Justice." Web Resources for Environmental Justice Activists. http://www.ejnet.org/ej/ principles.html.

Fischer, David Hackett. "Boston Common." *American Places: Encounters with History*, edited by William E. Leuchtenburg, 125–143. New York: Oxford University Press, 2000.

Fisher, Walter R. *Human Communication as Narration: Toward a Philosophy of Reason, Value, and Action*. Columbia: University of South Carolina Press, 1987.

Fitzgerald, Kathleen, and Diane Rodgers. "Radical Social Movement Organizations: A Theoretical Model." *Sociological Quarterly* 41 (2000): 573–592.

Flanagin, Andrew, Cynthia Stohl, and Bruce Bimber. "Reconceptualizing Collective Action in the Contemporary Media Environment." *Communication Theory* 15 (2005): 389–413.

Flemmons, Stephanie, and Liz McGathey. "'Green' Challenge to Residents, Businesses." *Plano Insider*, April 9, 2007. http://www.planoinsider.net/articles/2007/04/09/news/story3.txt.

Flores, Richard R. "Memory-Place, Meaning, and the Alamo." *American Literary History* 10, no. 3 (1998): 428–445.

Focus the Nation. "Focus the Nation: Global Warming Solutions for America." http://www.focusthenation.org.

Foot, Kirsten A., and Steven M. Schneider. "Online Action in Campaign 2000: An Exploratory Analysis of the U.S. Political Web Sphere." *Journal of Broadcasting and Electronic Media* 46, no. 2 (2002): 222–244.

Forsbert, Liz, Leah Burns, Clare O'Connor, Laura Reinsborough, and Jill Tomac. "Transforming Space into Place." *Women and Environments International Magazine* 72/73 (2006): 44–46.

Foss, Karen A., and Kathy L. Domenici. "Haunting Argentina: Synecdoche in the Protests of the Mothers of the Plaza de Mayo." *Quarterly Journal of Speech* 87 (2001): 237–258.

Foss, Sonja K. *Rhetorical Criticism: Exploration & Practice*. 2nd ed. Prospect Heights, IL: Waveland Press, 1996.

Fraser, Nancy. "Rethinking the Public Sphere: A Contribution to the Critique of Actually Existing Democracy." In *Habermas and the Public Sphere*, edited by Craig Calhoun, 109–142. Cambridge, MA: MIT Press, 1991.

Frederick, Howard. "Networks and the Emergence of Global Civil Society." In *Global Networks: Computers and International Communication*, edited Linda M. Harasim, 283–295. Cambridge, MA: MIT Press, 1993.

Frey, Lawrence, and Kevin Barge. *Managing Group Life: Communication in Decision-Making Groups*. Boston: Houghton Mifflin, 1998.

Fried, Amy. "US Environmental Interest Groups and the Promotion of Environmental Values: The Resounding Success and Failure of Earth Day." *Environmental Politics* 7 (1998): 1–22.

Gadette, Jamie. "Music Picks." *Salt Lake City Weekly*, April 12, 2007.

Gamson, William. "Constructing Social Protest." In *Social Movements and Culture*, edited by Hank Johnston and Bert Klandermans, 85–106. Minneapolis: University of Minnesota Press, 1995.

———. *Strategy of Social Protest*. Homewood, IL: Dorsey Press, 1975.

Gamson, William, David Croteau, William Hoynes, and Theodore Sasson. "Media Images and the Social Construction of Reality." *Annual Review of Sociology* 18 (1992): 373–393.

Ganesh, Shiv, Heather Zoller, and George Cheney. "Transforming Resistance, Broadening Our Boundaries: Critical Organizational Communication Meets Globalization From Below." *Communication Monographs* 72, no. 2 (2005): 169–191.

Garreau, Joel. *Edge City: Life on the Frontier*. New York: Anchor Books, 1991.

Geertz, Clifford. *The Interpretation of Cultures*. New York: Basic, 1973.

Gettier, Edmund L. "Is Justified True Belief Knowledge?" *Analysis* 23, no. 6 (1963): 121–123.

Gieryn, Thomas F. "Boundaries of Science." In *Handbook of Science and Technology Studies*, edited by Sheila Jasanoff, Gerald E. Markle, James C. Petersen, and Trevor Pinch, 393–443. Thousand Oaks, CA: Sage, 1995.

———. "Boundary-Work and the Demarcation of Science From Non-Science: Strains and Interests in Professional Ideologies of Science." *American Sociological Review* 48 (1983): 781–795.

Global Climate Campaign. "International Demonstrations on Climate Change." http://www.globalclimatecampaign.org.

Goffman, Erving. *Frame Analysis: An Essay on the Organization of Experience*. Cambridge, MA: Harvard University Press, 1974.

Goodnight, G. Thomas. "The Personal, Technical, and Public Spheres of Argument: A Speculative Inquiry into the Art of Public Deliberation." *Journal of the American Forensic Association* 18 (1982): 214–227.

Goodwyn, Lawrence. *Texas Oil, American Dreams: A Study of the Texas Independent Producers and Royalty Owners Association*. Austin: Texas State Historical Association, 1996.

Gottlieb, Roger S. "Religious Environmentalism in Action." In *The Oxford Handbook of Religion and Ecology*, edited by R. S. Gottlieb, 467–509. Cambridge: Oxford University Press, 2006.

Graham, Don. *Kings of Texas: The 150-Year Saga of an American Ranching Empire*. Hoboken, NJ: John Wiley and Sons, 2003.

Granzin, Kent L., and Janeen E. Olson. "Characterizing Participants in Activities Protecting the Environment: A Focus on Donating, Recycling, and Conservation Behaviors." *Journal of Public Policy & Marketing* 10 (1991): 1–27.

Gregg, Richard. "The Ego-Function of the Rhetoric of Protest." *Philosophy and Rhetoric* 4 (1971): 71–91.

Griffin, Leland M. "The Rhetorical Structure of the 'New Left' Movement: Part I." *Quarterly Journal of Speech* 50, no. 2 (1964): 113–135.

Grist. "One giant step for greenkind." *Grist Environmental News and Commentary*, June 19, 2007. www.grist.org/news/maindish/2007/04/19/stepitup.

Grønhøj, Alice. "Communication About Consumption: A Family Process Perspective on 'Green' Consumer Practices." *Journal of Consumer Behavior* 5 (2006): 491–503.

Grossberg, Larry. "History, Politics and Postmodernism: Stuart Hall and Cultural Studies." In *Critical Dialogues in Cultural Studies*, edited by David Morley and Kuan-Hsing Chen, 150–173. New York: Routledge, 2003.

Haack, Susan. *Defending Science—Within Reason: Between Scientism and Cynicism*. Amherst, NY: Prometheus Books, 2003.

Haack, Susan, and Robert Lane, eds. *Pragmatism, Old and New: Selected Writings*. Amherst, NY: Prometheus Books, 2006.

Habermas, Jürgen. "New Social Movements." *Telos* 49 (1981): 33–37.

———. "The Public Sphere: An Encyclopedia Article." *New German Critique* 1 (1974): 49–55.

———. *The Structural Transformation of the Public Sphere: An Inquiry into a Category of Bourgeois Society*. Translated by Thomas Burger. Cambridge, MA: MIT Press, 1989

———. *The Theory of Communicative Action, Volume 1: Reason and the Rationalization of Society*. Translated by T. McCarthy. Boston: Beacon, 1984.

Hafner, E. M., and Susan Presswood. "Strong Inference and Weak Interactions." *Science* 149, 2no. 3683 (1965): 503–510.

Haiman, Franklyn S. "Nonverbal Communication and the First Amendment: The Rhetoric of the Streets Revisited." *Quarterly Journal of Speech* 68 (1982): 371–383.

———. "The Rhetoric of the Streets Revisited: Some Legal and Ethical Considerations." *Quarterly Journal of Speech* 53 (1967): 99–114.

Haker, Hermann. "Narrative and Moral Identity in Paul Ricoeur." In *Creating Identity*, edited by Hermann Haring, Maureen Junker-Kenny, and Dietmar Mieth, 59–68. London: SCM Press, 2000.

Hariman, Robert, and John Louis Lucaites. "Public Identity and Collective Memory in U.S. Iconic Photography: The Image of 'Accidental Napalm.'" *Critical Studies in Media Communication* 20 (2003): 35–66.

Harré, Rom. "An Outline of the Social Constructionist Viewpoint." In *The Social Construction of Emotions*, edited by Rom Harré, 2–14. New York: Basil Blackwell, 1986.

Harrold, Christine. *Ourspace: Resisting the Corporate Control of Culture*. Minneapolis: University of Minnesota Press, 2007.

Hauser, Gerard A. "Demonstrative Displays of Dissident Rhetoric: The Case of Prisoner 885/63." In *Rhetorics of Display*, edited by Lawrence J. Prelli, 229–254. Columbia: University of South Carolina Press, 2006.

Hayden, Sara. "Family Metaphors and the Nation: Promoting a Politics of Care Through the Million Mom March." *Quarterly Journal of Speech* 89 (2003): 196–215.

Hilding, Tina. "The Heat is On: Community Discussion on Climate Change." Moscow Climate Change. http://www.moscowclimatechange.com/ (accessed November 30, 2007).

Hindman, Matthew. "'Open-Source Politics' Reconsidered: Emerging Patterns in Online Political Participation." In *Governance and*

Information Technology, edited by Viktor Mayer-Schonberger and David Lazer, 183–208. Cambridge, MA: MIT Press, 2006.

Hobson, Barbara. *Recognition Struggles and Social Movements: Contested Identities, Agency and Power*. Cambridge: Cambridge University Press, 2003.

Hubbard, Dean. "What Kind of Globalization? Organizing for Workers' Human Rights." *The Journal of Labor and Society* 9 (2006): 315–335.

Hybrid Cars. "Battery Toxicity." http://www.hybridcars.com/battery toxicity.html.

Hylton, Hilary, and Cam Rossie. *The Insider's Guide to Austin*. 3rd ed. Guilford, CT: Globe Pequot Press, 2001.

Intergovernmental Panel on Climate Change. *Climate Change 2007: Impacts, Adaptation and Vulnerability; Working Group II Contribution to the Intergovernmental Panel on Climate Change Fourth Assessment Report; Summary for Policymakers*. http://www.ipcc.ch/pdf/assessment-report/ar4/wg2/ar4-wg2-spm.pdf.

———. *Climate Change 2007: The Physical Science Basis: Contribution of Working Group I to the Fourth Assessment Report of the IPCC*. Cambridge: Cambridge University Press, 2007.

International Climate Justice Network. "Bali Principles of Climate Justice." Web Resources for Environmental Justice Activists. www.ejnet.org/ej/bali.pdf.

International Council for Local Environmental Initiatives. "Local Governments for Sustainability." http://www.iclei.org/.

Ivie, Robert L. *Democratic Dissent and the Trick of Rhetorical Critique*. Bloomington, IN: The Polynter Center for the Study of Ethics and American Institutions, 2005.

———. "Prologue to Democratic Dissent in America." *Jasnost—The Public* 11 (2004): 19–36.

Jamieson, Dale. "Ethics, Public Policy and Global Warming." *Science Technology Human Values* 17 (1992): 139–153.

———. "Justice: The Heart of Environmentalism." In *Environmental Justice and Environmentalism: The Social Justice Challenge to the Environmental Movement*, edited by Ronald Sandler and Phaedra Pezzullo, 85–101. Cambridge, MA: MIT Press, 2007.

———. "The Moral and Political Challenges of Climate Change." In *Creating a Climate for Change: Communicating Climate Change and Facilitating Social Change*, edited by Susanne C. Moser and Lisa Dilling, 475–482. Cambridge: Cambridge University Press, 2007.

Jasinoff, Sheila, Gerald E. Markle, James C. Petersen, and Trevor Pinch, eds. *Handbook of Science and Technology Studies*. Thousand Oaks, CA: Sage, 1995.

Jasinski, James. *Sourcebook on Rhetoric: Key Concepts in Contemporary Rhetorical Studies*. Thousand Oaks, CA: Sage, 2001.

Johnson, Chalmers A. *Revolutionary Change.* Stanford, CA: Stanford University Press, 1982.

Johnson, Davi. "Martin Luther King Jr.'s 1963 Birmingham Campaign as Image Event." *Rhetoric and Public Affairs* 10, no. 1 (2007): 1–26.

Johnson, Scott D., and L. Long. "'Being a Part and Being Apart': A Dialectical Perspective on Group Communication." In *New Directions in Group Communication Research*, edited by Larry Frey, 25–42. Thousand Oaks, CA: Sage, 2002.

Juris, Jeffery. "The New Digital Media and Activist Networking Within Anti-Corporate Globalization Movements." *The ANNALS of the American Academy of Political and Social Science* 597 (2005): 189–208.

Kahn, Richard, and Douglas Kellner. "New Media and Internet Activism: From the 'Battle in Seattle' to Blogging." *New Media & Society* 6 (2004): 87–95.

Kay, Jane. "The Plastic Garbage Pit of the Pacific." *San Francisco Chronicle*, November 6, 2006: 4.

Kellogg, W. "Community-Based Organizations and Neighborhood Environmental Problem Solving: A Framework for Adoption of Information Technologies." *Journal of Environmental Planning and Management* 42, no. 4 (1999): 445–469.

Kellogg, W., and A. Mathur. "Environmental Justice and Information Technologies: Overcoming the Information-Access Paradox in Urban Communities." *Public Administration Review* 63, no. 5 (2003): 573–585.

Killingsworth, M. Jimmie, and Jacqueline S. Palmer. *Ecospeak: Rhetoric and Environmental Politics in America.* Carbondale: Southern Illinois University Press, 1991.

———. "Millennial Ecology: The Apocalyptic Narrative from *Silent Spring* to Global Warming." In *Green Culture: Environmental Rhetoric in Contemporary America*, edited by Carl G. Herndl and Stuart C. Brown, 221–245. Madison: University of Wisconsin Press, 1996.

Kinsella, William J., and Jay Mullen. "Becoming Hanford Downwinders: Producing Community and Challenging Discursive Containment." In *Nuclear Legacies: Communication, Controversy, and the U.S. Nuclear Weapons Complex*, edited by Bryan C. Taylor, William J. Kinsella, Stephen P. Depoe, and Maribeth Metzler, 73–107. Lanham, MD: Lexington Books, 2007.

Klandermands, Bart. *The Social Psychology of Protest*. Oxford: Blackwell, 1997.

Kramer, Michael. "Toward a Communication Theory of Group Dynamics: An Ethnographic Study of a Community Theater Group." *Communication Monographs* 71 (2004): 311–322.

Krauss, Clifford. "Move Over, Oil, There's Money in Texas Wind. *New York Times*, February 23, 2008. http://www.nytimes.com/2008/02/23/business/ 23wind.html?pagewanted=1&_r=2&hp.

Kuhn, Thomas S. *The Structure of Scientific Revolutions*. Chicago: University of Chicago Press, 1962.

Lacitis, Erik. "Placid March Seeks Action on Warming." *Seattle Times*, April 15, 2007. http://seattletimes.nwsource.com/html/localnews/2003667134_stepitup15m.html (accessed August 27, 2009).

Lakatos, Imre. "Falsification and the Methodology of Scientific Research Programmes." In *Criticism and the Growth of Knowledge*, edited by I. Lakatos and A. Musgrave, 91–195. Cambridge: Cambridge University Press, 1970.

Lakoff, George. *Don't Think of an Elephant: Know Your Values and Frame the Debate*. New York: Chelsea Green Publishers, 2004.

Laraña, Enrique, Hank Johnston, and Joseph R. Gusfield, eds. *New Social Movements*. Philadelphia: Temple University Press, 1994.

Larson, John, and Robert Heilmayr. "Comparison of Legislative Climate Change Targets in the 110th Congress." World Resources Institute. http://www.wri.org/publications/ usclimatetargets.

Latour, Bruno. *We Have Never Been Modern*. Translated by C. Porter. Cambridge, MA: Harvard University Press, 1993.

Leiserowitz, Anthony. "Communicating the Risks of Global Warming: American Risk Perceptions, Affective Images, and Interpretive Communities." In *Creating a Climate for Change: Communicating Climate Change and Facilitating Social Change*, edited by Susanne C. Moser and Lisa Dilling, 44–63. Cambridge: Cambridge University Press, 2007.

Levenson, Michael. "Young, Old Urge Action on Global Warming—From Maine to Boston, Hundreds Rally on Dangers of Climate Change." *Boston Globe*, April 15, 2007. LexisNexis Academic database. http://www.lexisnexis.com.

Lichterman, Paul. *The Search for Political Community: American Activists Reinventing Commitment*. Cambridge: Cambridge University Press, 1996.

Lincoln, Yvonna S., and Egon G. Guba. *Naturalistic Inquiry*. Newbury Park, CA: Sage, 1985.

Lindberg, Anne. "Local Rally One Voice in Chorus of Protest." *St. Petersburg* [Florida] *Times*, April 11, 2007. LexisNexis *Academic* database. http://www.lexisnexis.com.

Livingston, David, and Richard Harrison. "Meaning through Metaphor: Analogy as Epistemology." *Annals of the Association of American Geographers* 71, no. 1 (1981): 95–107.

Lord, Clara, and Phillipe Keith. "Threatened Fishes of the World: Protogobius Attiti Watson & Pollabauer." *Environmental Biology of Fish* 77 (1998): 101–102.

Luhmann, Niklas. *Ecological Communication*. Translated by John Bednarz Jr. Chicago: University of Chicago Press, 1989.

Lutz, Catherine. *Unnatural Emotions: Everyday Sentiments on a Micronesian Atoll and Their Challenge to Western Theory*. Chicago: University of Chicago Press, 1988.

Mabry, Edward A. "The Systems Metaphor in Group Communication." In *The Handbook of Group Communication Theory & Research*, edited by Larry R. Frey, Dennis S. Gouran, and Marshall S. Poole, 71–91. Thousand Oaks, CA: Sage.

Manovich, Lee. *The Language of New Media*. Cambridge, MA: MIT Press, 2002.

Mason, Michael. "Transnational Environmental Obligations: Locating New Spaces of Accountability in a Post-Westphalian Global Order." *Transactions of the Institute of British Geographers* 26, no. 4 (2001): 407–429.

McAdam, Doug. *Political Process and the Development of Black Insurgency*. Chicago: University of Chicago Press, 1982.

McAdam, Doug, and W. Richard Scott. "Organizations and Movements." In *Social Movements and Organizational Theory*, edited by Gerald F.

Davis, Doug McAdam, W. Richard Scott, and Mayer N. Zald, 4–40. Cambridge: Cambridge University Press, 2006.

McKeon, Richard. "Discourse, Demonstration, Verification, Justification." In *Demonstration, Verification, Justification*, edited by Philippe Devaux, 37–55. Paris: Nauwelaerts, 1968.

———. "The Uses of Rhetoric in a Technological Age: Architectonic Productive Arts." In *The Province of Rhetoric*, edited by Lloyd F. Bitzer and Edwin Black, 44–63. Englewood Cliffs, NJ: Prentice-Hall, 1971.

McKibben, Bill. "Gandhi: A Man for All Seasons—What We Still Need to Learn From the Example of Gandhi." *Odemagazine.com*, January/February 2008. http://www.odemagazine.com/doc/50/what-the-world-needs-now.

———. "Hi, Tech: Who Needs Newspapers When the Web Can Do the Job?" *Grist*, February 26, 2007. http://www.grist.org/comments/dispatches/2007/01/08/mckibben/index7.html

(accessed June 9, 2008).

———. "Imagine That: What the Warming World Needs Now is Art Sweet Art." *Grist Magazine*, April 21, 2005. http://www.grist.org/comments/soapbox/2005/04/21/ mckibben-imagine/.

———. "A Letter from Bill McKibben." StepItUp.org. http://april.stepitup2007.org/ article.php?id=78.

———. "A Note From Bill McKibben." http://www.350.org.

———. "Step It Up: Introducing a Brand-New, Mass-Protest Climate Movement for 2007." *Grist: Environmental News and Commentary*, January 8, 2007. http://www.grist.org/comments/dispatches/2007/01/08/mckibben/index.html.

———. "We Stepped It Up! Day of Action Was a Raging Success, With 1,400 Climate Events around the U.S." *Grist*, April 15, 2007. http://www.grist.org/comments/dispatches/ 2007/01/08/mckibben/index14.html.

McKibben, Bill, Phil Aroneanu, Will Bates, May Boeve, Jamie Henn, Jeremy Osborn, and Jon Warnow. *Fight Global Warming Now: The Handbook for Taking Action in Your Community*. New York: Holt, 2007.

Mehra, B., C. Merkel, and A. P. Bishop. "The Internet for Empowerment of Minority and Marginalized Users." *New Media Society* 6 (2004): 718–802.

Melucci, Alberto. *Challenging Codes: Collective Action in the Information Age*. Cambridge: Cambridge University Press, 1996.

———. "The Process of Collective Identity." In *Social Movements and Culture*, edited by Hank Johnston and Bert Klandermans, 41–63. Minneapolis: University of Minnesota Press, 1995.

———. "A Strange Kind of Newness: What's 'New' in Social Movements?" In *New Social Movements: From Ideology to Identity*, edited by Enrique Larana, Hank Johnson, and Joseph R. Gunsfield, 101–130. Philadelphia: Temple University Press, 1994.

Menino, Thomas M. Remarks to the Greater Boston Chamber of Commerce, December 11, 2007. City of Boston. http://www.cityofboston.gov/mayor/pdfs/2007Chamber.pdf.

Middleton, Richard. *Studying Popular Music*. Philadelphia: Open University Press, 1990.

Mitchell, W. J. T. "What Do Pictures Really Want?" *October* 77 (1996): 71–82.

Moisander, Johanna. "Motivational Complexity of Green Consumerism." *International Journal of Consumer Studies* 31 (2007): 404–409.

Mol, Hans J. *Identity and the Sacred: A Sketch for a New Social-Scientific Theory of Religion*. New York: The Free Press, 1976.

Moore, Charles. "Trashed: Across the Pacific Ocean, Plastics, Plastics, Everywhere." *Natural History* 112 (2003): 46–51.

Morris, Charles E. III, and Stephen Howard Browne. Introduction in *Readings on the Rhetoric of Social Protest*, 2nd ed., edited by

Charles E. Morris III and Stephen Howard Browne, 1–3. State College, PA: Strata Publishing, 2006.

Morrison, Michael L., William M. Block, M. Dale Strickland, Bret A. Collier, and Markus J. Peterson. *Wildlife Study Design*. 2nd ed. New York: Springer, 2008.

Moser, Susanne C. "Communication Strategies." In *Ignition: What You Can Do to Fight Global Warming and Spark a Movement*, edited by Jonathan Isham and Sissel Waage, 11–21. Washington, DC: Island Press, 2007.

———. "Talk of the City: Engaging Urbanites on Climate Change." *Environmental Research Letters* 1 (2006): 1–10.

Moser, Susanne C., and Lisa Dilling. "Making Climate Hot: Communicating the Urgency and the Challenge of Global Climate Change." *Environment* 46 (2004): 32–46.

Moser, Susanne C., and Lisa Dilling, eds. *Creating a Climate for Change: Communicating Climate Change and Facilitating Social Change*. Cambridge: Cambridge University Press, 2007.

Mossberger, K., C. J. Tolbert, and M. Gilbert. "Race, Place and Information Technology." *Urban Affairs Review* 41 (2006): 583–620.

Mouffe, Chantal. *The Democratic Paradox*. New York: Verso, 2000.

———. *On the Political: Thinking in Action*. New York: Routledge, 2005.

———. "Schmitt's Vision of a Multipolar World Order." *South Atlantic Quarterly* 104, no. 2 (2005): 245–251.

Mumby, Dennis. "The Problem of Hegemony: Rereading Gramsci for Organizational Communication Studies." *Western Journal of Communication* 61, no. 4 (1997): 343–375.

———. "Theorizing Resistance in Organizational Studies: A Dialectical Approach." *Management Communication Quarterly* 19, no. 1 (2005): 19–44.

Murphy, Dennis D., and Barry D. Noon. "Coping with Uncertainty in Wildlife Biology." *Journal of Wildlife Management* 55, no. 4 (1991): 773–782.

Naess, Arne. "Intrinsic Value: Will the Defenders of Nature Please Rise?" In *Conservation Biology: The Science of Scarcity and Diversity*, edited by M. E. Soulé, 504–515. Sunderland, MA: Sinauer Associates, 1986.

———. "Shallow and Deep, Long-Range Ecology Movement: A Summary." *Inquiry—An Interdisciplinary Journal of Philosophy* 16, no. 1 (1973): 95–100.

National Energy Technology Lab. "Tracking New Coal-Fired Power Plants."

Coal Moratorium NOW! http://cmnow.org/NETL%20New%20Coal%205.2007.pdf.

National Telecommunications and Information Administration. U.S. Department of Commerce. "Falling through the Net, Toward Digital Inclusion." http://www.ntia.doc.gov.

———. "National Telecommunications and Information Administration Annual Report 2000." http://www.ntia.doc.gov.

———. "A Nation Online: How Americans Are Expanding Their Use of the Internet." http://www.ntia.doc.gov.

NBC News. *NBC Nightly News* transcript, April 24, 2007. LexisNexis *Academic* database. http://www.lexisnexis.com.

Nordhaus, Ted, and Michael Shellenberger. *Break Through: From the Death of Environmentalism to the Politics of Possibility*. New York: Houghton Mifflin, 2007.

Norton, Todd. "Property as Interorganizational Discourse: Rights in the Politics of Public Spaces." *Communication Theory* 18, no. 2 (2008): 210–239.

NPR (National Public Radio). "Rallies Encourage U.S. to Cut Carbon Emissions." *Talk of the Nation* transcript, April 13, 2007. LexisNexis *Academic* database. http://www.lexisnexis.com.

Olien, Roger M., and Diana Davids Olien. *Oil in Texas: The Gusher Age, 1895–1945*. Austin: University of Texas Press, 2002.

Orr, David. "The Politics of Poetry." *Poetry* 192 (2008): 408–418.

———. *Ecological Literacy: Education and the Transition to a Postmodern World.* Albany: State University of New York, 1992.

Ott, Brian L., and Eric Aoki. "The Politics of Negotiating Public Tragedy: Media Framing of the Matthew Shepard Murder." *Rhetoric & Public Affairs* 5, no. 3 (2002): 483–505.

Palczewski, Catherine Helen. "Cyber-Movements, New Social Movements, and Counterpublics." In *Counterpublics and the State*, edited by Robert Asen and Daniel C. Brouwer, 161–186. Albany: SUNY Press, 2001.

Papacharissi, Zizi. "Democracy Online: Civility, Politeness and the Democratic Potential of Online Political Discussion Groups." *New Media & Society* 6, no. 2 (2004): 259–283.

Pearce, W. Barnett. "A Sailing Guide for Social Constructionists." In *Social Approaches to Communication*, edited by W. Leeds-Hurwitz, 88–113. New York: Guilford Press, 1995.

Pearce, W. Barnett, and Stephen W. Littlejohn. *Moral Conflict: When Social Worlds Collide*. Newbury Park, CA: Sage, 1997.

Peters, John D. *Courting the Abyss: Free Speech and the Liberal Tradition*. Chicago: University of Chicago Press, 2005.

Peters, Stephanie M. "Global Warming Opponents Demand Cut in Carbon Emissions, Creation of 5m 'Green Jobs' Urged." *Boston Globe*, November 4, 2007. LexisNexis *Academic* database. http://www.lexisnexis.com.

Peterson, M. Nils, Markus J. Peterson, and Tarla R. Peterson. "Conservation and the Myth of Consensus." *Conservation Biology* 19 (2005): 762–767.

Peterson, M. Nils, Stacey A. Allison, Markus J. Peterson, Tarla R. Peterson, and Roel R. Lopez. "A Tale of Two Species: Habitat Conservation Plans as Bounded Conflict." *Journal of Wildlife Management* 68 (2004): 743–761.

Peterson, Markus J., and Tarla Rai Peterson. "Ecology: Scientific, Deep and Feminist." *Environmental Values* 5, no. 2 (1996): 123–146.

Peterson, Tarla Rai. "Jefferson's Yeoman Farmer as Frontier Hero: A Self-Defeating Mythic Structure." *Agriculture and Human Values* 7 (1990): 9–19.

———. "Telling the Farmers' Story: Competing Responses to Soil Conservation Rhetoric." *The Quarterly Journal of Speech* 77, no. 3 (1991): 289–308.

———. *Sharing the Earth: The Rhetoric of Sustainable Development.* Columbia: University of South Carolina Press, 1997.

Peterson, Tarla Rai, M. Nils Peterson, Markus J. Peterson, Stacey A. Allison, and David Gore. "To Play the Fool: Can Environmental Conservation and Democracy Survive Social Capital?" *Communication and Critical/Cultural Studies* 3 (2006): 116–140.

Peterson, Tarla Rai, and Todd Norton. "Discourses of Sustainability in Today's Public Sphere." In *The Debate over Corporate Social Responsibility*, edited by Steven May, George Cheney, and Juliet Roper, 786–819. Oxford: Oxford University Press, 2007.

Pew Reports. "An Even More Partisan Agenda for 2008." The Pew Research Center for the People and the Press. http://people-press.org/report/388/an-even-more-partisan-agenda-for-2008.

Pezzullo, Phaedra C. *Toxic Tourism: Rhetorics of Pollution, Travel, and Environmental Justice*. Tuscaloosa: University of Alabama Press, 2007.

Pezzullo, Phaedra, and Ronald Sandler. "Revisiting the Environmental Justice Challenge to Environmentalism." In *Environmental Justice and Environmentalism: The Social Justice Challenge to the Environmental Movement*, edited by Ronald Sandler and Phaedra Pezzullo, 1–24. Cambridge, MA: MIT Press, 2007.

———. "Working Together and Working Apart." In *Environmental Justice and Environmentalism: The Social Justice Challenge to the Environmental Movement*, edited by Ronald Sandler and Phaedra Pezzullo, 309–320. Cambridge, MA: MIT Press, 2007.

Pilkington, Tom. *State of Mind: Texas Literature and Culture*. College Station: Texas A&M University Press, 1998.

Pinker, Steven. "The Moral Instinct." *New York Times Magazine*, January 13, 2008. www.nytimes.com/pages/magazine/index.html.

Pirages, Paul R., and Dennis Ehrlich. *Ark II: Social Response to Environmental Imperatives*. San Francisco: W. H. Freeman, 1974.

Plato. *Theaetetus*. Translated by J. McDowell. Oxford: Clarendon Press, 1973.

Platt, John R. "Strong Inference: Certain Systematic Methods of Scientific Thinking May Produce Much More Rapid Progress Than Others." *Science* 146, no. 3642 (1964): 347–353.

Pol, Enrique. "The Theoretical Background of the City-Identity-Sustainability Network." *Environment and Behavior* 34 (2002): 8–25.

Pollan, Michael. "Why Bother?" *New York Times Magazine*, April 20, 2008. http://www.nytimes.com/indexes/2008/04/19/magazine/index.html.

Popper, Karl Raimund. *Conjectures and Refutations: The Growth of Scientific Knowledge*. New York: Basic Books, 1962.

———. *The Logic of Scientific Discovery*. London: Hutchinson, 1959.

———. *Logik der Forschung: Zur Erkenntnistheorie der Modernen Naturwissenschaft*. Wien, Österreich: Springer, 1934.

Poster, Mark. "Cyberdemocracy: Internet and the Public Sphere." In *Internet Culture*, edited by David Porter, 201–217. New York: Routledge, 1997.

———. *What's the Matter with the Internet?* Minneapolis: University of Minnesota Press, 2001.

Potter, Sulamith Heins. "The Cultural Construction of Emotion in Rural Chinese Social Life." *Ethos* 16, no. 2 (1988): 181–208.

Power Shift 2007. "About Power Shift 2007." http://powershift07.org/about.

Prelli, Lawrence J. "Rhetorics of Display: An Introduction." In *Rhetorics of Display*, edited by Lawrence J. Prelli, 1–38. Columbia: University of South Carolina Press, 2006.

Prothero, A. "Green Consumerism and the Societal Marketing Concept: Marketing Strategies for the 1990s." *Journal of Marketing Management* 6 (1990): 87–103.

Putnam, Linda L., and Cynthia Stohl. "Bona Fide Groups: A Reconceptualization of Groups in Context." *Communication Studies* 41 (1990): 248–265.

Rabe, Barry G. "Race to the Top: The Expanding Role of U.S. State Renewable Portfolio Standards." Pew Center on Global Climate Change. http://www.pewclimate.org/global-warming-in-depth/all_reports/race_to_the_top.

Raging Grannies. http://www.geocities.com/raginggrannies/.

Rasmussen, Matt. "Green Rage." *Orion*, January/February 2007. http://www.orionmagazine.org/ index.php/articles/article/6/ (accessed February 1, 2007).

Raymond, Eric. *The Cathedral and the Bazaar: Musing on Linux and Open Source by an Accidental Revolutionary*. Sebastopol, CA: O'Reilly and Associates, 1997.

ReDirect Guide. *Make Change with Every Dollar: 10 Easy Things You Can Do to Spend with Intention*. Handout.

Reese, Stephen D. "Prologue—Framing Public Life: A Bridging Model for Media Research." In *Framing Public Life: Perspectives on Media and Our Understanding of the Social World*, edited by Stephen D. Reese, Oscar H. Gandy Jr., and August E. Grant, 7–32. Mahwah, NJ: Lawrence Erlbaum, 2001.

Regan, Charles E. *Paul Ricoeur*. Chicago: University of Chicago Press, 1996.

Reinemann, C., and W. Schulz. "Introduction to the Special Issue: News Decisions and News Values." *Communications: The European Journal of Communication Research* 31, no. 1 (2006): 1–4.

Resnick, P. "Beyond Bowling Together: Sociotechnical Capital." In *Human Computer Interaction in the New Millennium*, edited by J. Carroll, 247–272. New York: Addison-Wesley, 2002.

Revkin, Andrew C. "Stuck on Coal and Stuck for Words in a High-Tech World." *New York Times*, December 4, 2007. http://www.nytimes.com/2007/12/04/science/earth/04comm.html?_r=1&scp=1&sq=Stuck%20on%20coal%20and%20stuck%20for%20words%20in%20a%20high-tech%20world&st=cse (accessed August 27, 2009).

Rheingold, Howard. *Smart Mobs: The Next Social Revolution*. Cambridge, MA: Perseus, 2002.

Ricoeur, Paul. "Humans as the Subject Matter of Philosophy." In *The Narrative Path*, edited by T. P. Kemp and D. Rasmussen, 80–101. Cambridge: Cambridge University Press, 1989.

———. "Life in Quest of Narrative." In *On Paul Ricoeur: Narrative and Interpretation*, edited by David Wood, 20–33. New York: Routledge, 1991.

———. "Narrative Identity." In *On Paul Ricoeur: Narrative and Interpretation*, edited by David Wood, 188–200. New York: Routledge, 1991.

———. *Oneself as Another*. Chicago: University of Chicago Press, 1992.

———. *Time and Narrative*. Vol. 1. Translated by K. McLaughlin and D. Pellauer. Chicago: University of Chicago Press, 1983.

Roberts, J. Timmons. "Globalizing Environmental Justice." In *Environmental Justice and Environmentalism: The Social Justice Challenge to the Environmental Movement*, edited by Ronald Sandler and Phaedra Pezzullo, 85–101. Cambridge, MA: MIT Press, 2007.

Rockett, Benne. "Green Wave: Austin Green Art." *Voices of Art Magazine*, 13, no. 2 (2006). http://www.voamag.com/index.php?option=com_content&task=view&id=71&Itemid=9.

Rogers, E. M. "The Digital Divide." *Convergence: The International Journal of Research into New Media Technologies* 7 (2001): 96.

Romesburg, H. Charles. "Wildlife Science: Gaining Reliable Knowledge." *Journal of Wildlife Management* 45, no. 2 (1981): 293–313.

Rosenberg, Alex. *Philosophy of Science: A Contemporary Introduction*. London: Routledge, 2000.

Rosenthal, Elizabeth. "U.N. Report Describes Risks of Inaction on Climate Change." *New York Times*, November 17, 2007. http://www.nytimes.com/2007/11/17/ science/earth/17cnd-climate.html?pagewanted=print.

Rothenbuhler, Eric D. "The Church of the Cult of the Individual." In *Media Anthropology*, edited by Eric D. Rothenbuhler and Mihai Coman, 91–100. Thousand Oaks, CA: Sage, 2005.

Saad, Lydia. 2007. "Did Hollywood's Glare Heat Up Public Concern About Global Warming?" *Gallup News Service*, March 21, 2007.

Salt Lake City. 2007. http://www.ci.slc.ut.us/.

Sanders, Bernie. "We Must Act Now on Global Warming." *States News Service*, April 16, 2007. LexisNexis *Academic* database. http://www.lexisnexis.com.

Sandler, Ronald, and Phaedra C. Pezzullo, eds. *Environmental Justice and Environmentalism: The Social Justice Challenge to the Environmental Movement*. Cambridge, MA: MIT Press, 2007.

Sayer, Kyeann. "Austin Green Art Temporary Recycled Artwork Installation." Treehugger. http://www.treehugger.com/files/2005/10/austin_green_ar.php.

Scher, Abbey. "Can the Arts Change the World? The Transformative Power of Community Arts." *New Directions for Adult and Continuing Education* 116 (2007): 3–11.

Schlosberg, David. *Environmental Justice and the New Pluralism*. New York: Oxford University Press, 1999.

Schnaiberg, A. *The Environment*. New York: Oxford University Press, 1980.

Schuler, D. "New Communities and New Community Networks." In *Community Informatics: Enabling Communications With Information and Communications Technologies*, edited by M. Gurstein, 174–189. Hershey, PA: Idea Group Publishing, 2000.

Schutten, Julie K. *Coming Out of the Coven: Contemporary Witches, Hidden Populations, and New Social Movements*. Saarbrücken, Germany: VDM Verlag Dr. Muller, 2008.

Scott, Robert L., and Donald K. Smith. "The Rhetoric of Confrontation." *Quarterly Journal of Speech* 55 (1969): 1–8.

Segal, Judy Z. "The Structure of Advocacy: A Study of Environmental Rhetoric." *Canadian Journal of Communication* 16 (1991): 409–415.

Seiter, Douglas. "The City of Austin Green Builder Program: Developing Sustainability Through Marketing and Education." Paper presented at the First Florida Sustainable Building Conference, Miami, 1997.

Selwyn, N. "Reconsidering Political and Popular Understandings of the Digital Divide." *New Media Society* 6 (2004): 341.

Shapin, Steven, and Simon Schaffer. *Leviathan and the Air-Pump: Hobbes, Boyle, and the Experimental Life*. Princeton, NJ: Princeton University Press, 1985.

Shellenberger, Michael, and Ted Nordhaus. "The Death of Environmentalism: Global Warming Politics in a Post-Environmental World." The Breakthrough Institute. http://www.thebreakthrough.org/PDF/Death_of_Environmentalism.pdf.

Shenhav, Shaul R. "Concise Narratives: A Structural Analysis of Political Discourse." *Discourse Studies* 7, no. 3 (2005): 315–335.

———. "Once Upon a Time There Was a Nation: Narrative Conceptualization Analysis, the Concept of 'Nation' in the Discourse of Israeli Likud Party Leaders." *Discourse and Society* 15, no. 1 (2004): 81–104.

———. "Political Narratives and Political Reality." *International Political Science Review* 27 (2006): 245–262.

Shils, Edward. *The Constitution of Society*. Chicago: University of Chicago Press, 1982.

Shrum, McCarty. "Buyer Characteristics of the Green Consumer and Their Implications for Advertising Strategy." *Journal of Advertising* 24 (1995): 71–83.

Shulman, S., J. Katz, C. Quinn, and P. Srivastava. "Empowering Environmental Burdened Communities in the US: A Primer on the Emerging Role for Information Technology." *Local Environment* 10 (2005): 501–512.

Sill, Randolph. "Polar Bear Sited in Waters off Myrtle Edwards Park Warns of Global Warming and Encourages Decrease of Carbon Emissions!" Step It Up. http://events.stepitup2007.org/april/reports/120.

Siskin, Sharon. "Drawing the Circle: From the Community to the Studio to the Classroom and Back Again." *Women & Environments International Magazine* 72/73 (2006): 6–9.

Snow, David A., and Doug McAdam. "Identity Work Processes in the Context of Social Movements: Clarifying the Movement/Movement Nexus." In *Self Identity and Social Movements*, edited by Sheldon Stryker, Timothy J. Owens, and Robert W. White, 41–67. Minneapolis: University of Minnesota Press, 2000.

Snow, David A., and Pamela E. Oliver. "Social Movements and Collective Behavior: Social Psychological Dimensions and Considerations." In *Sociological Perspectives on Social Psychology*, edited by Karen S. Cook, Gary Alan Fine, and James S. House, 571–599. Boston: Allyn and Bacon, 1995.

Soulé, Michael E., ed. *Conservation Biology: The Science of Scarcity and Diversity*. Sunderland, MA: Sinauer Associates, 1986.

Sowards, Stacey K. "Rhetoric of Perpetual Potential: A Case Study of the Environmentalist Movement to Protect Orangutans." In Vol. 3, *The Environmental Communication Yearbook*, edited by Steven P. Depoe, 115–136. Mahwah, NJ: Lawrence Erlbaum Associates, 2006.

Stein, Laura, and Nikhil Sinha. "New Global Media and the Role of the State." In *The Handbook of New Media*, edited by Leah A. Lievrouw and Sonia Livingstone, 415–442. Thousand Oaks, CA: Sage, 2006.

Stein, Linda. "The Marching Forest of Shoreline Takes It to the Streets." Step It Up. http://events.stepitup2007.org/april/events/show/1334.

Steinbeck, John. *Travels with Charley: In Search of America*. New York: The Viking Press, 1962.

Step It Up. "About." http://april.stepitup2007.org/section.php?id=8.

———. "Action Ideas." http://april.stepitup2007.org/article.php?id=70.

———. "April 14, 2007—National Day of Climate Action." http://april.stepitup2007.org/.

———. "Bannermaking 101." http://stepitup2007.org/article.php?id=686.

———. "The Big One: One Earth, One Climate Rally. Boston, MA." http://events.stepitup2007.org/events/show/573.

———. "Involving More People." http://april.stepitup2007.org/article.php?id=71.

———. "Media Attention." http://april.stepitup2007.org/article.php?id=72.

———. "No-Impact Man Features Step It Up on *The Colbert Report*." Step It Up 2007 blog. http://april.stepitup2007.org/article.php?list=class&class=20&offset=80.

———. "Organizer FAQs." http://april.stepitup2007.org/article.php?id=76.

———. "Our Story." http://stepitup2007.org/article.php?list=type&type=48.

———. "Photographing Your Action." http://april.stepitup2007.org/article.php?id=184.

———. "The Step It Up Team." http://april.stepitup2007.org/article.php?id=79.

———. "Tell a Friend." DemocracyInAction.org. http://www.democracyinaction.org/dia/ organizationsORG/SIU/tellafriend.jsp?tell_a_friend_KEY=1251.

Stewart, Charles J., Craig Allen Smith, and Robert E. Denton. *Persuasion and Social Movements*. 5th ed. Long Grove, IL: Waveland, 2007.

Stohl, Michael, and Cynthia Stohl. "Human Rights, Nation States and NGOs: Structural Holes and the Emergence of Global Regimes." *Communication Monographs* 72, no. 4 (2005): 442–467.

Stryker, Sheldon. "Identity Competition." In *Self Identity and Social Movements*, edited by Sheldon Stryker, Timothy J. Owens, and Robert W. White, 21–40. Minneapolis: University of Minnesota Press, 2000.

Stryker, Sheldon, Timothy J. Owens, and Robert W. White. "Social Psychology and Social Movements: Cloudy Past and Bright Future." In *Self Identity and Social Movements*, edited by Sheldon Stryker,

Timothy J. Owens, and Robert W. White, 1–17. Minneapolis: University of Minnesota Press, 2000.

Taylor, Charles. *Defining Science: A Rhetoric of Demarcation.* Madison: University of Wisconsin Press, 1996.

———. "The Politics of Recognition." In *Multiculturalism: Examining the Politics of Recognition*, edited by Amy Gutman, 25–73. Princeton, NJ: Princeton University Press, 1994.

Tesh, S. N. *Uncertain Hazards: Environmental Activists and Scientific Proof.* Ithaca, NY: Cornell University Press, 2000.

Teske, Nathan. *Political Activism in America: The Identity Construction Model of Political Participation.* Cambridge: Cambridge University Press, 1997.

Texas Commission on Environmental Quality. "Environmental quality." http://www.tceq.state.tx.us/nav/eq.

Texas State Historical Association. "Dallas, Texas." *The Handbook of Texas Online.* http://www.tshaonline.org/handbook/online/articles/DD/hdd1.html.

———. "Fort Worth Stockyards." *The Handbook of Texas Online.* http://www.tshaonline.org/handbook/online/articles/FF/dif4.html.

———. "Plano, Texas." *The Handbook of Texas Online.* http://www.tsha.utexas.edu/handbook/online/articles/PP/hdp4.html.

———. "Texas Revolution." *The Handbook of Texas Online.* http://www.tshaonline.org/handbook/online/articles/TT/qdt1.html.

Texas Water Development Board. *Water for Texas: A Consensus-Based Update to the State Water Plan.* Austin: Texas Water Development Board, 1997.

Thayer, Robert L. 2003. *LifePlace: Bioregional thought and practice.* Berkeley: University of California Press.

Thompson, Leigh L., and Richard Gonzales. "Environmental Disputes: Competition for Scarce Resources and Clashing of Values."

In *Environment, Ethics, and Behavior: The Psychology of Environmental Valuation and Degradation*, edited by Max Bazerman, David Messick, Anne Tenbrunsel, and Kimberly Wade-Benzoni, 75–104. San Francisco: New Lexington Books, 1997.

Thompson, Richard C., Ylva Olsen, Richard P. Mitchell, Anthony Davis, Steven J. Rowland, Anthony W. G. John, Daniel McGonigle, and Andrea E. Russell. "Lost at Sea: Where Is All the Plastic?" *Science* 304 (2004): 838.

Time's Up. "Bike Lane Liberation Ride." http://times-up.org/index.php?page=bike-lane-liberation.

Titscher, Stefan, Michael Meyer, Ruth Wodak, and Eva Vetter. *Methods of Text and Discourse Analysis*. London: Sage, 2000.

Torgerson, Doug. *The Promise of Green Politics: Environmentalism and the Public Sphere*. Durham, NC: Duke University Press, 1999.

Truini, Joe. "Can These Simply Be Thrown Away?" *Waste News* 12 (2007): 1–14.

Tuan, Yi-Fu. *Space and Place: The Perspectives of Experience*. Minneapolis: University of Minnesota Press, 1977.

UK Reuters. "Time Is Up for Coal." UK Reuters, February 14, 2008. http://uk.reuters.com/article/environmentNews/idUKN1445731320080214?pageNumber=2&virtualBrandChannel=0.

U.S. Bureau of the Census. "Quick Facts." http://quickfacts.census.gov/qfd/.

———. "State and County Quick Facts." http://quickfacts.census.gov/qfd/states/48000lk.html.

U.S. Federal Bureau of Investigation. "Interagency Domestic Terrorism Concepts of Operation Plan (CONPLAN)." http://www.fbi.gov/publications/conplan/conplan.pdf.

U.S. Senate. U.S. Senate Roll Call Votes 110th Cong., 2nd sess., June 6, 2008. http://www.senate.gov/legislative/LIS/roll_call_lists/ roll_call_vote_cfm.cfm?congress=110&session=2&vote=00145.

Van Dijk, Teun A. "Critical Discourse Analysis." In *The Handbook of Discourse Analysis*, edited by D. Schiffrin, D. Tannen and H. E. Hamilton, 352–357. Oxford: Blackwell, 2001.

Van Dijk, T., and K. Hacker. "The Digital Divide as a Complex and Dynamic Phenomenon." Paper presented at the 50th Annual Conference of the International Communication Association, Acapulco, June 2000.

Venema, Henry Isaac. *Identifying Selfhood, Imagination, Narrative, and Hermeneutics in the Thought of Paul Ricoeur*. New York: State University of New York Press, 2000.

Verba, Sidney, and Norman Nie. *Participation in America: Political Democracy and Social Equality*. New York: Harper and Row, 1972.

Walsh, Don. "The Plastic Ocean." *U.S. Naval Institute Proceedings* 131 (2005): 82.

Warschauer, M. *Technology and Social Inclusion: Rethinking the Digital Divide*. Cambridge, MA: MIT Press, 2003.

Wheaton, Belinda. "Identity, Politics, and the Beach: Environmental Activism in Surfers Against Sewage." *Leisure Studies* 26, no. 3 (2007): 279–302.

Whitacre, B. E., and B. F. Mills. "Infrastructure and the Rural-Urban Divide in High-Speed Residential Internet Access." *International Regional Science Review* 30 (2007): 249–273.

Witt, Howard. "Overheated Planet Seeking Solutions in Heart of Texas, Drumbeat for Green: Austin Attacks the Problems of Climate Change—Right at Home." *Chicago Tribune*, September 28, 2007. http://www.ci.austin.tx.us/council/downloads/mw_planet_trib.pdf.

Worster, Donald. *Nature's Economy: A History of Ecological Ideas*. 2nd ed. Cambridge: Cambridge University Press, 1994.

Wright, B. H., P. Bryant, and R. D. Bullard. "Coping with Poisons in Cancer Alley." In *Unequal Protection: Environmental Justice and Communities of Color*, edited by R. D. Bullard, 110–129. San Francisco: Sierra Club Books, 1994.

List of Contributors

Leigh Bernacchi is a PhD student in wildlife and fisheries sciences, Texas A&M University. Leigh's research interests include public participation for public-land management, animal-human relationships (especially with birds), and performance and visual art as effective environmental communication. While her MS in environmental humanities research focused on the relationships of water and family in Yosemite National Park's Tuolumne River watershed and Hetch Hetchy Valley, Leigh has returned to the park as a naturalist ranger. She also worked in Denali National Park and Preserve and Sequoia National Park after receiving her BA in environmental science from Willamette University.

Deborah Cox Callister is a doctoral student and teaching fellow in the Department of Communication at the University of Utah and the 2009–2010 recipient of the David C. Williams Memorial Graduate Fellowship. Her research and teaching interests engage environmental communication, conflict theory, participatory communication, and critical rhetoric. Deborah has more than ten years of experience in environmental conflict mediation. Prior to graduate school, she worked as a senior project coordinator in a statewide Utah BLM wilderness-dispute-resolution project and the Community and Wild Lands Futures Project in Emery County, Utah. She holds a BA in biological sciences from the University of California, Davis, and an MS in communication from the University of Utah.

Tracylee Clarke is an assistant professor in the Communication Department at California State University, Channel Islands. Her current research and teaching interests include environmental communication, environmental policy development, identity politics, and the role of dialogue in environmental conflict prevention and resolution. Prior to joining the faculty of CSUCI, Tracylee worked as an environmental mediator and policy analyst and has more than fourteen years of experience in environmental conflict management and collaborative policy development. Tracylee received her PhD in communication and environmental conflict management from the University of Utah, her MA in communication from Washington State University, and her BA in communication from Southern Utah University.

Robert Cox is professor of communication studies and the curriculum for the environment and ecology at the University of North Carolina at Chapel Hill. He has served as president of the Sierra Club three times (2007–08; 2000–01; 1994–96), and has been on the board of directors for the Sierra Club for fourteen years. His principal research and teaching foci are rhetorical theory, environmental communication, and critical study of the discourse of change agents and social movements. He is considered one of the nation's leading scholars in environmental communication. Robert has received numerous awards for his research and environmental service, including the Christine L. Oravec Research Award in Environmental Communication, the American Communication Association's Outstanding Service to the Public Award, and "Best Monograph" award (American Forensic Association) for "The Die Is Cast: Topical and Ontological Dimensions of the Locus of the Irreparable," *Quarterly Journal of Speech* 68 (1982). He has published *Environmental Communication and the Public Sphere* (SAGE 2006, 2009); other work appears in *Quarterly Journal of Speech* and *Environmental Communication: A Journal of Nature and Culture*. In addition to performing research, he advises environmental groups on strategy in their advocacy campaigns and is called upon regularly to participate in and comment on numerous national initiatives concerning the environment. While serving as Sierra Club president in 2000, he campaigned with former vice president Al Gore in the U.S. presidential election. Robert received his PhD and MA degrees from the University of Pittsburgh and his BA from the University of Richmond.

Danielle Endres is an assistant professor of communication and faculty in the environmental humanities masters program at the University of Utah. Danielle's research and teaching interests lie in the intersection of rhetoric, environmental communication, social movements, and Native American cultures. She is currently working on several projects, including a book on the rhetoric of nuclear-waste-siting decisions, documenting oral histories of nuclearism in the American West, and tracing the intersections between environmental justice and energy policy in the United States. Danielle's work can be seen in *Communication and Critical Cultural Studies*, *Environmental Communication: A Journal of Nature and Culture*, *Western Journal of Communication*, and *Local Environment*. Danielle received her PhD in communication from the University of Washington, her MA in communication from San Diego State University, and her BA in history from the University of Oregon.

Andrea M. Feldpausch-Parker is a PhD candidate at Texas A&M University in the Department of Wildlife and Fisheries Sciences, where she is studying science communication regarding technologies for mitigating climate change. Andrea also coordinates outreach and education efforts for the Southwest Regional Partnership on Carbon Sequestration (SWP). Andrea has a BS in environmental biology/zoology and a BS in fisheries and wildlife from Michigan State University, and an MS in wildlife and fisheries science from Texas A&M University. In addition to her current research, Andrea has had experience in a wide range of scientific and management-oriented wildlife-research projects as well as public outreach and education programs in both Michigan and Texas.

Autumn Garrison is a PhD student and teaching fellow at the University of Utah. She considers herself a praxis-based research scholar and currently uses her research to inform her role as the cochair of communication for the Utah Society for Environmental Education (http://www.usee.org), an environmental education nonprofit in Salt Lake City. Autumn applies theories and concepts from mass communication, integrated marketing communication (IMC), and environmental communication to help USEE make choices about its communicative strategies and tactics. Autumn is also the 2008–2009 editor of the *Rocky Mountain Communication Review* (*RMCR*), an online journal based at the University of Utah and made for graduate students by graduate students. Prior to accepting her fellowship at the University of Utah, Autumn earned her MA in communication from the University of Cincinnati, where she cofounded and advised the multidisciplinary student group LEAP (Leaders for Environmental Awareness and Protection). While she earned her BA in communication and marketing from Northern Kentucky University, she also worked full-time in the sales and marketing industries.

Damon M. Hall is a PhD candidate and Boone and Crockett Fellow in wildlife and conservation policy in the Department of Wildlife and Fisheries Sciences at Texas A&M University in College Station. His dissertation research is on the discourses of place and coupled human-natural systems to improve environmental decision making among landowners, managers, and publics of the Yellowstone River (North Dakota, Montana, and Wyoming). He has a BS in forestry and natural resources and an MA in communication from Purdue University.

Cristi Choat Horton is a PhD student in the Department of Wildlife and Fisheries Sciences at Texas A&M University. Her research interests include how landowners negotiate the multiple demands made on them when faced with loss of space or land fragmentation due to urban development or land purchased for recreational use. In particular, she is interested in discovering and facilitating dimensions of conservation policy that provide opportunities for property owners to contribute to wildlife and other conservation measures as a fundamental aspect of their land-management portfolio. She received her BA and

MA in speech communication from Texas A&M University. Currently, she is employed at Tarleton State University (Stephenville, Texas) in the Department of Communication Studies.

William J. Kinsella is an associate professor of communication at North Carolina State University and a faculty member in two interdisciplinary programs: communication, rhetoric, and digital media and science, technology, and society. His work addresses the intersections among organizational communication, environmental communication, rhetoric of science and technology, and rhetoric of public policy. He is especially interested in the role of communication in scientific research, technology development, and public-policy debate on issues with technical dimensions, and on the communication relationships among experts, government, and citizens. His publications include articles in *Communication Yearbook*, *Environmental Communication: A Journal of Nature and Culture*, *Environmental Communication Yearbook*, *Management Communication Quarterly*, *New Jersey Journal of Communication*, *Qualitative Research Reports in Communication*, *Rhetoric Society Quarterly*, *Science as Culture*, *Technical Communication Quarterly*, and chapters in four academic books. He is a coeditor of *Nuclear Legacies: Communication, Controversy, and the U.S. Nuclear Weapons Complex* (Lexington Books, 2007). His work has received five awards from the (U.S.) National Communication Association and a Fulbright Fellowship award for research and lecturing based at the University of Stuttgart in 2010. He earned a bachelor's degree in physics from Manhattan College, did graduate work in astronomy and physics at New Mexico State University, and worked as a science educator before receiving master's and doctoral degrees in communication from Rutgers University.

Bill McKibben is an American environmentalist and writer who frequently writes about global warming, alternative energy, and the risks associated with human genetic engineering. In January 2007, he founded Step It Up to demand that Congress enact curbs on carbon emissions that would cut global-warming pollution 80 percent by 2050. Bill's first book, *The End of Nature*, was published in 1989 by Random House after being serialized in the *New Yorker*. It is regarded as the first book for a general audience about climate change and has been printed in more than twenty languages. Subsequent books include *The Age of Missing Information*; *Hope, Human and Wild*; *Maybe One*; *Long Distance: A Year of Living Strenuously*; and *Wandering Home*; and *Deep Economy*: *Fight Global Warming Now*, a guide to help people initiate environmental activism in their community, was published in late 2007. Bill has been awarded Guggenheim and Lyndhurst Fellowships and the Lannan Prize for nonfiction writing. He has honorary degrees from Green Mountain College, Unity College, Lebanon Valley College, and Sterling College. He is a scholar in residence at Middlebury College.

Trudy Milburn is an associate professor in communication at the California State University, Channel Islands. Her most recent book, *Nonprofit Organizations: Creating Membership Through Communication* (Hampton Press, 2009), explores the way that group members create a sense of organizational identity. Her next book, *Speaking of Superfunds: Clean-ups, Communities, and the Conversations That Bind*, coauthored with the geographer Susan Gilbertz, is due out from Cambria Press in 2010.

Tema Oliveira Milstein is an assistant professor in the Communication and Journalism Department at the University of New Mexico. Her research focuses on the intersections of nature and culture. You can learn more about her work at http://www.unm.edu/~tema/.

Jodi Minion is a wildlife biologist for People for the Ethical Treatment of Animals (PETA) in their Cruelty Investigations Department. Prior to taking a position with PETA, Jodi was a graduate research assistant and Boone and Crockett Fellow in the Department of Wildlife and Fisheries Sciences at Texas A&M University. Her research focused on the relationship between new media and democracy—in the context of climate-change action—as a way to improve our understanding of contemporary activism. Jodi has also organized and implemented direct actions for environmental organizations. Prior to her graduate work, she graduated from Western Washington University's Huxley College of the Environment, magna cum laude, with a degree in environmental policy and planning.

Anna Maria Munoz is a doctoral candidate in the Department of Wildlife and Fisheries Sciences at Texas A&M University. Her current research interests include environmental conflict management, public participation in conservation planning, and environmental policy development. Her dissertation research project focuses on the benefits and constraints associated with the use of collaboration to address the Edwards Aquifer endangered-species conflict in Texas. Prior to entering the graduate program at Texas A&M University, Anna worked as a wildlife and fisheries biologist for the U.S. Fish and Wildlife Service in New Mexico and an environmental protection specialist for Region 10 of the U.S. Environmental Protection Agency. She has an MS in wildlife and fisheries sciences from Texas A&M University and a BS in wildlife science from New Mexico State University.

Todd Norton is an assistant professor in the Edward R. Murrow College of Communication at Washington State University. A native of the Great Lakes region, Todd obtained his bachelor's and master's degrees from the University of Wisconsin—Stevens Point. Moving west to the University of Utah, Todd earned the distinction of the David C. Williams Memorial Graduate Fellowship from the College of Social and Behavioral Sciences for his dissertation, "Grand Staircase-Escalante National Monument: Organizing Land Politics." Todd's ongoing research focuses on the organizational dimensions of public-sector policy processes, with particular interest in multistakeholder environmental conflicts as evident in contestation over rights to public land, community planning, and control/resistance among stakeholders. His research appears in *Management Communication Quarterly*, *Communication Theory*, *Environmental Communication: A Journal of Nature and Culture*, and *Southern Communication Journal*.

Chad O'Neil was a PhD student in communication, rhetoric, and digital media at North Carolina State University in the College of Humanities and Social Sciences while collaborating on a chapter for the Step It Up book. Chad's research interests encompass a wide range of community uses of technology. He received his MA in digital rhetoric and professional writing from the Rhetoric and Writing Program at Michigan State University and his BA in English from the University of Michigan.

Israel D. Parker is a PhD candidate in the Department of Wildlife and Fisheries Sciences at Texas A&M University. His interests are in wildlife ecology in general and endangered-species ecology in particular. He is also interested in the evolving human-nature relationship and the critical role of environmental communication in the future of wildlife ecology. He completed a BS in fishery and wildlife science at New Mexico State University and an MS in wildlife and fisheries sciences at Texas A&M University. He has ten years of experience in wildlife ecology, working with a variety of state and federal agencies.

Markus J. Peterson is an associate professor of wildlife ecology and conservation policy at Texas A&M University. Markus' research program addresses biodiversity and environmental conservation. From an ecological perspective, his research emphasizes processes influencing wild-animal abundance, such as parasitism/disease, predation, weather, habitat conditions, and human exploitation. From a social perspective, his research emphasizes processes influencing environmental policy formation and implementation, such as environmental democracy, public processes used in environmental decision making, and how the legal and economic social systems influence environmental decision making.

In addition to several book chapters, Markus coauthored *Wildlife Study Design* and is published in numerous refereed journals, including *Communication and Critical/Cultural Studies*, *Conservation Biology*, *Ecography*, *Ecological Modeling*, *Environmental Communication: A Journal of Nature and Culture*, *Environmental Values*, *Landscape and Urban Planning*, *Journal of Wildlife Diseases*, and *Journal of Wildlife Management*. Markus previously worked as an administrator for a state natural-resource agency, and he serves on several state and national advisory boards related to controversial wildlife disease and conservation issues. Markus received his PhD and MS degrees in wildlife ecology and conservation from Texas A&M University, his DVM from Washington State University, and his BS from the University of Idaho.

M. Nils Peterson is an assistant professor in the Department of Forestry and Environmental Resources at North Carolina State University. His current research and teaching interests lie in exploring the intersections between human and natural systems. Nils uses land use, household dynamics, and democracy as nexuses between people and the environment. His research interests are diverse and include policy analysis, environmental attitude and behavior assessment, and assessing impacts of human behaviors on endangered wildlife populations. His research is published in journals such as *Conservation Biology*, *Journal of Wildlife Management*, *Environmental Communication: A Journal of Nature and Culture*, and *Landscape and Urban Planning*. Nils received his PhD in fisheries and wildlife from Michigan State University and his MS and BS in wildlife and fisheries science from Texas A&M University.

Tarla Rai Peterson holds the Boone & Crocket Chair of Wildlife and Conservation Policy at Texas A&M University. She is a professor of environmental communication at the Swedish University of Agricultural Sciences and an adjunct professor of communication at the University of Utah. Dr. Peterson received an interdisciplinary PhD (emphasis in environmental rhetoric) and her MA from Washington State University, and a BA in American history from the University of Idaho. She is associate editor for *North America at Environmental Communication: A Journal of Nature and Culture. Her publications include Green Talk in the White House: The Rhetorical Presidency Encounters Ecology, Sharing the Earth: The Rhetoric of Sustainable Development*, and numerous articles.

Phaedra C. Pezzullo is an associate professor in the Department of Communication and Culture at the University of Indiana. Her research, teaching, and advocacy interests include environmental communication, environmental justice, social movements, tourism, and rhetoric and public culture. Phaedra has published two books: *Toxic Tourism: Rhetorics of Travel, Pollution, and Environmental Justice* (University of Alabama Press, 2007), and *Environmental Justice and Environmentalism* (coedited with Ronald Sandler, MIT Press, 2007). In addition to her books and several chapters, Phaedra has published articles in *Tourism and Cultural Change*, *Cultural Studies*, *Environmental Communication Yearbook*, *Text and Performance Quarterly*, *Quarterly Journal of Speech*, and *Western Journal of Communication*. Currently, she serves as the chair of the city of Bloomington's Environmental Commission.

Travis Paveglio is a PhD student in the Department of Natural Resource Sciences at Washington State University. His research and teaching interests include wildland fire policy and management, natural resource conflict/collaboration in the wildland-urban interface, community hazard resilience, and the sociology of natural-resource management. Travis' current research is focused on alternatives to evacuation during wildland fire and the implications of social diversity on hazard management. His work can be seen in *The Journal of Forestry*, *Environmental Management*, *Society & Natural Resources*, and *Environmental Communication: A Journal of Nature and Culture*. He received his MA in communication and BA in journalism from the Edward R. Murrow College of Communication at Washington State University.

Lawrence J. Prelli is a professor of communication and chair in the Department of Communication and an affiliate professor in the Department of Natural Resources at the University of New Hampshire. He teaches courses and conducts research in rhetorical theory and criticism, with special attention to examining rhetoric in relation to science, environmental discourses, and display. In addition to journal articles and book chapters, he is editor of *Rhetorics of Display* (2006) and author of *A Rhetoric of Science: Inventing Scientific Discourse* (1989), which won the 1990 Eastern Communication Association's Everett Lee Hunt Award for distinguished scholarship in rhetoric and public address. He also is recipient of that association's 2004 Donald H. Ecroyd and Carolyn Drummond Ecroyd Teaching Excellence Award and the 2007 National Communication Association's Golden Anniversary Monograph Award. He received a PhD in speech communication from the Pennsylvania State University, an MS in natural resources from the University of New Hampshire, an MA in rhetoric and communication from the State University of New York at Albany, and a BS in political science from the State University of New York College at Brockport.

Julie Kalil Schutten is a visiting assistant professor of communication at Northern Arizona University in Flagstaff, Arizona. Her research focuses on the discursive practices of new social movements, environmental communication, and gender/feminist studies. Her dissertation has been published as a book titled *Coming Out of the Coven: Contemporary Witches, Hidden Populations and New Social Movements*. More of Julie's work can be found in the journals *Environmental Communication: A Journal of Nature and Culture, Western Journal of Communication*, and *The Communication Review*. Julie earned her PhD in communication from the University of Utah in 2007, her MA in communication studies from the University of Nevada, Las Vegas, and her BA in speech communication from Northern Arizona University.

Samantha Senda-Cook is a doctoral candidate at the University of Utah. Samantha's research investigates intersections of rhetoric, identity, perceptions of the environment, consumerism, and authenticity. She received the David C. Williams Memorial Graduate Fellowship for the 2008–2009 year and the Floyd O'Neil Scholarship in Western American Studies for the 2009–2010 year. In teaching, her goals are to encourage and create spaces for students to critically engage cultural phenomena. Samantha received her MA in rhetoric from Colorado State University and her BA in communication and journalism from Shippensburg University.

Leah Sprain is an assistant professor of communication at Colorado State University, where she also works with the Center for Public Deliberation. Leah's research and teaching draw on cultural communication perspectives to study deliberation, environmental activism, conflict, and social movements. She has extended ethnographic fieldwork in Nicaragua and the United States. Her work can be found in the *International Journal of Public Participation, Environmental Communication: A Journal of Nature and Culture*, and the *Handbook of Communication Ethics*. Leah received her PhD and MA in communication from the University of Washington, and her BA in environmental studies and BFA in communication from Pacific Lutheran University.

Micheal R. Vickery is a professor of communication and co-director of the Center for Responsible Leadership at Alma College. Micheal teaches courses in environmental communication, comparative environmental policy, health communication, and the interaction of media and political discourse. He has long been interested in the interdependent rhetorics of energy and progress and is currently working on projects related to the rhetoric of alternative and renewable energy, sustainable communities, and public health. Micheal received his PhD in communication from the University of Texas at Austin and his BS in psychology from the University of Alabama.

Illustration Credits

Figure 1. Image of Sustainable Ballard's community art project at the Seattle, Washington, SIU event: Sustainable Ballard participant.

Figure 2. Image of two ceremony kettles form the Gathering of the Waters SIU event in Salt Lake City, Utah: Leigh A. Bernacchi.

Figure 3. Image of people gathered for the Gathering of the Waters SIU event in Salt Lake City, Utah: Leigh A. Bernacchi.

Figure 4. Image of an art installation at the Austin, Texas, Step It Up event: Damon M. Hall.

Figure 5. Close-up image of art installation at the Austin, Texas, Step It Up event: Damon M. Hall.

Figure 6. Photo of "Free Concert" sign at Step It Up event in downtown Salt Lake City, Utah: Danielle Endres.

Figure 7. Picture of an SIU "Eight Ways You Can Reduce You Impact Today" poster at Gathering of the Waters SIU event in Salt Lake City, Utah: Craig O. Rich.

Figure 8. Picture of a banner signed by participants at the Gathering of the Waters SIU event in Salt Lake City, Utah: Craig O. Rich.

Figure 9. Screen shot of the Step It Up 2007 National Day of Climate Action Web site: Step It Up 2007.

Figure 10. Photo of Raging Grannies holding compact florescent light-bulbs at the Step It Up event in Raleigh, North Carolina: William J. Kinsella.

Name Index

Subject Index

CPSIA information can be obtained at www.ICGtesting.com
Printed in the USA
BVOW040744270812

298684BV00003B/22/P

9 781604 976410